AF739952

THE NATURE OF LIGHT

What Is a Photon?

OPTICAL SCIENCE AND ENGINEERING

Founding Editor
Brian J. Thompson
University of Rochester
Rochester, New York

1. Electron and Ion Microscopy and Microanalysis: Principles and Applications, *Lawrence E. Murr*
2. Acousto-Optic Signal Processing: Theory and Implementation, *edited by Norman J. Berg and John N. Lee*
3. Electro-Optic and Acousto-Optic Scanning and Deflection, *Milton Gottlieb, Clive L. M. Ireland, and John Martin Ley*
4. Single-Mode Fiber Optics: Principles and Applications, *Luc B. Jeunhomme*
5. Pulse Code Formats for Fiber Optical Data Communication: Basic Principles and Applications, *David J. Morris*
6. Optical Materials: An Introduction to Selection and Application, *Solomon Musikant*
7. Infrared Methods for Gaseous Measurements: Theory and Practice, *edited by Joda Wormhoudt*
8. Laser Beam Scanning: Opto-Mechanical Devices, Systems, and Data Storage Optics, *edited by Gerald F. Marshall*
9. Opto-Mechanical Systems Design, *Paul R. Yoder, Jr.*
10. Optical Fiber Splices and Connectors: Theory and Methods, *Calvin M. Miller with Stephen C. Mettler and Ian A. White*
11. Laser Spectroscopy and Its Applications, *edited by Leon J. Radziemski, Richard W. Solarz, and Jeffrey A. Paisner*
12. Infrared Optoelectronics: Devices and Applications, *William Nunley and J. Scott Bechtel*
13. Integrated Optical Circuits and Components: Design and Applications, *edited by Lynn D. Hutcheson*
14. Handbook of Molecular Lasers, *edited by Peter K. Cheo*
15. Handbook of Optical Fibers and Cables, *Hiroshi Murata*
16. Acousto-Optics, *Adrian Korpel*
17. Procedures in Applied Optics, *John Strong*
18. Handbook of Solid-State Lasers, *edited by Peter K. Cheo*
19. Optical Computing: Digital and Symbolic, *edited by Raymond Arrathoon*
20. Laser Applications in Physical Chemistry, *edited by D. K. Evans*
21. Laser-Induced Plasmas and Applications, *edited by Leon J. Radziemski and David A. Cremers*

UNIVERSITY OF
BIRMINGHAM

Library Services

This item must be returned promptly if recalled by
…her borrower, otherwi…

22. Infrared Technology Fundamentals, *Irving J. Spiro and Monroe Schlessinger*
23. Single-Mode Fiber Optics: Principles and Applications, Second Edition, Revised and Expanded, *Luc B. Jeunhomme*
24. Image Analysis Applications, *edited by Rangachar Kasturi and Mohan M. Trivedi*
25. Photoconductivity: Art, Science, and Technology, *N. V. Joshi*
26. Principles of Optical Circuit Engineering, *Mark A. Mentzer*
27. Lens Design, *Milton Laikin*
28. Optical Components, Systems, and Measurement Techniques, *Rajpal S. Sirohi and M. P. Kothiyal*
29. Electron and Ion Microscopy and Microanalysis: Principles and Applications, Second Edition, Revised and Expanded, *Lawrence E. Murr*
30. Handbook of Infrared Optical Materials, *edited by Paul Klocek*
31. Optical Scanning, *edited by Gerald F. Marshall*
32. Polymers for Lightwave and Integrated Optics: Technology and Applications, *edited by Lawrence A. Hornak*
33. Electro-Optical Displays, *edited by Mohammad A. Karim*
34. Mathematical Morphology in Image Processing, *edited by Edward R. Dougherty*
35. Opto-Mechanical Systems Design: Second Edition, Revised and Expanded, *Paul R. Yoder, Jr.*
36. Polarized Light: Fundamentals and Applications, *Edward Collett*
37. Rare Earth Doped Fiber Lasers and Amplifiers, *edited by Michel J. F. Digonnet*
38. Speckle Metrology, *edited by Rajpal S. Sirohi*
39. Organic Photoreceptors for Imaging Systems, *Paul M. Borsenberger and David S. Weiss*
40. Photonic Switching and Interconnects, *edited by Abdellatif Marrakchi*
41. Design and Fabrication of Acousto-Optic Devices, *edited by Akis P. Goutzoulis and Dennis R. Pape*
42. Digital Image Processing Methods, *edited by Edward R. Dougherty*
43. Visual Science and Engineering: Models and Applications, *edited by D. H. Kelly*
44. Handbook of Lens Design, *Daniel Malacara and Zacarias Malacara*
45. Photonic Devices and Systems, *edited by Robert G. Hunsberger*
46. Infrared Technology Fundamentals: Second Edition, Revised and Expanded, *edited by Monroe Schlessinger*
47. Spatial Light Modulator Technology: Materials, Devices, and Applications, *edited by Uzi Efron*
48. Lens Design: Second Edition, Revised and Expanded, *Milton Laikin*
49. Thin Films for Optical Systems, *edited by Francoise R. Flory*
50. Tunable Laser Applications, *edited by F. J. Duarte*
51. Acousto-Optic Signal Processing: Theory and Implementation, Second Edition, *edited by Norman J. Berg and John M. Pellegrino*

52. Handbook of Nonlinear Optics, *Richard L. Sutherland*
53. Handbook of Optical Fibers and Cables: Second Edition, *Hiroshi Murata*
54. Optical Storage and Retrieval: Memory, Neural Networks, and Fractals, *edited by Francis T. S. Yu and Suganda Jutamulia*
55. Devices for Optoelectronics, *Wallace B. Leigh*
56. Practical Design and Production of Optical Thin Films, *Ronald R. Willey*
57. Acousto-Optics: Second Edition, *Adrian Korpel*
58. Diffraction Gratings and Applications, *Erwin G. Loewen and Evgeny Popov*
59. Organic Photoreceptors for Xerography, *Paul M. Borsenberger and David S. Weiss*
60. Characterization Techniques and Tabulations for Organic Nonlinear Optical Materials, *edited by Mark G. Kuzyk and Carl W. Dirk*
61. Interferogram Analysis for Optical Testing, *Daniel Malacara, Manuel Servin, and Zacarias Malacara*
62. Computational Modeling of Vision: The Role of Combination, *William R. Uttal, Ramakrishna Kakarala, Spiram Dayanand, Thomas Shepherd, Jagadeesh Kalki, Charles F. Lunskis, Jr., and Ning Liu*
63. Microoptics Technology: Fabrication and Applications of Lens Arrays and Devices, *Nicholas Borrelli*
64. Visual Information Representation, Communication, and Image Processing, *edited by Chang Wen Chen and Ya-Qin Zhang*
65. Optical Methods of Measurement, *Rajpal S. Sirohi and F. S. Chau*
66. Integrated Optical Circuits and Components: Design and Applications, *edited by Edmond J. Murphy*
67. Adaptive Optics Engineering Handbook, *edited by Robert K. Tyson*
68. Entropy and Information Optics, *Francis T. S. Yu*
69. Computational Methods for Electromagnetic and Optical Systems, *John M. Jarem and Partha P. Banerjee*
70. Laser Beam Shaping, *Fred M. Dickey and Scott C. Holswade*
71. Rare-Earth-Doped Fiber Lasers and Amplifiers: Second Edition, Revised and Expanded, *edited by Michel J. F. Digonnet*
72. Lens Design: Third Edition, Revised and Expanded, *Milton Laikin*
73. Handbook of Optical Engineering, *edited by Daniel Malacara and Brian J. Thompson*
74. Handbook of Imaging Materials: Second Edition, Revised and Expanded, *edited by Arthur S. Diamond and David S. Weiss*
75. Handbook of Image Quality: Characterization and Prediction, *Brian W. Keelan*
76. Fiber Optic Sensors, *edited by Francis T. S. Yu and Shizhuo Yin*
77. Optical Switching/Networking and Computing for Multimedia Systems, *edited by Mohsen Guizani and Abdella Battou*
78. Image Recognition and Classification: Algorithms, Systems, and Applications, *edited by Bahram Javidi*

79. Practical Design and Production of Optical Thin Films: Second Edition, Revised and Expanded, *Ronald R. Willey*
80. Ultrafast Lasers: Technology and Applications, *edited by Martin E. Fermann, Almantas Galvanauskas, and Gregg Sucha*
81. Light Propagation in Periodic Media: Differential Theory and Design, *Michel Nevière and Evgeny Popov*
82. Handbook of Nonlinear Optics, Second Edition, Revised and Expanded, *Richard L. Sutherland*
83. Polarized Light: Second Edition, Revised and Expanded, *Dennis Goldstein*
84. Optical Remote Sensing: Science and Technology, *Walter Egan*
85. Handbook of Optical Design: Second Edition, *Daniel Malacara and Zacarias Malacara*
86. Nonlinear Optics: Theory, Numerical Modeling, and Applications, *Partha P. Banerjee*
87. Semiconductor and Metal Nanocrystals: Synthesis and Electronic and Optical Properties, edited by *Victor I. Klimov*
88. High-Performance Backbone Network Technology, *edited by Naoaki Yamanaka*
89. Semiconductor Laser Fundamentals, *Toshiaki Suhara*
90. Handbook of Optical and Laser Scanning, *edited by Gerald F. Marshall*
91. Organic Light-Emitting Diodes: Principles, Characteristics, and Processes, *Jan Kalinowski*
92. Micro-Optomechatronics, *Hiroshi Hosaka, Yoshitada Katagiri, Terunao Hirota, and Kiyoshi Itao*
93. Microoptics Technology: Second Edition, *Nicholas F. Borrelli*
94. Organic Electroluminescence, *edited by Zakya Kafafi*
95. Engineering Thin Films and Nanostructures with Ion Beams, *Emile Knystautas*
96. Interferogram Analysis for Optical Testing, Second Edition, *Daniel Malacara, Manuel Sercin, and Zacarias Malacara*
97. Laser Remote Sensing, *edited by Takashi Fujii and Tetsuo Fukuchi*
98. Passive Micro-Optical Alignment Methods, *edited by Robert A. Boudreau and Sharon M. Boudreau*
99. Organic Photovoltaics: Mechanism, Materials, and Devices, *edited by Sam-Shajing Sun and Niyazi Serdar Saracftci*
100. Handbook of Optical Interconnects, *edited by Shigeru Kawai*
101. GMPLS Technologies: Broadband Backbone Networks and Systems, *Naoaki Yamanaka, Kohei Shiomoto, and Eiji Oki*
102. Laser Beam Shaping Applications, *edited by Fred M. Dickey, Scott C. Holswade and David L. Shealy*
103. Electromagnetic Theory and Applications for Photonic Crystals, *Kiyotoshi Yasumoto*
104. Physics of Optoelectronics, *Michael A. Parker*
105. Opto-Mechanical Systems Design: Third Edition, *Paul R. Yoder, Jr.*
106. Color Desktop Printer Technology, *edited by Mitchell Rosen and Noboru Ohta*
107. Laser Safety Management, *Ken Barat*

108. Optics in Magnetic Multilayers and Nanostructures, *Štefan Višňovský*
109. Optical Inspection of Microsystems, *edited by Wolfgang Osten*
110. Applied Microphotonics, *edited by Wes R. Jamroz, Roman Kruzelecky, and Emile I. Haddad*
111. Organic Light-Emitting Materials and Devices, *edited by Zhigang Li and Hong Meng*
112. Silicon Nanoelectronics, *edited by Shunri Oda and David Ferry*
113. Image Sensors and Signal Processor for Digital Still Cameras, *Junichi Nakamura*
114. Encyclopedic Handbook of Integrated Circuits, *edited by Kenichi Iga and Yasuo Kokubun*
115. Quantum Communications and Cryptography, *edited by Alexander V. Sergienko*
116. Optical Code Division Multiple Access: Fundamentals and Applications, *edited by Paul R. Prucnal*
117. Polymer Fiber Optics: Materials, Physics, and Applications, *Mark G. Kuzyk*
118. Smart Biosensor Technology, *edited by George K. Knopf and Amarjeet S. Bassi*
119. Solid-State Lasers and Applications, *edited by Alphan Sennaroglu*
120. Optical Waveguides: From Theory to Applied Technologies, *edited by Maria L. Calvo and Vasudevan Lakshiminarayanan*
121. Gas Lasers, *edited by Masamori Endo and Robert F. Walker*
122. Lens Design, Fourth Edition, *Milton Laikin*
123. Photonics: Principles and Practices, *Abdul Al-Azzawi*
124. Microwave Photonics, *edited by Chi H. Lee*
125. Physical Properties and Data of Optical Materials, *Moriaki Wakaki, Keiei Kudo, and Takehisa Shibuya*
126. Microlithography: Science and Technology, Second Edition, *edited by Kazuaki Suzuki and Bruce W. Smith*
127. Coarse Wavelength Division Multiplexing: Technologies and Applications, *edited by Hans Joerg Thiele and Marcus Nebeling*
128. Organic Field-Effect Transistors, *Zhenan Bao and Jason Locklin*
129. Smart CMOS Image Sensors and Applications, *Jun Ohta*
130. Photonic Signal Processing: Techniques and Applications, *Le Nguyen Binh*
131. Terahertz Spectroscopy: Principles and Applications, *edited by Susan L. Dexheimer*
132. Fiber Optic Sensors, Second Edition, *edited by Shizhuo Yin, Paul B. Ruffin, and Francis T. S. Yu*
133. Introduction to Organic Electronic and Optoelectronic Materials and Devices, *edited by Sam-Shajing Sun and Larry R. Dalton*
134. Introduction to Nonimaging Optics, *Julio Chaves*
135. The Nature of Light: What Is a Photon?, *edited by Chandrasekhar Roychoudhuri, A. F. Kracklauer, and Katherine Creath*

THE NATURE OF LIGHT

What Is a Photon?

Edited by
CHANDRASEKHAR ROYCHOUDHURI
A. F. KRACKLAUER
KATHERINE CREATH

CRC Press
Taylor & Francis Group
Boca Raton London New York

CRC Press is an imprint of the
Taylor & Francis Group, an **informa** business

UNIVERSITY OF BIRMINGHAM LIBRARY

CRC Press
Taylor & Francis Group
6000 Broken Sound Parkway NW, Suite 300
Boca Raton, FL 33487-2742

© 2008 by Taylor & Francis Group, LLC
CRC Press is an imprint of Taylor & Francis Group, an Informa business

No claim to original U.S. Government works
Printed in the United States of America on acid-free paper
10 9 8 7 6 5 4 3 2 1

International Standard Book Number-13: 978-1-4200-4424-9 (Hardcover)

This book contains information obtained from authentic and highly regarded sources Reasonable efforts have been made to publish reliable data and information, but the author and publisher cannot assume responsibility for the validity of all materials or the consequences of their use. The Authors and Publishers have attempted to trace the copyright holders of all material reproduced in this publication and apologize to copyright holders if permission to publish in this form has not been obtained. If any copyright material has not been acknowledged please write and let us know so we may rectify in any future reprint

Except as permitted under U.S. Copyright Law, no part of this book may be reprinted, reproduced, transmitted, or utilized in any form by any electronic, mechanical, or other means, now known or hereafter invented, including photocopying, microfilming, and recording, or in any information storage or retrieval system, without written permission from the publishers.

For permission to photocopy or use material electronically from this work, please access www.copyright.com (http://www.copyright.com/) or contact the Copyright Clearance Center, Inc. (CCC) 222 Rosewood Drive, Danvers, MA 01923, 978-750-8400. CCC is a not-for-profit organization that provides licenses and registration for a variety of users. For organizations that have been granted a photocopy license by the CCC, a separate system of payment has been arranged.

Trademark Notice: Product or corporate names may be trademarks or registered trademarks, and are used only for identification and explanation without intent to infringe.

Library of Congress Cataloging-in-Publication Data

Roychoudhuri, Chandrasekhar.
The nature of light : what is a photon? / Chandra Roychoudhuri, A.F. Kracklauer, Kathy Creath.
p. cm.
Includes bibliographical references and index.
ISBN 978-1-4200-4424-9 (alk. paper)
1. Photons. 2. Light. I. Kracklauer, Al F. II. Creath, Kathy. III. Title.

QC793.5.P427R69 2008
539.7'217--dc22 2008002446
23544678

Visit the Taylor & Francis Web site at
http://www.taylorandfrancis.com

and the CRC Press Web site at
http://www.crcpress.com

Contents

Preface

This book is an attempt to rekindle active interest by both aspiring scientists (senior and graduate students) and practicing scientists in the nature of light—an unresolved issue in the field of physics. Many fundamental issues pertaining to light persist; they should be explored and understood, hopefully *inter alia* opening up many new applications.

The deeply enigmatic nature of light (groups of photons) can be appreciated from the long history of controversy starting with Newton and Huygens in the early 1700s. Newton claimed that light had a "corpuscular" nature. Huygens asserted that it had a "wave" nature. In the early 1800s, Thomas Young tried to resolve the issue by his famous double slit experiment. He demonstrated the generation of sinusoidal fringes under a common single-slit diffraction pattern in a far-field location. His experiment was overridden a century later by Einstein's heuristic hypothesis that light beams consist of indivisible quanta of electromagnetic energy, $h\nu$. Einstein was inspired by Planck's successful representation of measured blackbody spectra. This hypothesis successfully explained the observed phenomenon of photoelectron emission. Now, however, more than another hundred years later, we still are experiencing conceptual conundrums.

Most of the active physics community is comfortable with claims that quantum computers, quantum communication systems, and quantum encryption techniques can be developed by generating, manipulating, propagating, and detecting a single photon that, according to Dirac's view, "interferes only with itself." On the other hand, others claim that light beams do not "interfere" (interact) with each other to produce a redistribution of field energy (fringes of superposition) unless photodetecting molecules are physically present within the volume in which superposition occurs to facilitate energy redistribution. The first group relies on conceptual premises such as non-locality in superposition effects and teleportation as a physical possibility. The second group actively attempts to bridge classical and quantum physics by innovatively using various semiclassical methods and concepts to restore "reality" and "locality" to physics. Their key premise is that all measurable transformation *processes* require energy exchanges among interactants as allowed by a natural force law that is practically effective only within a finite range. This implies that each interactant must be within another's sphere of influence to generate a detectable transformation. Our view is that "if nobody understands quantum mechanics" in spite of its very useful formalism, an attempt should be made to revisit both the interpretation and the formalism. We must discover the real origin of our failures to understand quantum mechanics and imagine and visualize the physical processes behind these light-matter interaction *processes.*

This book has three sections. The first one contains five articles from well known quantum optics groups. These articles originally published by OSA in *Optics and Photonics News* are written for senior level college students who plan to specialize in quantum optics. Scientists and engineers from fields other than quantum optics can also use these articles to understand mainstream views and the state of knowledge of the nature of light and photons. The second section contains two articles. Their purpose is to prepare the audience for the diverse out-of-the-box photon models presented in the third section summarizing the paradoxes, contradictions, and confusions arising from the currently accepted definition of a photon as a monochromatic Fourier mode of vacuum. The epistemology article also offers a novel methodology of organizing incomplete information and framing it into a theory using human logics and helping to redefine physics as discovering realities of nature rather than trying to invent them. The third section consists of articles characterized as out-of-the-box thinking. The last four chapters of this section present diverse experimental results and viewpoints. Collectively they underscore that the semi-classical model for photons as space and time finite wave packets allows one to conceptualize and visualize a causal model for photons à la Planck's original version and as further developed by E.T. Jaymes.

We thank the Taylor & Francis editorial team for their work in publishing this compilation as a book, thereby promoting accessibility of these articles to a broader audience. We earnestly hope that this book will inspire the next generation of scientists and engineers in quantum optics to explore the nature of light and originate many new ideas to elucidate light–matter interaction processes with many practical new applications. Only real applications can firmly validate the reality of the proposed hypotheses.

Chandrasekhar Roychoudhuri

A. F. Kracklauer

Katherine Creath

Acknowledgments

This book is, essentially, a compilation of edited and selected articles already published in a special issue magazine of the Optical Society of America (OSA), several conference proceedings of the Society of Photo-Optical Instrumentation Engineers (SPIE) and the journal *Science.* Our sincere thanks to these three organizations for extending the permission to re-publish them as a single book to serve our community better. We would also like to thank Nippon Sheet Glass Co. Ltd. of Japan for being a consistent financial supporter for both the OSA publication and the SPIE conferences.

Section 1. Chapters 1 through 5: Edited and re-printed from the magazine of October 2003 Special Issue of Optics and Photonics News, "The nature of light: What is a Photon?" Eds. Chandrasekhar Roychoudhuri and Rajarshi Roy.

Section 2. Chapter 6: Previously unpublished.

Section 2. Chapter 7: Edited and re-printed from SPIE Proc. Vol. **5866** (2005), *The nature of light: What is a photon?*

Section 3. Chapters 8 through 25: Edited and re-printed from SPIE Proc. Vol. **5866** (2005), *The nature of light: What is a photon?*

Section 3. Chapter 26: Edited and re-printed from SPIE Proc. Vol. **6372** (2006).

Section 3. Chapter 27: Edited and re-printed from Science **305,** 1267 (2004).

Editors

Professor Chandrasekhar Roychoudhuri is a member of the faculty of the physics department, University of Connecticut in Storrs. His current research interests are exploration of the fundamental nature of light and photons and the principle of superposition. Professor Roychoudhuri focuses on applications requiring miniaturization and integration of various optical and photonics sensors exploiting spectral super resolution and other techniques. He worked for major United States corporations such as TRW, Perkin-Elmer, and United Technologies for 14 years and developed advanced optical systems for space and other applications. He has worked in academia for some 20 years in India, Mexico, and the United States.

Professor Roychoudhuri made pioneering contributions to laser multiplexing (20-channel WDM) and nonlinear optics for satellite and satellite–submarine communications technologies at TRW. He led the high power semiconductor laser phase locking program for Perkin–Elmer. At United Technologies and during his early years at University of Connecticut, he promoted various experimental concepts for laser machining, nonlinear optics, and two-photon fluorescence using phase-locked and directly pulsed diode lasers. Working with DARPA and the U.S. Air Force, he facilitated the spin-off of Infinite Photonics, a high power diode company, now restructured as Radiant Energy.

He served on the boards of directors of both SPIE and OSA. He is a fellow of SPIE and OSA, a member of IEEE-LEOS, and a life member of APS. He served as a key organizing chairperson for a special biannual conference series on "The Nature of Light: What Are Photons?" He was also the key motivator and cost-defraying fund raiser behind a special issue of *Optics and Photonics News,* dedicated by OSA in October 2003 to the education of senior level students.

Dr. A.F. Kracklauer was employed as a software engineer for McDonald-Douglas (NASA) and had a career as a technology development specialist and export control foreign service officer with the U.S. Government before becoming a private research consultant. He has had a longstanding interest in the foundations of physics and in using numerical simulation to study foundations issues in optics, quantum mechanics, and relativity. He is a member of the American Physics Society and has been published in various professional journals, including *Physics Review, Journal of Optics B,* and *Foundations of Physics Letters.*

Katherine Creath has PhDs in optical sciences and music from the University of Arizona. Her professional career began in industry, developing optical measurement instrumentation at Wyco Corporation where she is best known for her developments in optical metrology and a seminal monograph on phase-measuring interferometry.

She began teaching classes at the University of Arizona in the Optical Sciences Center (now known as the College of Optical Sciences) in 1986 and entered academia full time in 1991. She has taught classes in fundamentals of applied optics, optical testing, interferometry and holography, and currently teaches optical fabrication and testing.

While back in school to earn degrees in music in 1995 she began an optical engineering consultancy practice called Optineering and has since been active as a consultant. In the last twenty years she has been an internationally-recognized expert in optical measurement and, more recently, low-light level imaging. In the past decade her research interests have been focused on the development of instrumentation for energy and medicine research for the development of bioassays and therapeutic modalities in complimentary and alternative medicine.

Dr. Creath is a fellow of the Optical Society of America (OSA) and SPIE—the International Society for Optical Engineering. She is the author of more than 125 technical publications including 12 book chapters, 5 encyclopedia articles, and editor of 13 books, the best known of which is the *Encyclopedia of Optics*. She has always been fascinated with light and helps to organize and foster discussion at the philosophical and, especially, the experimental and empirical levels.

Contributors

Shahriar S. Afshar
Harvard University
Cambridge, Massachusetts, USA

Camil Alexandrescu
Physics Department
York University
Toronto, Ontario, Canada

David L. Andrews
Nanostructures and Photomolecular Systems
School of Chemical Sciences
University of East Anglia
Norwich, United Kingdom

Marco Bellini
Department of Physics
University of Florence
Florence, Italy

Yannick Bidel
Institut Nonlinéare de Nice
Valbonne, France

Thierry Chaneliére
Institut Nonlinéare de Nice
Valbonne, France

Katherine Creath
University of Arizona
Tucson, Arizona, USA

Edward Henry Dowdye, Jr.
National Aeronautics and Space Administration
Houston, Texas, USA

David Finkelstein
School of Physics
Georgia Institute of Technology
Atlanta, Georgia, USA

Geoffrey Hunter
Chemistry Department
York University
Toronto, Ontario, Canada

Robin Kaiser
Institut Nonlinéare de Nice
Valbonne, France

R. M. Kiehn
Physics Department
University of Houston
Houston, Texas, USA

Bruce Klappauf
Institut Nonlinéare de Nice
Valbonne, France

B. P. Kosyakov
Russian Federal Nuclear Center
Sarov, Russia

Marian Kowalski
Optech Inc.
Toronto, Ontario, Canada

A. F. Kracklauer
Weimar, Germany

Rodney Loudon
University of Essex
Colchester, United Kingdom

Holger Mack
Institute of Quantum Physics
University of Ulm
Ulm, Germany

Christian Miniatura
Institut Nonlinéare de Nice
Valbonne, France

Michael J. Mobley
The Biodesign Institute
Arizona State University
Tempe, Arizona, USA

Ashok Muthukrishnan
Institute for Quantum Studies
Department of Physics
Texas A&M University
College Station, Texas, USA

John M. Myers
Gordon McKay Laboratory
Harvard University
Cambridge, Massachusetts, USA

Emilio Panarella
Physics Essays
Ottawa, Ontario, Canada

C. Rangacharyulu
Department of Physics and Engineering Physics
University of Saskatchewan
Saskatoon, Saskatchewan, Canada

M. G. Raymer
Oregon Center for Optics
Department of Physics
University of Oregon
Eugene, Oregon, USA

Chandrasekhar Roychoudhuri
Photonics Laboratory
Physics Department
University of Connecticut
Storrs, Connecticut, USA

Emilio Santos
Department of Physics
University of Cantabria
Santander, Spain

Wolfgang P. Schleich
Institute of Quantum Physics
University of Ulm
Ulm, Germany

Marlan O. Scully
Departments of Chemistry and Aerospace and Mechanical Engineering
Princeton University
Princeton, New Jersey, USA

Brian J. Smith
Oregon Center for Optics
Department of Physics
University of Oregon
Eugene, Oregon, USA

Tuomo Suntola
Suntola Consulting Ltd.
Tampere University of Technology
Tampere, Finland

Silvia Viciani
Department of Physics
University of Florence
Florence, Italy

David Wilkowki
Institut Nonlinéare de Nice
Valbonne, France

Arthur Zajonc
Physics Department
Amherst College
Amherst, Massachusetts, USA

Alessandro Zavatta
Department of Physics
University of Florence
Florence, Italy

M. Suhail Zubairy
Department of Electronics
Quaid-i-Azam University
Islamabad, Pakistan

Section 1

Critical Reviews of Mainstream Photon Model

1

Light Reconsidered

Arthur Zajonc
Physics Department, Amherst College

CONTENTS

> I therefore take the liberty of proposing for this hypothetical new atom, which is not light but plays an essential part in every process of radiation, the name photon.[1]
>
> **Gilbert N. Lewis, 1926**

© 2003 Optical Society of America

Light is an obvious feature of everyday life, and yet light's true nature has eluded us for centuries. Near the end of his life Albert Einstein wrote, "All the fifty years of conscious brooding have brought me no closer to the answer to the question: What are light quanta? Of course today every rascal thinks he knows the answer, but he is deluding himself." We are today in the same state of "learned ignorance" with respect to light as was Einstein.

In 1926 when the chemist Gilbert Lewis suggested the name "photon," the concept of the light quantum was already a quarter of a century old. First introduced by Max Planck in December of 1900 in order to explain the spectral distribution of blackbody radiation, the idea of concentrated atoms of light was suggested by Einstein in his 1905 paper to explain the photoelectric effect. Four years later on September 21, 1909 at Salzburg, Einstein delivered a paper to the Division of Physics of German Scientists and Physicians on the same subject. Its title gives a good sense of its content: "On the development of our views concerning the nature and constitution of radiation."[2]

Einstein reminded his audience how great had been their collective confidence in the wave theory and the luminiferous ether just a few years earlier.

Now they were confronted with extensive experimental evidence that suggested a particulate aspect to light and the rejection of the ether outright. What had seemed so compelling was now to be cast aside for a new if as yet unarticulated view of light. In his Salzburg lecture he maintained "that a profound change in our views on the nature and constitution of light is imperative," and "that the next stage in the development of theoretical physics will bring us a theory of light that can be understood as a kind of fusion of the wave and emission theories of light." At that time Einstein personally favored an atomistic view of light in which electromagnetic fields of light were "associated with singular points just like the occurrence of electrostatic fields according to the electron theory." Surrounding these electromagnetic points he imagined fields of force that superposed to give the electromagnetic wave of Maxwell's classical theory. The conception of the photon held by many if not most working physicists today is, I suspect, not too different from that suggested by Einstein in 1909.

Others in the audience at Einstein's talk had other views of light. Among those who heard Einstein's presentation was Max Planck himself. In his recorded remarks following Einstein's lecture we see him resisting Einstein's hypothesis of atomistic light quanta propagating through space. If Einstein were correct, Planck asked, how could one account for interference when the length over which one detected interference was many thousands of wavelengths. How could a quantum of light interfere with itself over such great distances if it were a point object? Instead of quantized electromagnetic fields Planck maintained that "one should attempt to transfer the whole problem of the quantum theory to the area of *interaction* between matter and radiation energy." That is, only the exchange of energy between the atoms of the radiating source and the classical electromagnetic field is quantized. The exchange takes place in units of Planck's constant times the frequency, but the fields remain continuous and classical. In essence, Planck was holding out for a semi-classical theory in which only the atoms and their interactions were quantized while the free fields remained classical. This view has had a long and honorable history, extending all the way to the end of the 20th century. Even today we often use a semi-classical approach to handle many of the problems of quantum optics, including Einstein's photoelectric effect.[3]

The debate between Einstein and Planck as to the nature of light was but a single incident in the four thousand year inquiry concerning the nature of light.[4] For the ancient Egyptian light was the activity of their god Ra seeing. When Ra's eye (the Sun) was open, it was day. When it was closed, night fell. The dominant view in ancient Greece focused likewise on vision, but now the vision of human beings instead of the gods. The Greeks and most of their successors maintained that inside the eye a pure ocular fire radiated a luminous stream out into the world. This was the most important factor in sight. Only with the rise of Arab optics do we find strong arguments advanced against the extromissive theory of light expounded by the Greeks. For example around 1000 A.D. Ibn al-Haytham (Alhazen in the West) used his invention of the *camera obscura* to advocate for a view of light in which rays streamed from luminous sources traveling in straight lines to the screen or the eye.

By the time of the scientific revolution the debate as to the physical nature of light had divided into the two familiar camps of waves and particles. In broad strokes Galileo and Newton maintained a corpuscular view of light, while Huygens, Young and Euler advocated a wave view. The evidence supporting these views is well known.

1.1 The Elusive Single Photon

One might imagine that with the more recent developments of modern physics the debate would finally be settled and a clear view of the nature of light attained. Quantum electro-dynamics (QED) is commonly treated as the most successful physical theory ever invented, capable of predicting the effects of the interaction between changed particles and electro-magnetic radiation with unprecedented precision. While this is certainly true, what view of the photon does the theory advance? And how far does it succeed in fusing wave and particle ideas. In 1927 Dirac, one of the inventors of QED, wrote confidently of the new theory that, "There is thus a complete harmony between the wave and quantum descriptions of the interaction."[5] While in some sense quantum field theories do move beyond wave particle duality, the nature of light and the photon remains elusive. In order to support this I would like to focus on certain fundamental features of our understanding of photons and the philosophical issues associated with quantum field theory.[6]

In QED the photon is introduced as the unit of excitation associated with a quantized mode of the radiation field. As such it is associated with a plane wave of precise momentum, energy and polarization. Because of Bohr's principle of complementarity we know that a state of definite momentum and energy must be completely indefinite in space and time. This points to the first difficulty in conceiving of the photon. If it is a particle, then in what sense does it have a location? This problem is only deepened by the puzzling fact that, unlike other observables in quantum theory, there is no Hermetian operator that straightforwardly corresponds to position for photons. Thus while we can formulate a well-defined quantum-mechanical concept of position for electrons, protons and the like, we lack a parallel concept for the photon and similar particles with integer spin. The simple concept of spatio-temporal location must therefore be treated quite carefully for photons.

We are also accustomed to identifying an object by a unique set of attributes. My height, weight, shoe size, etc. uniquely identify me. Each of these has a well-defined value. Their aggregate is a full description of me. By contrast the single photon can, in some sense, take on multiple directions, energies and polarizations. Single-photon spatial interference and quantum beats require superpositions of these quantum descriptors for single photons. Dirac's refrain "photons interfere with themselves" while not universally true is a reminder of the importance of superposition. Thus the single photon should *not* be thought of as like a simple plane wave having a unique direction, frequency or polarization. Such states are rare special cases. Rather the superposition state for single photons is the common situation. Upon detection, of course, light appears as if discrete and indivisible possessing well-defined attributes. In transit things are quite otherwise.

Nor is the single photon state itself easy to produce. The anti-correlation experiments of Grangier, Roger and Aspect provide convincing evidence that with suitable care one can prepare single-photon states of light.[7] When sent to a beam splitter such photon states display the type of statistical correlations we would expect of particles. In particular the single photons appear to go one way or the other. Yet such single-photon states can interfere with themselves, even when run in "delayed choice."[8]

1.2 More Than One Photon

If we consider multiple photons the conceptual puzzles multiply as well. As spin one particles, photons obey Bose-Einstein statistics. The repercussions of this fact are very significant both for our conception of the photon and for technology. In fact Planck's law for the distribution of blackbody radiation makes use of Bose-Einstein statistics. Let us compare the statistics suited to two conventional objects with that of photons. Consider two marbles that are only distinguished by their colors: red (R) and green (G). Classically, four distinct combinations exist: RR, GG, RG and GR. In writing this we presume that although identical except for color, the marbles are, in fact, distinct because they are located at different places. At least since Aristotle we have held that two objects cannot occupy exactly the same location at the same time and therefore the two marbles, possessing distinct locations, are two distinct objects.

Photons by contrast are defined by the three quantum numbers associated with momentum, energy and polarization; position and time do not enter into consideration. This means that if two photons possess the same three values for these quantum numbers they are indistinguishable from one another. Location in space and in time is no longer a means for theoretically distinguishing photons as elementary particles. In addition, as bosons, any number of photons can occupy the same state, which is unlike the situation for electrons and other fermions. Photons do not obey the Pauli Exclusion Principle. This fact is at the foundation of laser theory because laser operation requires many photons to occupy a single mode of the radiation field.

To see how Bose-Einstein statistics differ from classical statistics consider the following example. If instead of marbles we imagine we have two photons in our possession which are distinguished by one of their attributes, things are quite different. For consistency with the previous example I label the two values of the photon attribute R and G. As required by Bose-Einstein

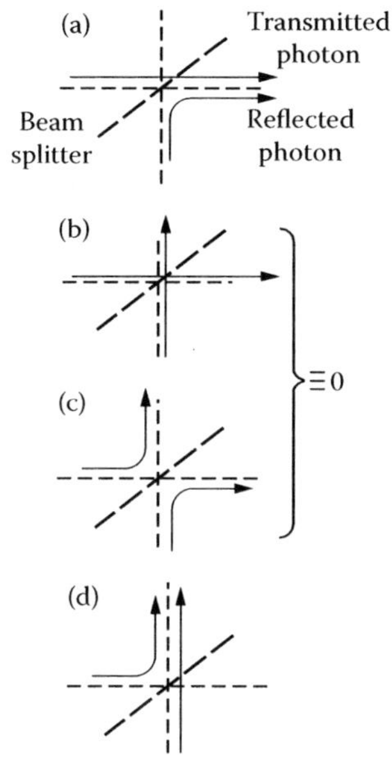

FIGURE 1.1
Copyright permission granted by *Nature*.[9]

statistics, the states available to the two photons are those that are symmetric states under exchange: RR, GG and ½(RG + GR). The states RG and GR are non-symmetric, while the combination ½(RG – GR) is anti-symmetric. These latter states are not suitable for photons. All things being equal we expect equal occupation for the three symmetric states with 1/3 as the probability for finding a pair of photons in each of the three states, instead of ¼ for the case of two marbles. This shows that is makes no sense to continue to think of photons as if they were "really" in classical states like RG and GR.

Experimentally we can realize the above situation by sending two photons onto a beam splitter. From a classical perspective there are four possibilities. They are sketched out in Fig. 1.1. We can label them RR for two right-going photons, UR for up and right, RU for right and up, and UU for the two photon going up. The quantum amplitudes for the UR and RU have opposite signs due the reflections which the photons undergo in Fig. 1.1c, which leads to destructive interference between these two amplitudes. The signal for one photon in each direction therefore vanishes. Surprisingly both photons are always found together. Another way of thinking about the experiment is in terms of the bosonic character of photons. Instead of thinking of the photons as having individual identities we should really think of there being three ways of pairing the two photons: two up (UU), two right (RR) and the symmetric combination (1/2(UR + RU)). All things being equal, we would expect the experiment to show an even distribution between the three options, 1/3 for each. But the experiment does not show this; why not? The answer is found in the opposite signs associated with UR and RU due to reflections. As a consequence the proper way to write the state for combination of b and c is ½(UR – RU). But this is anti-symmetric and therefore forbidden for photons which must have a symmetric state.

From this example we can see how Bose statistics confounds our conception of the identity of individual photons and rather treats them as aggregates with certain symmetry properties. These features are reflected in the treatment of photons in the formal mathematical language of Fock space. In this representation we only specify how many quanta are to found in each mode. All indexing of individual particles disappears.

1.3 Photons and Relativity

In his provocatively titled paper "Particles Do not Exist," Paul Davies advances several profound difficulties for any conventional particle conception of the photon, or for that matter for particles in general as they appear in relativistic quantum field theory.[10] One of our deepest tendencies is to reify the features that appear in our theories. Relativity confounds this habit of mind, and many of the apparent paradoxes of relativity arise because of our erroneous expectations due to this attitude. Every undergraduate is confused when, having mastered the electromagnetic theory of Maxwell he or she learns about Einstein treatment of the electrodynamics of moving bodies. The foundation of Einstein's revolutionary 1905 paper was his recognition that the values the electric and magnetic fields take on are always relative to the observer. That is, two observers in relative motion to one another will record on their measuring instruments different values of E and B for the same event. They will, therefore, give different casual accounts for the event. We habitually reify the electromagnetic field so that particular values of E and B are imagined as truly extent in space independent of any observer. In relativity we learn that in order for the laws of electromagnetism to be true in different inertial frames the values of the electric and magnetic fields (among other things) must change for different inertial frames. Matters only become more subtle when we move to accelerating frames.

Davies gives special attention to the problems that arise for the photon and other quanta in relativistic quantum field theory. For example, our concept of reality has, at its root, the notion that either an object exists or it does not. If the very existence of a thing is ambiguous, in what sense is it real? Exactly this is challenged by quantum field theory. In particular the quantum vacuum is the state in which no photons are present in any of the modes of the radiation field. However the vacuum only remains empty of particles for inertial observers. If instead we posit an observer in a uniformly accelerated frame of reference, then what was a vacuum state becomes a thermal bath of photons for the accelerated observer. And what is true for accelerated observers is similarly true for regions of space-time curved by gravity.

Davies uses these and other problems to argue for a vigorous Copenhagen interpretation of quantum mechanics that abandons the idea of a "particle as a really existing thing skipping between measuring devices."

To my mind, Einstein was right to caution us concerning light. Our understanding of it has increased enormously in the 100 years since Planck, but I suspect light will continue to confound us, while simultaneously luring us to inquire ceaselessly into its nature.

References

1. Gilbert N. Lewis, *Nature,* vol. 118, Part 2, December 18, 1926, pp. 874–875. [What Lewis meant by the term photon was quite different from our usage.]
2. *The Collected Papers of Albert Einstein,* vol. 2, translated by Anna Beck (Princeton, NJ: Princeton University Press, 1989), pp. 379–98.
3. George Greenstein and Arthur Zajonc, *The Quantum Challenge, Modern Research on the Foundations of Quantum Mechanics,* 2nd ed. (Boston, MA: Jones & Bartlett, 2007); T. H. Boyer, Scientific American, "The Classical Vacuum" August 1985, 253(2) pp. 56–62.
4. For a full treatment of the history of light see Arthur Zajonc, *Catching the Light, the Entwined History of Light and Mind* (NY: Oxford University Press, 1993).
5. P. A. M. Dirac, *Proceedings of the Royal Society (London)* A**114** (1927) pp. 243–65.
6. See Paul Teller, *An Interpretive Introduction to Quantum Field Theory* (Princeton, NJ: Princeton University Press, 1995).
7. P. Grangier, G. Roger and A. Aspect, *Europhysics Letters,* vol. 1, (1986) pp. 173–179.
8. T. Hellmuth, H. Walther, A. Zajonc, and W. Schleich, Phys. Rev. A, vol. 35, (1987) pp. 2532–41.
9. Figure is from Philippe Grangier, "Single Photons Stick Together," *Nature* 419, p. 577 (10 Oct 2002).
10. P. C. W. Davies, *Quantum Theory of Gravity,* edited by Steven M. Christensen (Bristol: Adam Hilger, 1984).

2

What Is a Photon?

Rodney Loudon
University of Essex, Colchester CO4 3SQ, United Kingdom

CONTENTS

The concept of the photon is introduced by discussion of the process of electromagnetic field quantization within a closed cavity or in an open optical system. The nature of a single-photon state is clarified by consideration of its behavior at an optical beam splitter. The importance of linear superposition or entangled states in the distinctions between quantum-mechanical photon states and classical excitations of the electromagnetic field is emphasized. These concepts and the ideas of wave–particle duality are illustrated by discussions of the effects of single-photon inputs to Brown–Twiss and Mach–Zehnder interferometers. Both the theoretical predictions and the confirming experimental observations are covered. The defining property of the single photon in terms of its ability to trigger one, and only one, photodetection event is discussed.

The development of theories of the nature of light has a long history, whose main events are well reviewed by Lamb[1]. The history includes strands of argument in favor of either a particle or a wave view of light. The realm of *classical optics* includes all of the phenomena that can be understood and interpreted on the basis of classical wave and particle theories. The conflicting views of the particle or wave essence of light were reconciled by the establishment of the quantum theory, with its introduction of the idea that all excitations simultaneously have both particle-like and wave-like properties. The demonstration of this dual behavior in the real world of experimental physics is, like so many basic quantum-mechanical phenomena, most readily achieved in optics. The fundamental properties of the photon, particularly the discrimination of its particle-like and wave-like properties, are most clearly illustrated by observations based on the use of beam splitters. The realm of *quantum optics* includes all of the phenomena that are not embraced by classical optics and require the quantum theory for their understanding and interpretation. The aim of the present article is to try to clarify the nature of the photon by considerations of electromagnetic fields in optical cavities or in propagation through free space.

2.1 Single Photons and Beam Splitters

A careful description of the nature of the photon begins with the electromagnetic field inside a *closed* optical resonator, or perfectly-reflecting cavity. This is the system usually assumed in textbook derivations of Planck's radiation law[2]. The field excitations in the cavity are limited to an infinite discrete set of spatial modes determined by the boundary conditions at the cavity walls. The allowed standing-wave spatial variations of the electromagnetic field in the cavity are identical in the classical and quantum theories. However, the time dependence of each mode is governed by the equation of motion of a harmonic oscillator, whose solutions take different forms in the classical and quantum theories.

Unlike its classical counterpart, a quantum harmonic oscillator of angular frequency ω can only be excited by energies that are integer multiples of $\hbar\omega$. The integer n thus denotes the number of energy quanta excited in the oscillator. For application to the electromagnetic field, a single spatial mode whose associated harmonic oscillator is in its nth excited state unambiguously contains n photons, each of energy $\hbar\omega$. Each photon has a spatial distribution within the cavity that is proportional to the square modulus of the complex field amplitude of the mode function. For the simple, if unrealistic, example of a one-dimensional cavity bounded by perfectly reflecting mirrors, the spatial modes are standing waves and the photon may be found at any position in the cavity except the nodes. The single-mode photons are said to be *delocalized*.

These ideas can be extended to *open* optical systems, where there is no identifiable cavity but where the experimental apparatus has a finite extent determined by the sources, the transverse cross sections of the light beams, and the detectors. The discrete standing-wave modes of the closed cavity are replaced by discrete travelling-wave modes that propagate from sources to detectors. The simplest system to consider is the optical beam splitter, which indeed is the central component in many of the experiments that study the quantum nature of light. Fig. 2.1 shows a representation of a lossless beam splitter, with two input arms denoted 1 and 2 and two output arms denoted 3 and 4. An experiment to distinguish the classical and quantum natures of

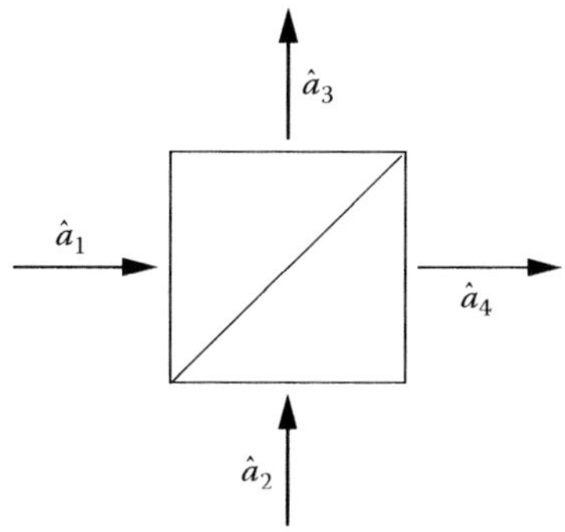

FIGURE 2.1
Schematic representation of an optical beam splitter showing the notation for the field operators in the two input and two output arms. In practice the beam-splitter cube is often replaced by a partially reflecting plate at 45° or a pair of optical fibers in contact along a fused section.

light consists of a source that emits light in one of the input arms and which is directed by the beam splitter to detectors in the two output arms. The relevant spatial modes of the system in this example include a joint excitation of the selected input arm and both output arms.

The operators $\hat{a}_i$ in Fig. 2.1 are the *photon destruction operators* for the harmonic oscillators associated with the two input $(i = 1, 2)$ and two output $(i = 3, 4)$ arms. These destruction operators essentially represent the amplitudes of the quantum electromagnetic fields in the four arms of the beam splitter, analogous to the complex classical field amplitudes. The real electric-field operators of the four arms are proportional to the sum of $\hat{a}_i \exp(-i\omega t)$ and the Hermitean conjugate operators $\hat{a}_i^\dagger \exp(i\omega t)$. The proportionality factor includes Planck's constant $\hbar$, the angular frequency ω, and the permittivity of free space ε_0, but its detailed form does not concern us here. For the sake of brevity, we refer to $\hat{a}_i$ as the *field* in arm i. The operator $\hat{a}_i^\dagger$ is the *photon creation operator* for arm i and it has the effect of generating a single-photon state $|1\rangle_i$ in arm i, according to

$$\hat{a}_i^\dagger |0\rangle = |1\rangle_i. \tag{2.1}$$

Here $|0\rangle$ is the *vacuum state* of the entire input–output system, which is defined as the state with no photons excited in any of the four arms.

The relations of the output to the input fields at a symmetric beam splitter have forms equivalent to those of classical theory,

$$\hat{a}_3 = R\hat{a}_1 + T\hat{a}_2 \quad \text{and} \quad \hat{a}_4 = T\hat{a}_1 + R\hat{a}_2, \tag{2.2}$$

where R and T are the reflection and transmission coefficients of the beam splitter. These coefficients are generally complex numbers that describe the amplitudes and phases of the reflected and transmitted light relative to those of the incident light. They are determined by the boundary conditions for the electromagnetic fields at the partially transmitting and partially reflecting interface within the beam splitter. The boundary conditions are the same for classical fields and for the quantum-mechanical field operators $\hat{a}_i$. It follows that the coefficients satisfy the standard relations[3]

$$|R|^2 + |T|^2 = 1 \quad \text{and} \quad RT^* + TR^* = 0. \tag{2.3}$$

It can be shown[2] that these beam-splitter relations ensure the conservation of optical energy from the input to the output arms, in both the classical and quantum forms of beam-splitter theory.

The essential property of the beam splitter is its ability to convert an input photon state into a *linear superposition* of output states. This is a basic quantum-mechanical manipulation that is less easily achieved and studied in other physical systems. Suppose that there is one photon in input arm 1 and no photon in input arm 2. The beam splitter converts this joint input state to the output state determined by the simple calculation

$$|1\rangle_1 |0\rangle_2 = \hat{a}_1^\dagger |0\rangle = (R\hat{a}_3^\dagger + T\hat{a}_4^\dagger)|0\rangle = R|1\rangle_3 |0\rangle_4 + T|0\rangle_3 |1\rangle_4, \tag{2.4}$$

where $|0\rangle$ is again the vacuum state of the entire system. The expression for $\hat{a}_1^\dagger$ in terms of output arm operators is obtained from the Hermitean conjugates of the relations in eqn 2.2 with the use of eqn 2.3. In words, the state on the right is a superposition of the state with one photon in arm 3 and nothing in arm 4, with probability amplitude R, and the state with one photon in arm 4 and nothing in arm 3, with amplitude T. This conversion of the input state to a linear superposition of the two possible output states is the basic quantum-mechanical process performed by the beam splitter. In terms of travelling-wave modes, this example combines the input-arm excitation on the left of eqn 2.4 with the output-arm excitation on the right of eqn 2.4 to form a joint single-photon excitation of a mode of the complete beam-splitter system.

Note that the relevant spatial mode of the beam splitter, with light incident in arm 1 and outputs in arms 3 and 4, is the same in the classical and quantum theories. What is quantized in the latter theory is the energy content of the electromagnetic field in its distribution over the complete spatial extent of the mode. In the classical theory, an incident light beam of intensity I_1 excites the two outputs with intensities $|R|^2 I_1$ and $|T|^2 I_1$, in contrast to the excitation of the quantum state shown on the right of eqn 2.4 by a single incident photon. A state of this form, with the property that each contribution to the superposition is a product of states of different subsystems (output arms), is said to be *entangled*. Entangled states form the basis of many of the applications of quantum technology in information transfer and processing[4].

2.2 Brown–Twiss Interferometer

The experiment described in essence by eqn 2.4 above is performed in practice by the use of a kind of interferometer first constructed by Brown and Twiss in the 1950s. They were not able to use a single-photon input but their apparatus was essentially that illustrated in Fig. 2.1 with light from a mercury arc incident in arm 1. Their interest was in measurements of the angular diameters of stars by interference of the intensities of starlight[5] rather than the interference of field amplitudes used in traditional classical interferometers. The techniques they developed work well with the random multiphoton light emitted by arcs or stars.

However, for the study of the quantum entanglement represented by the state on the right of eqn 2.4, it is first necessary to obtain a single-photon input state, and herein lies the main difficulty of the experiment. It is true, of course, that most sources emit light in single-photon processes but the sources generally contain large numbers of emitters whose emissions occur at random times, such that the experimenter cannot reliably isolate a single photon. Even when an ordinary light beam is heavily attenuated, statistical analysis shows that single-photon effects cannot be detected by the apparatus in Fig. 2.1. It is necessary to find a way of identifying the presence of one and only one photon. The earliest reliable methods of single-photon generation depended on optical processes that generate photons in pairs. Thus, for example, the nonlinear optical process of parametric down conversion[6] replaces a single incident photon by a pair of photons whose frequencies sum to that of the incident photon to ensure energy conservation. Again, two-photon cascade emission is a process in which an excited atom decays in two steps, first to an intermediate energy level and then to the ground state, emitting two photons in succession with a delay determined by the lifetime of the intermediate state[7]. If one of the photons of the pair produced by these processes is detected, it is known that the other photon of the pair must be present more-or-less simultaneously. For a two-photon source sufficiently weak that the time separation between one emitted pair and the next is longer than the resolution time of the measurement, this second photon can be used as the input to a single-photon experiment. More versatile single-photon light sources are now available[8].

The arrangement of the key single-photon beam-splitter experiment[9] is represented in Fig. 2.2. Here, the two photons came from cascade emission in an atomic Na light source S and one of them activated photodetector D. This first detection opened an electronic gate that activated the recording of the responses of two detectors in output arms 3 and 4 of the Brown–Twiss beam splitter. The gate was closed again after a period of time sufficient for the photodetection. The experiment was repeated many times and the results were processed to determine the average values of the mean photocounts $\langle n_3 \rangle$ and $\langle n_4 \rangle$ in the two arms and the average value $\langle n_3 n_4 \rangle$ of their correlation product. It is convenient to work with the normalized correlation

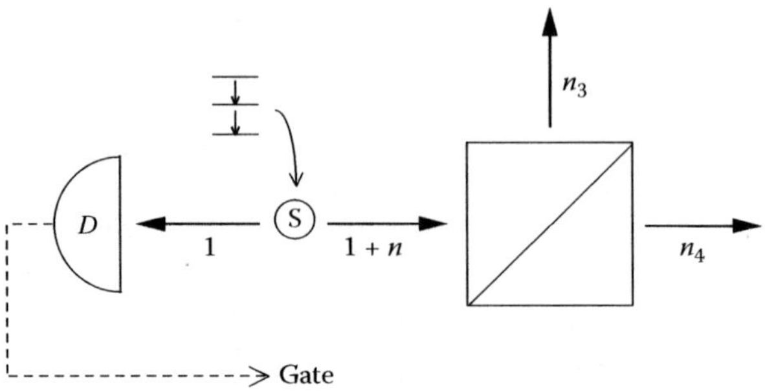

FIGURE 2.2
Brown–Twiss interferometer using a single-photon input obtained from cascade emission with an electronic gate.

$\langle n_3 n_4 \rangle / \langle n_3 \rangle \langle n_4 \rangle$, which is independent of the detector efficiencies and beam splitter reflection and transmission coefficients. In view of the physical significance of the entangled state in eqn 2.4, the single-photon input should lead to a single photon either in arm 3 or arm 4 but never a photon in both output arms. The correlation $\langle n_3 n_4 \rangle$ should therefore ideally vanish.

However, in the real world of practical experiments, a purely single-photon input is difficult to achieve. In addition to the twin of the photon that opens the gate, n additional 'rogue' photons may enter the Brown–Twiss interferometer during the period that the gate is open, as represented in Fig. 2.2. These rogue photons are emitted randomly by other atoms in the cascade light source and their presence allows two or more photons to pass through the beam splitter during the detection period. Fig. 2.3 shows experimental results for the normalized correlation, with its dependence on the average number $\langle n \rangle$ of additional photons that enter the interferometer for different gate periods. The continuous curve shows the calculated value of the correlation in the presence of the additional rogue photons. It is seen that both experiment

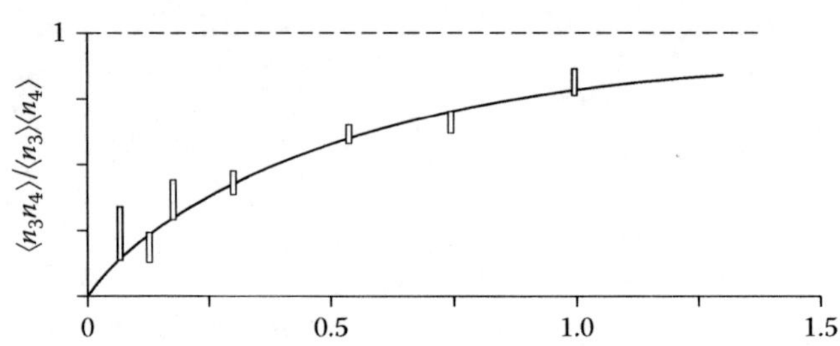

FIGURE 2.3
Normalized output correlation as a function of the average additional photon number $\langle n \rangle$, as measured in the experiment represented in Fig. 2.2 (after ref. 9).

and theory agree on the tendency of the correlation to zero as $\langle n \rangle$ becomes very small, in confirmation of the quantum expectation of the particle-like property of the output photon exciting only one of the output arms.

2.3 Mach–Zehnder Interferometer

The excitation of one photon in a single travelling-wave mode is also frequently considered in the discussion of the quantum theory of the traditional classical amplitude-interference experiments, for example Young's slits or the Michelson and Mach–Zehnder interferometers. Each classical or quantum spatial mode in these systems includes input light waves, *both* paths through the interior of the interferometer, and output waves appropriate to the geometry of the apparatus. A one-photon excitation in such a mode again carries an energy quantum $\hbar\omega$ distributed over the entire interferometer, including both internal paths. Despite the absence of any localization of the photon, the theory provides expressions for the distributions of light in the two output arms, equivalent to a determination of the interference fringes.

The arrangement of a Mach–Zehnder interferometer with a single-photon input is represented in Fig. 2.4. The two beam splitters are assumed to be symmetric and identical, with the properties given in eqn 2.3. The complete interferometer can be regarded as a composite beam splitter, whose two output fields are related to the two input fields by

$$\hat{a}_3 = R_{MZ}\hat{a}_1 + T_{MZ}\hat{a}_2 \quad \text{and} \quad \hat{a}_4 = T_{MZ}\hat{a}_1 + R'_{MZ}\hat{a}_2, \tag{2.5}$$

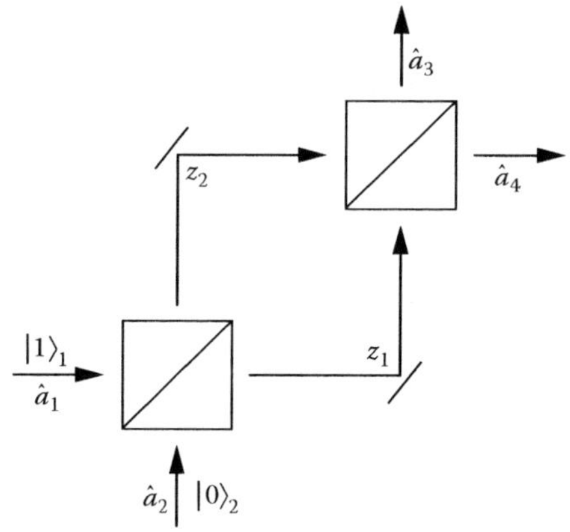

FIGURE 2.4
Representation of a Mach–Zehnder interferometer showing the notation for input and output field operators and the internal path lengths.

similar to eqn 2.2 but with different reflection coefficients in the two relations. Without going into the details of the calculation[2], we quote the quantum result for the average number of photons in output arm 4 when the experiment is repeated many times with the same internal path lengths z_1 and z_2,

$$\langle n_4 \rangle = |T_{MZ}|^2 = |RT(e^{i\omega z_1/c} + e^{i\omega z_2/c})|^2 = 4\,|R|^2|T|^2 \cos^2[\omega(z_1 - z_2)/2c]. \quad (2.6)$$

The fringe pattern is contained in the trigonometric factor, which has the same dependence on frequency and relative path length as found in the classical theory. Fig. 2.5 shows the fringe pattern measured with the same techniques as used for the Brown–Twiss experiment of Figs. 2.2 and 2.3. The average photon count $\langle n_4 \rangle$ in output arm 4 was determined[9] by repeated measurements for each relative path length. The two parts of Fig. 2.5 show the improvements in fringe definition gained by a fifteenfold increase in the number of measurements for each setting.

The existence of the fringes seems to confirm the wave-like property of the photon and we need to consider how this behavior is consistent with the particle-like properties that show up in the Brown–Twiss interferometer. For the Mach–Zehnder interferometer, each incident photon must propagate through the apparatus in such a way that the probability of its leaving

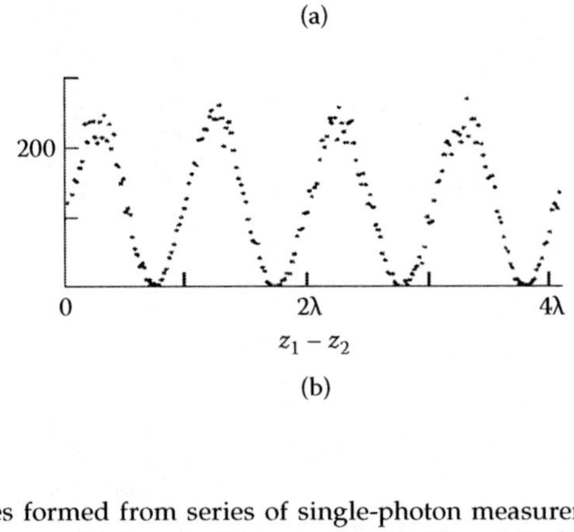

FIGURE 2.5
Mach–Zehnder fringes formed from series of single-photon measurements as a function of the path difference expressed in terms of the wavelength. The vertical axis shows the number of photodetections in arm 4 for (a) 1 sec and (b) 15 sec integration times per point. The latter fringes have 98% visibility (after ref. 9).

the interferometer by arm 4 is proportional to the calculated mean photon number in eqn 2.6. This is achieved only if each photon excites both internal paths of the interferometer, so that the input state at the second beam splitter is determined by the complete interferometer geometry. This geometry is inherent in the entangled state in the output arms of the first beam splitter from eqn 2.4, with the output labels 3 and 4 replaced by internal path labels, and in the propagation phase factors for the two internal paths shown in T_{MZ} in eqn 2.6. The photon in the Mach–Zehnder interferometer should thus be viewed as a composite excitation of the appropriate input arm, internal paths and output arms, equivalent to the spatial field distribution produced by illumination of the input by a classical light beam. The interference fringes are thus a property not so much of the photon itself as of the spatial mode that it excites.

The internal state of the interferometer excited by a single photon is the same as that investigated by the Brown–Twiss experiment. There is, however, no way of performing both kinds of interference experiment simultaneously. If a detector is placed in one of the output arms of the first beam splitter to detect photons in the corresponding internal path, then it is not possible to avoid obscuring that path, with consequent destruction of the interference fringes. A succession of suggestions for more and more ingenious experiments has failed to provide any method for simultaneous fringe and path observations. A complete determination of the one leads to a total loss of resolution of the other, while a partial determination of the one leads to an accompanying partial loss of resolution of the other[10].

2.4 Detection of Photon Pulses

The discussion so far is based on the idea of the photon as an excitation of a single traveling-wave mode of the complete optical system considered. Such an excitation is independent of the time and it has a nonzero probability over the whole system, apart from isolated interference nodes. This picture of delocalized photons gives reasonably correct results for the interference experiments treated but it does not provide an accurate representation of the physical processes in real experiments. The typical light source acts by spontaneous emission and this is the case even for the two-photon emitters outlined above. The timing of an emission is often determined by the random statistics of the source but, once initiated, it occurs over a limited time span Δt and the light is localized in the form of a *pulse* or *wavepacket*. The light never has a precisely defined angular frequency and ω is distributed over a range of values $\Delta\omega$ determined by the nature of the emitter, for example by the radiative lifetime for atoms or by the geometry of the several beams involved in a nonlinear-optical process. The minimum values of pulse duration and frequency spread are related by Fourier transform theory such that their product $\Delta t \Delta\omega$ must have a value at least of order unity.

The improved picture of the photon thus envisages the excitation of a pulse that is somewhat localized in time and involves several traveling-wave modes of the optical system. These modes are exactly the same as the collection of those used in single-mode theory and they are again the same as the spatial modes of classical theory. Their frequency separation is often small compared to the wavepacket frequency spread $\Delta\omega$, and it is convenient to treat their frequency ω as a continuous variable. The theories of optical interference experiments based on these single-photon continuous-mode wavepackets are more complicated than the single-mode theories but they provide more realistic descriptions of the measurements. For example, the frequency spread of the wavepacket leads to a blurring of fringe patterns and its limited time span may lead to a lack of simultaneity in the arrival of pulses by different paths, with a destruction of interference effects that depend on their overlap.

The good news is that the single-mode interference effects outlined above survive the change to a wavepacket description of the photon for optimal values of the pulse parameters. The discussions of the physical significances of the Brown–Twiss and Mach–Zehnder interference experiments in terms of particle-like and wave-like properties thus remain valid. However, some of the concepts of single-mode theory need modification. Thus, the single-mode photon creation operator $\hat{a}^{\dagger}$ is replaced by the *photon wavepacket creation operator*

$$\hat{a}_{\xi}^{\dagger} = \int d\omega \xi(\omega) \hat{a}^{\dagger}(\omega), \tag{2.7}$$

where $\xi(\omega)$ is the spectral amplitude of the wavepacket and $\hat{a}^{\dagger}(\omega)$ is the continuous-mode creation operator. The integration over frequencies replaces the idea of a single energy quantum $\hbar\omega$ in a discrete mode by an average quantum $\hbar\omega_0$, where ω_0 is an average frequency of the wavepacket spectrum $|\xi(\omega)|^2$.

The main change in the description of experiments, however, lies in the theory of the optical detection process[2]. For the detection of photons by a phototube, the theory must allow for its switch-on time and its subsequent switch-off time; the difference between the two times is the *integration time*. The more accurate theory includes the need for the pulse to arrive during an integration time in order for the photon to be detected. More importantly, it shows that the single-photon excitation created by the operator defined in eqn 2.7 can at most trigger a single detection event. Such a detection only occurs with certainty, even for a 100% efficient detector, in conditions where the integration time covers essentially all of the times for which the wavepacket has significant intensity at the detector. Of course, this feature of the theory merely reproduces some obvious properties of the passage of a photon wavepacket from a source to a detector but it is nevertheless gratifying to have a realistic representation of a practical experiment. Real phototubes miss some fraction of the incident wavepackets, but the effects of detector efficiencies of less than 100% are readily included in the theory[2].

2.5 So What Is a Photon?

The question posed by this chapter has a variety of answers, which hopefully converge to a coherent picture of this somewhat elusive object. The present article reviews a series of three physical systems in which the spatial distribution of the photon excitation progresses from a single discrete standing-wave mode in a closed cavity to a single discrete traveling-wave mode of an open optical system to a traveling pulse or wavepacket. The first two excitations are spread over the complete optical system but the wavepacket is localized in time and contains a range of frequencies. All of these spatial distributions of the excitation are the same in the classical and quantum theories. What distinguishes the quantum theory from the classical is the limitation of the energy content of the discrete-mode systems to integer multiples of the $\hbar\omega$ quantum. The physically more realistic wavepacket excitation also carries a basic energy quantum $\hbar\omega_0$, but ω_0 is now an average of the frequencies contained in its spectrum. The single-photon wavepacket has the distinguishing feature of causing at most a single photodetection and then only when the detector is in the right place at the right time.

It cannot be emphasized too strongly that the spatial modes of the optical system, classical and quantum, include the combinations of all routes through the apparatus that are excited by the light sources. In the wavepacket picture, a single photon excites this complete spatial distribution, however complicated, and what is measured by a detector is determined both by its position within the complete system and by the time dependence of the excitation. The examples outlined here show how particle-like and wave-like aspects of the photon may appear in suitable experiments, without any conflict between the two.

Acknowledgment

Figures 2.1, 2.2, and 2.4 are reproduced from reference 2 by permission of Oxford University Press and Figures 2.3 and 2.5 from reference 9 by permission of EDP Sciences.

References

1. W.E. Lamb, Jr., Anti-photon, *Appl. Phys. B 60,* 77–84 (1995).
2. R. Loudon, *The Quantum Theory of Light,* 3rd edn (University Press, Oxford, 2000).

3. M. Mansuripur, *Classical Optics and its Applications* (University Press, Cambridge, 2002).
4. I. Walmsley and P. Knight, Quantum information science, OPN 43–9 (November 2002).
5. R.H. Brown, *The Intensity Interferometer* (Taylor & Francis, London, 1974).
6. D.C. Burnham and D.L. Weinberg, Observation of simultaneity in parametric production of optical photon pairs, *Phys. Rev. Lett.* 25, 84–7 (1970).
7. J.F. Clauser, Experimental distinction between the quantum and classical field-theoretic predictions for the photoelectric effect, *Phys. Rev. D9*, 853–60 (1974).
8. P. Grangier and I. Abram, Single photons on demand, *Physics World 31–5* (February 2003).
9. P. Grangier, G. Roger and A. Aspect, Experimental evidence for a photon anti-correlation effect on a beam splitter: a new light on single-photon interferences, *Europhys. Lett. 1*, 173–9 (1986).
10. M.O. Scully and M.S. Zubairy, *Quantum Optics* (University Press, Cambridge, 1997).

3

What Is a Photon?

David Finkelstein
School of Physics, Georgia Institute of Technology, Atlanta, Georgia 30032

CONTENTS

Modern developments in the physicist's concept of nature have expanded our understanding of light and the photon in ever more startling directions. We take up expansions associated with the established physical constants c, $\hbar$, G, and two proposed "transquantum" constants $\hbar'$, $\hbar''$.

© 2003 Optical Society of America

From the point of view of experience, "What is a photon?" is not the best first question. We never experience a photon as it "is." For example, we never see a photon in the sense that we see an apple, by scattering diffuse light off it and forming an image of it on our retina. What we experience is what photons do. A better first question is "What do photons do?" After we answer this we can define what photons are, if we still wish to, by what they do.

Under low resolution the transport of energy, momentum and angular momentum by electromagnetic radiation often passes for continuous but under sufficient resolution it breaks down into discrete jumps, quanta. Radiation is not the only way that the electromagnetic field exerts forces; there are also Coulomb forces, say, but only the radiation is quantized. Even our eyes, when adapted sufficiently to the dark, see any sufficiently dim light as a succession of scintillations. What photons do is couple electric charges and electric or magnetic multipoles by discrete irreducible processes of photon emission and absorption connected by continuous processes of propagation. All electromagnetic radiation resolves into a flock of flying photons, each carrying its own energy, momentum and angular momentum.

Francis Bacon and Isaac Newton were already certain that light was granular in the 17th century but hardly anyone anticipated the radical conceptual expansions in the physics of light that happened in the 20th century. Now a simple extrapolation tells us to expect more such expansions.

These expansions have one basic thing in common: Each revealed that the resultant of a sequence of certain processes depends unexpectedly on their order. Processes are said to *commute* when their resultant does *not* depend on their order, so what astounded us each time was a non-commutativity. Each such discovery was made without connection to the others, and the phenomenon of non-commutativity was called several things, like non-integrability, inexactness, anholonomy, curvature, or paradox (of two twins, or two slits). These aliases must not disguise this underlying commonality. Moreover the prior commutative theories are unstable relative to their non-commutative successors in the sense that an arbitrarily small change in the commutative commutation relations can change the theory drastically,[9] but not in the non-commutative relations.

Each of these surprising non-commutativities is proportional to its own small new fundamental constant. The expansion constants and non-commutativities most relevant to the photon so far have been k (Boltzmann's constant, for the kinetic theory of heat) c (light speed, for special relativity), G (gravitational constant, for general relativity), h (Planck's constant, for quantum theory), e (the electron charge, for the gauge theory of electromagnetism), g (the strong coupling constant) and W (the mass of the W particle, for the electroweak unification). These constants are like flags. If we find a c in an equation, for instance, we know we are in the land of special relativity. The historic non-commutativities introduced by these expansions so far include those of reversible thermodynamic processes (for k), boosts (changes in the velocity of the observer, for c), filtration or selection processes (for h), and space-time displacements (of different kinds of test-particles for G, e, and g).

Each expansion has its inverse process, a *contraction* that reduces the fundamental constant to 0, recovering an older, less accurate theory in which the processes commute.[6] Contraction is a well-defined mathematical process. Expansion is the historical creative process, not a mathematically well-posed problem. When these constants are taken to 0, the theories "contract" to their more familiar forms; but in truth the constants are not 0, and the expanded theory is more basic than the familiar one, and is a better starting point for further exploration.

Einstein was the magus of these expansions, instrumental in raising the flags of k, c, G and h. No one comes close to his record. In particular he brought the photon back from the grave to which Thomas Young's diffraction studies had consigned it, though he never accommodated to the h expansion.

Each expansion establishes a reciprocity between mutually coupled concepts that was lacking before it, such as that between space and time in special relativity. Each thereby dethroned a false absolute, an unmoved mover, what Frances Bacon called an "idol," usually an "idol of the theater." Each made physics more relativistic, more processual, less mechanical.

There is a deeper commonality to these expansions. Like earthquakes and landslides, they stabilize the region where they occur, specifically against small changes in the expansion constant itself.

Each expansion also furthered the unity of physics in the sense that it replaced a complicated kind of symmetry (or group) by a simple one.

Shifting our conceptual basis from the familiar idol-ridden theory to the strange expanded theory has generally led to new and deeper understanding. The Standard Model, in particular, gives the best account of the photon we have today, combining expansions of quantum theory, special relativity, and gauge theory, and it shows signs of impending expansions as drastic as those of the past. Here we describe the photon as we know it today and speculate about the photon of tomorrow.

1. ***c*** The expansion constant c of special relativity, the speed of light, also measures how far the photon flouts Euclid's geometry and Galileo's relativity. In the theory of space-time that immediately preceded the c expansion, associated with the relativity theory of Galileo, reality is a collection of objects or fields distributed over space at each time, with the curious codicil that different observers in uniform relative motion agree about simultaneity – having the same time coordinate – but not about colocality – having the same space coordinates. One could imagine history as a one-dimensional stack of three-dimensional slices. If V is a boost vector, giving the velocity of one observer O' relative to another O, then in Galileo relativity: $x' = x - Vt$ but $t' = t$. The transformation $x' = x - Vt$ couples time into space but the transformation $t' = t$ does not couple space into time. O and O' slice history the same way but stack the slices differently.

Special relativity boosts couple time into space and space back into time, restoring reciprocity between space and time. The very constancy of c implies this reciprocity. Relatively moving observers may move different amounts during the flight of a photon and so may disagree on the distance Δx covered by a photon, by an amount depending on Δt. In order to agree on the speed $c = \Delta x/\Delta t$, they must therefore disagree on the duration Δt as well, and by the same factor. They slice history differently.

We could overlook this fundamental reciprocity for so many millennia because the amount by which space couples into time has a coefficient $1/c^2$ that is small on the human scale of the second, meter, and kilogram. When $c \to \infty$ we recover the old relativity of Galileo.

The c non-commutativity is that between two boosts B, B' in different directions. In Galileo relativity $BB' = B'B$; one simply adds the velocity vectors v and v' of B and B' to compute the resultant boost velocity $v + v' = v' + v$ of BB' or $B'B$. In special relativity BB' and $B'B$ differ by a rotation in the plane of the two boosts, called Thomas precession, again with a coefficient $1/c^2$.

The reciprocity between time and space led to a parallel one between energy and momentum, and to the identification of mass and energy. The photon has both. The energy and momentum of a particle are related to the rest-mass m_0 in special relativity by $E^2 - c^2p^2 = (m_0c^2)^2$. The parameter m_0 is 0 for the photon, for which $E = cp$. When we say that the photon "has mass 0," we speak elliptically. We mean that it has rest-mass 0. Its mass is actually E/c^2.

Some say that a photon is a bundle of energy. This statement is not meaningful enough to be wrong. In physics, energy is one property of a system among many others. Photons have energy as they have spin and momentum and cannot *be* energy any more than they can be spin or momentum. In the late 1800's some thinkers declared that all matter is made of one philosophical stuff that they identified with energy, without much empirical basis. The theory is dead but its words linger on.

When we speak of a reactor converting mass into energy, we again speak elliptically and archaically. Strictly speaking, we can no more heat our house by converting mass into energy than by converting Centigrade into Fahrenheit. Since the c expansion, mass *is* energy. They are the same conserved stuff, mass-energy, in different units. Neither ox-carts nor nuclear reactors convert mass into energy. Both convert rest mass-energy into kinetic mass-energy.

2. *G* In special relativity the light rays through the origin of space-time form a three-dimensional cone in four dimensions, called the light cone, whose equation is $c^2t^2 - x^2 - y^2 - z^2 = 0$. Space-time is supposed to be filled with such light cones, one at every point, all parallel, telling light where it can go. This is a reciprocity failure of special relativity: Light cones influence light, light does not influence light cones. The light-cone field is an idol of special relativity.

In this case general relativity repaired reciprocity. An acceleration $\mathbf{a}$ of an observer is equivalent to a gravitational field $\mathbf{g} = -\mathbf{a}$ in its local effects. Even in the presence of gravitation, special relativity still describes correctly the infinitesimal neighborhood of each space-time point. Since an acceleration clearly distorts the field of light cones, and gravity is locally equivalent to acceleration, Einstein identified gravity with such a distortion. In his G expansion, which is general relativity, the light-cone field is as much a dynamical variable as the electromagnetic field, and the two fields influence each other reciprocally, to an extent proportional to Newton's gravitational constant G.

The light-cone directions dx at one point x can be defined by the vanishing of the norm $d\tau^2 = \Sigma_{\mu\nu} g_{\mu\nu}(x) dx^\mu dx^\nu = 0$; since Einstein, one leaves such summation signs implicit. General relativity represents gravity in each frame by the coefficient matrix $g_{..}$, which now varies with the space-time point. To have the light cones uniquely determine the matrix g, one may posit $\det g = 1$. The light cones guide photons and planets, which react back on the light cones through their energy and momentum. Newton's theory of gravity survives as the linear term in a series expansion of Einstein's theory of gravity in powers of G under certain physical restrictions.

The startling non-commutativity introduced by the G expansion is space-time curvature. If T, T' are infinitesimal translations along two orthogonal coordinate axes then in special relativity $TT' = T'T$ and in a gravitational field $TT' \neq T'T$. The differences $TT' - T'T$ define curvature. The Einstein gravitational equations describe how the flux of momentum-energy – with coefficient G – curves space-time. When $G \to 0$ we recover the flat space-time of special relativity.

Photons are the main probes in two of the three classic tests of general relativity, which provided an example of a successful gauge theory that ultimately inspired the gauge revolution of the Standard Model. The next expansion that went into the Standard Model is the h expansion.

3. $\hbar$ Before quantum mechanics, the theory of a physical system split neatly into two phases. *Kinematics* tells about all the complete descriptions of the system or of reality, called states. *Dynamics* tells about how states change in dynamical processes. Operationally speaking, kinematics concerns filtration processes, which select systems of one kind, and dynamics concerns propagation processes, which change systems of one kind into another. Filtration processes represent predicates about the system. Such "acts of election" seem empirically to commute, Boole noted in 1847, as he was laying the foundations of his laws of thought.[4] But dynamical processes represent actions on the system and need not commute.

In h-land, quantum theory, filtrations no longer commute. This is what we mean operationally when we say that observation changes the system observed.

Such non-commuting filtrations were first used practically by Norse navigators who located the cloud-hidden sun by sighting clouds through beam-splitting crystals of Iceland spar. This phenomenon, like oil-slick colors and partial specular reflection, was not easy for Newton's granular theory of light. Newton speculated that some kind of invisible transverse guide wave accompanied light corpuscles and controlled these phenomena, but he still argued for his particle theory of light, declaring that light did not "bend into the shadow," or diffract, as waves would. Then Thomas Young exhibited light diffraction in 1804, and buried the particle theory of light.

Nevertheless Étienne-Louis Malus still applied Newton's photon theory to polarization studies in 1805. Malus was truer than Newton to Newton's own experimental philosophy and anticipated modern quantum practice. He did not speculate about invisible guide waves but concerned himself with experimental predictions, specifically the transition probability P – the probability that a photon passing the first filter will pass the second. For liner polarizers with polarizing axes along the unit vectors **a** and **b** normal to the light ray, $P = |\mathbf{a} \cdot \mathbf{b}|^2$, the Malus law. Malus may have deduced his law as much from plausible principles of symmetry and conservation as from experiment.

Write $f' < f$ to mean that all f' photons pass f but not conversely, a relation schematized in Figure 3.1.

A filtration process f is called *sharp* (homogeneous, pure) if it has no proper refinement $f' < f$.

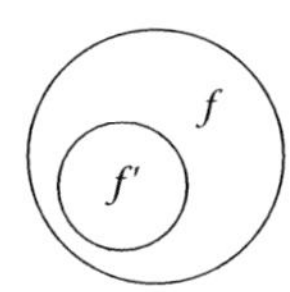

FIGURE 3.1
If no such f' exists, f is sharp.

In mechanics one assumed implicitly that if 1 and 2 are two sharp filtration processes, then the transition probability for a particle from 1 to pass 2 is either 0 (when 1 and 2 filter for different kinds of particle) or 1 (when they filter for the same kind); briefly put, that all sharp filtrations are *non-dispersive*. (Von Neumann 1934 spoke of pure ensembles rather than sharp filtrations; the upshot is the same.) The successive performance of filtration operations, represented by P_2P_1, to be read from right to left, is a kind of AND combination of predicates and their projectors, though the resultant of two filtrations may not be a filtration.

The Malus law, applied to two sharp filtrations in succession implies that even sharp filtrations are dispersive, and that photon filtrations do not commute, confirming Boole's uncanny premonition. Since we do not directly perceive polarization, we need three polarizing filters to verify that two do not commute. Let the polarization directions of P_1 and P_2 be obliquely oriented, neither parallel nor orthogonal. Compare experiments $P_1P_2P_1$ and $P_1P_1P_2 = P_1P_2$. Empirically, and in accord with the Malus law, all photons from P_1P_2 pass through P_1 but not all from P_2P_1 pass through P_1. Therefore empirically $P_1P_2P_1 \neq P_1P_1P_2$, and so $P_2P_1 \neq P_1P_2$.

This non-commutativity revises the logic that we use for photons.

If we generalize **a** and **b** to vectors of many components, representing general ideal filtration processes, Malus' Law becomes the fundamental Born statistical principle of quantum physics today. The guide wave concept of Newton has evolved into the much less object-like wave-function concept of quantum theory. The traditional boundary between commutative kinematical processes of information and non-commutative dynamical processes of transformation has broken down.

One reasons today about photons, and quantum systems in general, with a special quantum logic and quantum probability theory. One represents quantum filtrations and many other processes by matrices, and expresses quantum logic with matrix addition and multiplication; hence the old name "matrix mechanics."

We can represent any photon source by a standard perfectly white source o followed by suitable processes, and any photon counter by a standard perfect counter • preceded by suitable processes. This puts experiments into a convenient standard form

$$\bullet \leftarrow P_n \leftarrow \cdots \leftarrow P_1 \leftarrow \mathrm{o} \tag{3.1}$$

of a succession of physical processes between a source and a target.

Quantum theory represents all these intermediate processes by square matrices, related to experiment by the generalized Malus-Born law: For unit incident flux from o, the counting rate P at • for this experiment is determined by the matrix product

$$T = T_n \ldots T_1 \tag{3.2}$$

and its Hermitian conjugate T^* (the complex-conjugate transpose of T) as the trace

$$P = \frac{\mathrm{Tr}T^*T}{\mathrm{Tr}1}. \tag{3.3}$$

This is the unconditioned probability for transmission. A photon that stops in the first filter contributes 0 to the count at the counter but 1 to the count at the source. The vectors **a** and **b** of the Malus law are column vectors on which these quantum matrices act.

The physical properties of the quantum process determine the algebraic properties of its quantum matrix. For example a filtration operation P for photon polarizations becomes a 2×2 projection matrix or projector, one obeying $P^2 = P = P^\dagger$.

Heisenberg introduced quantum non-commutativity through the (non-) commutation relation

$$xp - px = i\hbar, \tag{3.4}$$

for the observables of momentum p and position x, not for filtrations ($\hbar \equiv h/2\pi$ is a standard abbreviation). But all observables are linear combinations of projectors, even in classical thought, and all projectors are functions of observables, polynomials in the finite-dimensional cases. So Heisenberg's non-commutativity of observables is equivalent to the non-commutativity of filtration processes, and so leads to a quantum logic.

The negation of the predicate P is $1 - P$ for quantum logic as for Boole logic. Quantum logic reduces to the Boole logic for diagonal filtration matrices, with elements 0 or 1. Then Boole rules. The classical logic also works well for quantum experiments with many degrees of freedom. Two directions chosen at random in a space of huge dimensionality are almost certainly almost orthogonal, and then Boole's laws almost apply. Only in low-dimensional playgrounds like photon polarization do we easily experience quantum logic.

Quantum theory represents the passage of time in an isolated system by a unitary matrix $U = U^{-1\dagger}$ obeying Heisenberg's Equation, the first-order differential equation $i\hbar\, dU/dt = HU$. It does not give a complete description of what evolves, but only describes the process. H is called the Hamiltonian operator and historically was at first constructed from the Hamiltonian of a classical theory. U appears as a block in eqn 3.1 and a factor in eqn 3.2 for every time-lapse t between operations.

$U(t)$ transforms any vector $\psi(0)$ to a vector $\psi(t)$ that obeys the Schrödinger Equation $i\hbar\, d\psi/dt = H\psi$ during the transformation U. A quantum vector ψ is not a dynamical variable or a complete description of the system but represents an irreversible operation of filtration, and so the Schrödinger Equation does not describe the change of a dynamical variable. The Heisenberg Equation does that. The Schrödinger Equation describes a coordinate-transformation that solves the Heisenberg Equation. The pre-quantum correspondent of the Heisenberg Equation is the Hamiltonian equation of motion, giving the rate of change of all observables. In the correspondence between quantum and pre-quantum concepts as $\hbar \to 0$, the Heisenberg Equation is the quantum equation of motion. The pre-quantum correspondent of the Schrödinger Equation is the Hamilton-Jacobi Equation, which is an equation for a coordinate transformation that solves the equation of motion, and is not the equation of motion.

As has widely been noted, starting with the treatises of Von Neumann and Dirac on the fundamental principles of quantum theory, the input wave-function for a transition describes a sharp input filtration *process*, not a system variable. Common usage nevertheless calls the input wave-function of an experiment the "state of the photon."

There are indeed systems whose states are observable wave-functions. They are called waves. But a quantum wave-function is not the state of some wave. Calling it the "quantum state" is a relic of early failed attempts at a wave theory of the atom. The "state-vector" is not the kind of thing that can be a system observable in quantum theory. Each observable is a fixed operator or matrix.

The state terminology, misleading as it is, may be too widespread and deep-rooted to up-date. After all, we still speak of "sunrise" five centuries after Copernicus. One must read creatively and let context determine the meaning of the word "state." In spectroscopy it usually refers to a sharp input or output operation.

It is problematical to attribute absolute values even to true observables in quantum theory. Consider a photon in the middle of an experiment that begins with a process of linear polarization along the x axis and ends with a right-handed circular polarization around the z axis, given that the photon passes both polarizers. Is it polarized along the x axis or y axis? If we reason naively forward from the first filter, the polarization between the two filters is certainly along the x axis, since the photon passed the first filter. If we reason naively backward from the last filter, the intermediate photon polarization must be circular and right-handed, since it is going to pass the last filter; it has probability 1/2 of being along the x axis. If we peek – measure the photon polarization in the middle of the experiment – we only answer a different question, concerning an experiment that ends with our new measurement. Measurements on a photon irreducibly and unpredictably change the photon, to an extent measured by h, so the question of the value between measurements has no immediate experimental meaning.

Common usage conventionally assigns the input properties to the photon. Assigning the output properties would work as well. Either choice breaks the time symmetry of quantum theory unnecessarily. The most operational procedure is to assign a property to the photon not absolutely but only relative to an experimenter who ascertains the property, specifying in particular whether the experimenter is at the input or output end of the optical bench. Quantum logic thus requires us to put some of our pre-quantum convictions about reality on holiday, but they can all come back to work when h can be neglected.

The photon concept emerges from the combination of the Maxwell equations with the Heisenberg non-commutativity eqn 3.4. Pre-quantum physicists recognized that by a Fourier analysis into waves $\sim e^{ik \cdot x}$ one can present the free electromagnetic field in a box as a collection of infinitely many linear harmonic oscillators, each with its own canonical coordinate q, canonical momentum p, and Hamiltonian

$$H = \frac{1}{2}(p^2 + \omega^2 q^2). \tag{3.5}$$

When the coefficient of p is scaled to unity in this way, the coefficient of q^2 is the square of the natural frequency ω of the oscillator. The Fourier analysis associates a definite wave-vector k with each oscillator. The energy spectrum of each oscillator is the set of roots E of the equation $HX = EX$ with arbitrary non-zero "eigenoperator" X.

The energy spectrum is most elegantly found by the ladder method. One seeks a linear combination a of q and p that obeys $Ha = a(H - E_1)$. This means that a lowers E (and therefore H) by steps of E_1 in the sense that if $HX = EX$ then $H(aX) = (E - E_1)(aX)$, unless $aX = 0$. Such an a, if it exists, is called a ladder operator, therefore. It is easy to see that a ladder operator exists for the harmonic oscillator, namely $a = 2^{-1/2}(p - iq)$, with energy step $E_1 = \hbar\omega$. One scales a so that H takes the form

$$H = \hbar\omega\left(n + \frac{1}{2}\right), \tag{3.6}$$

$n = a^\dagger a$, and a lowers n by steps of 1: $na = a(n - 1)$. Then n counts "excitation quanta" of the harmonic oscillator, each contributing an energy $E_1 = \hbar\omega$ to the total energy, and a momentum $\hbar k$ to the total momentum. The excitation quanta of the electromagnetic field oscillators are photons. The operator a is called an annihilation operator or annihilator for the photon because it lowers the photon count by 1. By the same token its adjoint is a photon creator.

The term 1/2 in H contributes a zero-point energy that is usually arbitrarily discarded, primarily because any non-zero vacuum energy would violate Lorentz invariance and so disagree somewhat with experiment. One cannot deduce that the vacuum energy is zero from the present dynamical theory, and astrophysicists are now fairly sure that it is not zero.

A similar process leads to the excitation quanta of the field oscillators of other fields. Today one accounts for all allegedly fundamental quanta as excitation quanta of suitably designed field oscillators.

Now we can say what a photon is. Consider first what an apple is. When I move it from one side of the table to the other, or turn it over, it is still the same apple. So the apple is not its state, not what we know about the apple. Statistical mathematicians formulate the concept of a constant object with varying properties by identifying the object—sometimes called a random variable – not with one state but with the space of all its possible states. This works just as well for quantum objects as for random objects, once we replace states by more appropriate actions on the quantum object. The object is defined, for example, by the processes it can undergo. For example, the sharp filtration processes for one photon, relative to a given observer, form a collection with one structural element, the transition probability between two such processes. For many purposes we can identify a photon with this collection of processes.

The filtration processes mentioned are usually represented by lines through the origin in a Hilbert space. If we are willing to start from a Hilbert space, we can define a photon by its Hilbert space; not by one wave-function, which just says one way to produce a photon, but the collection of them all. This gives preference to input over output and spoils symmetry a bit. One restores time symmetry by using the algebra of operators rather than the Hilbert space to define the photon. In words, the photon is the creature on which those operations can act.

From the current viewpoint the concept of photon is not as fundamental as that of electromagnetic field. Not all electromagnetic interactions are photon-mediated. There are also static forces, like the Coulomb force. Different observers may split electromagnetic interactions into radiation and static forces differently. Gauge theory leads us to quantum fields, and photons arise as quantum excitations of one of these fields.

Quantum theory has a non-Boolean logic in much the sense that general relativity has a non-Euclidean geometry: it renounces an ancient commutativity. A Boolean logic has non-dispersive predicates called states, common to all observers; a quantum logic does not. Attempting to fit the quantum non-commutativity of predicates into a classical picture of an object with absolute states is like attempting to fit special relativity into a space-time with absolute time. Possibly we can do it but probably we shouldn't. If we accept that the expanded logic contracts to the familiar one when $\hbar$*to*0, we can go on to the next expansion.

4. $\hbar'\hbar''$ In this section I describe a possible future expansion suggested by Segal[9] that might give a simpler and more finite structure to the photon and other quanta. There are clear indications, both experimental and structural, that quantum theory is still too commutative. Experiment indicates limits to the applicability of the concept of time both in the very small and the very large, ignored by present quantum theory. The theoretical assumption that all feasible operations commute with the imaginary i makes i a prototypical idol. The canonical commutation relations are unstable.

To unseat this idol and stabilize this instability, one first rewrites the defining relations for a photon oscillator in terms of antisymmetric operators $\hat{q} := iq$, $\hat{p} = -ip$:

$$\begin{aligned} \hat{q}\hat{p} - \hat{p}\hat{q} &= \hbar i, \\ i\hat{q} - \hat{q}i &= 0, \\ i\hat{p} - \hat{p}i &= 0. \end{aligned} \tag{3.7}$$

One stabilizing variation, for example, is

$$\begin{aligned} \hat{q}\hat{p} - \hat{p}\hat{q} &= \hbar i, \\ i\hat{q} - \hat{q}i &= \hbar'\hat{p}, \\ i\hat{p} - \hat{p}i &= -\hbar''\hat{q} \end{aligned} \tag{3.8}$$

with Segal constants $\hbar'$, $\hbar'' > 0$ supplementing the Planck quantum constant $\hbar$.[9] No matter how small the Segal constants are, if they have the given sign the expanded oscillator commutation relations can be rescaled to angular momentum relations[2]

$$\begin{aligned} \hat{L}_x\hat{L}_y - \hat{L}_y\hat{L}_x &= \hat{L}_z, \\ \hat{L}_y\hat{L}_z - \hat{L}_z\hat{L}_y &= \hat{L}_x, \\ \hat{L}_z\hat{L}_x - \hat{L}_x\hat{L}_z &= \hat{L}_y. \end{aligned} \tag{3.9}$$

by a scaling

$$\begin{aligned} \breve{q} &= Q\hat{L}_1, \\ \breve{p} &= P\hat{L}_2, \\ \breve{i} &= J\hat{L}_3, \end{aligned} \tag{3.10}$$

with

$$\begin{aligned} J &= \sqrt{\hbar'\hbar''} = \frac{1}{l}, \\ Q &= \sqrt{\hbar\hbar'}, \\ P &= \sqrt{\hbar\hbar''}. \end{aligned} \tag{3.11}$$

As customary we have designated the maximum eigenvalue of $|\hat{L}|_z$ by l. This theory is now stabilized by its curvature against further small changes in $\hbar$, $\hbar'$, $\hbar''$; just as a small change in curvature turns any straight line into a circle but leaves almost all circles circular; and just as quantum theory is stable against small changes in $\hbar$.

To be sure, when $\hbar'$, $\hbar'' \to 0$ we recover the quantum theory. As in all such expansions of physical theory, the quantum theory with c-number i is a case of probability zero in an ensemble of more likely expanded theories with operator i's. The canonical commutation relations might be right, but that would be a miracle of probability 0. Data always have some error bars, so an exactly zero commutator is never based entirely on experiment and usually incorporates faith in some prior absolute: here i. Renouncing that absolute makes room for a more stable kind of theory, based more firmly on experiment and at least as consistent with the existing data. Which one of these possibilities is in better agreement with experiment than the canonical theory can only be learned from experiment.

The most economical way to stabilize the Heisenberg relations is to close them on themselves as we have done here. A more general stabilization might also couple each oscillator to others. In the past the stabilizations that worked have usually been economical but not always.

These transquantum relations describe a rotator, not an oscillator. What we have thought were harmonic oscillators are more likely to be quantum rotators. It has been recognized for some time that oscillators can be approximated by rotators and conversely.[1,2,7] In particular, photons too are infinitely more likely to be quanta of a kind of rotation than of oscillation. If so, they can still have exact ladder operators, but their ladders now have a top as well as a bottom, with $2l + 1$ rungs for rotational transquantum number l.

In the most intense lasers, there can be as many as 10^{13} photons in one mode at one time.[8] Then $2l \geq 10^{13}$ and $\hbar'\hbar'' \leq 10^{-26}$ in order of magnitude.

When we expand the commutation relations for time and energy in this way, the two new transquantum constants that appear indeed limit the applicability of these concepts both in the small and the large. They make the photon advance step by quantum step. We will probably never be able to visualize a photon but we might soon be able to choreograph one; to describe the process rather than the object.

References

1. Arecchi, F. T., E. Courtens, R. Gilmore, and H. Thomas, *Physical Review A* **6**, 2211 (1972).
2. Atakishiyev, N. M., G. S. Pogosyan, and K. B. Wolf. Contraction of the finite one-dimensional oscillator. *International Journal of Modern Physics* **A18** (2003) 317–327.
3. Baugh, J., D. Finkelstein, A. Galiautdinov, and H. Saller, Clifford algebra as quantum language. *J. Math. Phys.* **42** (2001) 1489–1500.
4. Boole, G. (1847). The mathematical analysis of logic; being an essay towards a calculus of deductive reasoning. Cambridge. Reprinted, Philosophical Library, New York, 1948.
5. Galiautdinov A. A., and D. R. Finkelstein. Chronon corrections to the Dirac equation. hep-th/0106273. *J. Math. Phys.* **43**, 4741 (2002).
6. Inönü, E. and E. P. Wigner. On the contraction of groups and their representations. *Proceedings of the National Academy of Sciences* **39**(1952) 510–525.

7. Kuzmich, A., N. P. Bigelow, and L. Mandel, *Europhysics Letters* **A 42**, 481 (1998). Kuzmich, A., L. Mandel, J. Janis, Y. E. Young, R. Ejnisman, and N. P. Bigelow, *Physical Review A* **60**, 2346 (1999). Kuzmich, A., L. Mandel, and N. P. Bigelow, *Physical Review Letters* **85**, 1594 (2000).
8. A. Kuzmich, Private communication. Cf. e.g., A. Siegman, *Lasers*, University Science Books, (1986).
9. Segal, I. E. A class of operator algebras which are determined by groups. *Duke Mathematics Journal* **18** (1951) 221.

4

The Concept of the Photon—Revisited

Ashok Muthukrishnan,[1] Marlan O. Scully,[1,2] and M. Suhail Zubairy[1,3]

CONTENTS

The photon concept is one of the most debated issues in the history of physical science. Some thirty years ago, we published an article in *Physics Today* entitled "The Concept of the Photon,"[1] in which we described the "photon" as a classical electromagnetic field plus the fluctuations associated with the vacuum. However, subsequent developments required us to envision the photon as an intrinsically quantum mechanical entity, whose basic physics is much deeper than can be explained by the simple "classical wave plus vacuum fluctuations" picture. These ideas and the extensions of our conceptual understanding are discussed in detail in our recent quantum optics book.[2] In this

[1] Institute for Quantum Studies and Department of Physics, Texas A&M University, College Station, TX 77843
[2] Departments of Chemistry and Aerospace and Mechanical Engineering, Princeton University, Princeton, NJ 08544
[3] Department of Electronics, Quaid-i-Azam University, Islamabad, Pakistan

chapter we revisit the photon concept based on examples from these sources and more.

© 2003 Optical Society of America

The "photon" is a quintessentially twentieth-century concept, intimately tied to the birth of quantum mechanics and quantum electrodynamics. However, the root of the idea may be said to be much older, as old as the historical debate on the nature of light itself—whether it is a wave or a particle—one that has witnessed a seesaw of ideology from antiquity to present. The transition from classical to quantum descriptions of light presents yet another dichotomy, one where the necessity of quantizing the electromagnetic field (over and above a quantization of matter) has been challenged. The resolution lies in uncovering key behavior of quantum light fields that are beyond the domain of the classical, such as vacuum fluctuations and quantum entanglement, which necessitate a quantum theory of radiation.[2–5] Nevertheless, a precise grasp of the "photon" concept is not an easy task, to quote Albert Einstein:

> "These days, every Tom, Dick and Harry thinks he knows what a photon is, but he is wrong."

We ought to proceed with diligence and caution. In the words of Willis Lamb:[6]

> "What do we do next? We can, and should, use the Quantum Theory of Radiation. Fermi showed how to do this for the case of Lippmann fringes. The idea is simple, but the details are somewhat messy. A good notation and lots of practice makes it easier. Begin by deciding how much of the universe needs to be brought into the discussion. Decide what normal modes are needed for an adequate treatment. Decide how to model the light sources and work out how they drive the system."

We proceed to elucidate the photon concept by specific experiments (real and gedanken) which demonstrate the need for and shed light on the meaning of the "photon." Specifically, we will start by briefly reviewing the history of the wave-particle debate and then giving seven of our favorite examples, each clarifying some key aspect of the quantum nature of light. The two facets of the photon that we focus on are vacuum fluctuations (as in our earlier article[1]), and aspects of many-particle correlations (as in our recent book[2]). Examples of the first are spontaneous emission, Lamb shift, and the scattering of atoms off the vacuum field at the entrance to a micromaser. Examples of the second facet include quantum beats, quantum eraser, and photon correlation microscopy. Finally, in the example of two-site downconversion interferometry, the essence of both facets is combined and elucidated.

In the final portions of the article, we return to the basic questions concerning the nature of light in the context of the wave-particle debate: What is a photon and where is it? To the first question, we answer in the words of Roy Glauber:

"A photon is what a photodetector detects."

To the second question (on the locality of the photon), the answer becomes: "A photon is *where* the photodetector detects it." In principle, the detector could be a microscopic object such as an atom. Guided by this point of view, we address the much debated issue of the existence of a photon wave function Ψ $(\mathbf{r},t)$.[2,7,8] Arguments to the contrary notwithstanding, we show that the concept of the photon wave function arises naturally from the quantum theory of photodetection (see Ref. [2], ch. 1). A wealth of insight is gained about the interference and entanglement properties of light by studying such one-photon, and related two-photon, "wave functions".[2]

4.1 Light—Wave or Particle?

The nature of light is a very old issue in the history of science. For the ancient Greeks and Arabs, the debate centered on the connection between light and vision. The tactile theory, which held that our vision was initiated by our eyes reaching out to "touch" or feel something at a distance, gradually lost ground to the emission theory, which postulated that vision resulted from illuminated objects emitting energy that was sensed by our eyes. This paradigm shift is mainly due to the eleventh-century Arab scientist Abu Ali Hasan Ibn Al-Haitham (or "Alhazen") who laid the groundwork for classical optics through investigations into the refraction and dispersion properties of light. Later Renaissance thinkers in Europe envisioned light as a stream of particles, perhaps supported by the ether, an invisible medium thought to permeate empty space and all transparent materials.

In the seventeenth century, Pierre de Fermat introduced the *principle of least time* to account for the phenomenon of refraction. Equivalently, his principle states that a ray of light takes the path that minimizes the optical path length between two points in space:

$$\delta \int_{\mathbf{r}_0}^{\mathbf{r}} n\,ds = 0, \tag{4.1}$$

where $n = c/v$ is the (spatially varying) refractive index that determines the velocity of the light particle, and δ denotes a variation over all paths connecting $\mathbf{r}_0$ and $\mathbf{r}$. Fermat's principle is the foundation for geometrical optics, a theory based on the view that light is a particle that travels along well-defined geometrical rays. The idea of light as particle (or "corpuscle") was of course

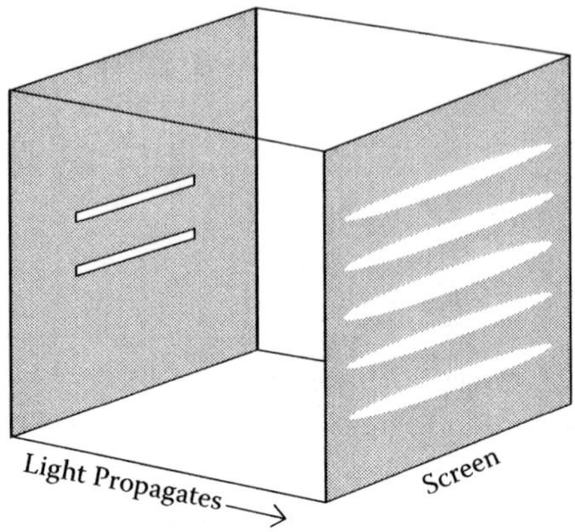

FIGURE 4.1
Young's two-slit experiment—Light incident on two slits in a box propagates along two pathways to a given point on the screen, displaying constructive and destructive interference. When a single photon is incident on the slits, it is detected with highest probability at the interference peaks, but never at the interference nodes.

adopted by Isaac Newton, who bequeathed the weight of his scientific legacy, including the bearing of his laws of mechanics, on the nature of light.

Christian Huygens on the other hand, a contemporary of Newton, was a strong advocate of the wave theory of light. He formulated a principle (that now bears his name) which describes wave propagation as the interference of secondary wavelets arising from point sources on the existing wave-front. It took the mathematical genius of Augustin Fresnel, 150 years later, to realize the consequences of this discovery, including a rigorous development of the theory of wave diffraction. Light does not form sharp, geometrical shadows that are characteristic of a particle, but bends around obstacles and apertures.

The revival of the wave theory in the early nineteenth century was initiated by Thomas Young. In 1800, appearing before the Royal Society of London, Young spoke for an analogy between light and sound, and declared later that a two-slit interference experiment would conclusively demonstrate the wave nature of light (see Figure 4.1). It is hard for the modern reader to visualize how counter-intuitive this suggestion was at the time. The idea that a screen uniformly illuminated by a single aperture could develop dark fringes with the introduction of a second aperture—that the addition of *more* light could result in *less* illumination—was hard for Young's audience to digest.

Likewise, Fresnel's diffraction theory was received with skepticism by the judges on the 1819 prize committee in Paris. In particular, the esteemed Pierre Simon de Laplace was very skeptical of the wave theory. His protégé, Siméon-Denis Poisson, highlighted the seemingly absurd fact that the theory implied a bright spot at the center of the shadow of an illuminated opaque disc, now known as Poisson's spot. The resistance to switch from a particle description to a wave description for light by these pre-eminent scientists of the early nineteenth century gives an indication of the great disparity between these two conceptions. It was a precursor of the struggle to come a hundred years later with the advent of quantum mechanics.

The wave theory really came into its own in the late nineteenth century in the work of James Clerk Maxwell. His four equations, known to all students of undergraduate physics, is the first self-contained theory of radiation. Receiving experimental confirmation by Heinrich Hertz, the Maxwell theory unified the disparate phenomena of electricity and magnetism, and gave physical meaning to the transverse polarizations of light waves. The far-reaching success of the theory explains the hubris of late nineteenth century physicists, many of whom believed that there were really only two "clouds" on the horizon of physics at the dawn of the twentieth century. Interestingly enough, both of these involved light.

The first cloud, namely the null result of the Michelson-Morley experiment, led to special relativity, which is the epitome of classical mechanics, and the logical capstone of classical physics. The second cloud, the Rayleigh-Jeans ultraviolet (UV) catastrophe and the nature of blackbody radiation, led to the advent of quantum mechanics, which of course was a radical change in physical thought. While both of these problems involved the radiation field, neither (initially) involved the concept of a photon. That is, neither Albert Einstein and Hendrik Lorentz in the first instance, nor Max Planck in the second, called upon the particulate nature of light for the explanation of the observed phenomena. Relativity is strictly classical, and Planck quantized the energies of the oscillators in the walls of his cavity, not the field.[9]

The revival of the particle theory of light, and the beginning of the modern concept of the photon, was due to Einstein. In his 1905 paper on the photoelectric effect,[10] the emission of electrons from a metallic surface irradiated by UV rays, Einstein postulated that light comes in discrete bundles, or quanta of energy, borrowing Planck's five-year old hypothesis: $E = \hbar V$, where V is the circular frequency and $\hbar$ is Planck's constant divided by 2π. This re-introduced the particulate nature of light into physical discourse, not as localization in space in the manner of Newton's corpuscles, but as discreteness in energy. But irony upon irony, it is a historical curiosity that Einstein got the idea for the photon from the physics of the photoelectric effect. In fact, it can be shown that the essence of the photoelectric effect does not require the quantization of the radiation field,[11] a misconception perpetuated by the mills of textbooks, to wit, the following quote from a mid-century text:[12]

"Einstein's photoelectric equation played an enormous part in the development of the modern quantum theory. But in spite of its generality and of the many successful applications that have been made of it in physical theories, the equation:

$$\hbar V = E + \Phi \tag{4.2}$$

is, as we shall see presently, based on a concept of radiation – the concept of "light quanta" – completely at variance with the most fundamental concepts of the classical electromagnetic theory of radiation."

We will revisit the photoelectric effect in the next section and place it properly in the context of radiation theory.

Both the Planck hypothesis and the Einstein interpretation follow from considerations of how energy is *exchanged* between radiation and matter. Instead of an electromagnetic wave continuously driving the amplitude of a classical oscillator, we have the discrete picture of light of the right frequency absorbed or emitted by a quantum oscillator, such as an atom in the walls of the cavity, or on a metallic surface. This seemingly intimate connection between energy quantization and the interaction of radiation with matter motivated the original coining of the word "photon" by Gilbert Lewis in 1926:[13]

"It would seem inappropriate to speak of one of these hypothetical entities as a particle of light, a corpuscle of light, a light quantum, or light quant, if we are to assume that it spends only a minute fraction of its existence as a carrier of radiant energy, while the rest of the time it remains as an important structural element within the atom ... I therefore take the liberty of proposing for this hypothetical new atom, which is not light but plays an essential part in every process of radiation, the name *photon*."

Energy quantization is the essence of the old quantum theory of the atom proposed by Niels Bohr. The electron is said to occupy discrete orbitals with energies E_i and E_j, with transitions between them caused by a photon of the right frequency: $v = (E_i - E_j)/\hbar$. An ingenious interpretation of this quantization in terms of matter waves was given by Louis de Broglie, who argued by analogy with standing waves in a cavity, that the wavelength of the electron in each Bohr orbital is quantized—an integer number of wavelengths would have to fit in a circular orbit of the right radius. This paved the way for Erwin Schrödinger to introduce his famous wave equation for matter waves, the basis for (non-relativistic) quantum mechanics of material systems.

Quantum mechanics provides us with a new perspective on the wave-particle debate, vis á vis Young's two-slit experiment (Figure 4.1). In the paradigm of quantum interference, we add the probability amplitudes associated with different pathways through an interferometer. Light (or matter) is neither wave nor particle, but an intermediate entity that obeys the superposition principle. When a single photon goes through the slits, it registers as a point-like event on the screen (measured, say, by a CCD array). An accumulation of such events over repeated trials builds up a probabilistic fringe pattern that is characteristic of classical wave interference. However, if we arrange to acquire information about which slit the photon went through, the interference nulls disappear. Thus, from the standpoint of complementarity, both wave and particle perspectives have equal validity. We will return to this issue later in the chapter.

4.2 The Semiclassical View

The interaction of radiation and matter is key to understanding the nature of light and the concept of a photon. In the semiclassical view, light is treated classically and only matter is quantized. In other words, both are treated on an *equal* footing: a wave theory of light (the Maxwell equations) is combined self-consistently with a wave theory of matter (the Schrödinger equation). This yields a remarkably accurate description of a large class of phenomena, including the photoelectric effect, stimulated emission and absorption, saturation effects and nonlinear spectroscopy, pulse propagation phenomena, "photon" echoes, etc. Many properties of laser light, such as frequency selectivity, phase coherence, and directionality, can be explained within this framework.[14]

The workhorse of semiclassical theory is the two-level atom, specifically the problem of its interaction with a sinusoidal light wave.[15] In reality, real atoms have lots of levels, but the two-level approximation amounts to isolating a particular transition that is nearly resonant with the field frequency ν. That is, the energy separation of the levels is assumed to be $E_a - E_b = \hbar\omega \approx \hbar v$. Such a comparison of the atomic energy difference with the field frequency is in the spirit of the Bohr model, but note that this already implies a discreteness in light energy, $\Delta E = \hbar v$. That a semiclassical analysis is able to bring out this discreteness—in the form of *resonance*—is a qualitative dividend of this approach.

Schrödinger's equation describes the dynamics of the atom, but how about the dynamics of the radiation field? In the semiclassical approach, one assumes that the atomic electron cloud $\psi^* \psi$, which is polarized by the incident field, acts like an oscillating *charge* density, producing an ensemble dipole moment that re-radiates a classical Maxwell field. The effects of radiation reaction, i.e., the back action of the emitted field on the atom, are taken into account by requiring the coupled Maxwell-Schrödinger equations to be self-consistent with respect to the *total* field. That is, the field that the atoms see should be consistent with the field radiated. In this way, semiclassical theory becomes a self-contained description of the dynamics of a quantum mechanical atom interacting with a classical field. As we have noted above, its successes far outweigh our expectations.

Let us apply the semiclassical analysis to the photoelectric effect, which provided the original impetus for the quantization of light. There are three observed features of this effect that need accounting. First, when light shines on a photo-emissive surface, electrons are ejected with a kinetic energy E equal to $\hbar$ times the frequency v of the incident light less some work function Φ, as in Eq. 4.2. Second, it is observed that the rate of electron ejection is proportional to the square of the incident electric field E_0. Third, and more subtle, there is not necessarily a time delay between the instant the field is turned on, and the time when the photoelectron is ejected, contrary to classical expectations.

All three observations can be nominally accounted for by applying the semiclassical theory to lowest order in perturbation of the atom-field interaction $V(t) = -eE_0r$.[11] This furnishes a Fermi Golden Rule for the probability of transition of the electron from the ground state g of the atom to the kth excited state in the continuum:

$$P_k = [2\pi(e|r_{kg}|E_0/2\hbar)^2\, t]\, \delta[\nu - (E_k - E_g)/\, \hbar], \tag{4.3}$$

where er_{kg} is the dipole matrix element between the initial and final states. The δ-function (which has units of time) arises from considering the frequency response of the surface, and assuming that t is at least as long as several optical cycles: $\nu t \gg 1$. Now, writing energy $E_k - E_g$ as $E + \Phi$, we see that the δ-function immediately implies Eq. 4.2. The second fact is also clearly contained in Eq. 4.3 since P_k is proportional to E^2_0. The third fact of photoelectric detection, the finite time delay, is explained in the sense that P_k is linearly proportional to t, and there is a finite probability of the atom being excited even at infinitesimally small times.

Thus, the experimental aspects of the photoelectric effect are completely understandable from a semiclassical point of view. Where we depart from a classical intuition for light is a subtle issue connected with the third fact, namely that there is negligible time delay between the incidence of light and the photoelectron emission. While this is understandable from an *atomic* point of view—the electron has finite probability of being excited even at very short times—the argument breaks down when we consider the implications for the field. That is, if we persist in thinking about the field classically, energy is not conserved. Over a time interval t, a classical field E_0 brings in a flux of energy $\varepsilon_0\, E^2_0\, At$ to bear on the atom, where A is the atomic cross-section. For short enough times t, this energy is negligible compared to $\hbar v$, the energy that the electron supposedly absorbs (instantaneously) when it becomes excited. We just do not have the authority, within the Maxwell formalism, to affect a similar *quantum jump* for the field energy.

For this and other reasons (see next section), it behooves us to supplement the epistemology of the Maxwell theory with a quantized view of the electromagnetic field that fully accounts for the probabilistic nature of light and its inherent fluctuations. This is exactly what Paul Dirac did in the year 1927, when the photon concept was, for the first time, placed on a logical foundation, and the quantum theory of radiation was born.[16] This was followed in the 1940s by the remarkably successful theory of quantum electrodynamics (QED)—the quantum theory of interaction of light and matter—that achieved unparalleled numerical accuracy in predicting experimental observations. Nevertheless, a short twenty years later, we would come back full circle in the saga of semiclassical theory, with Ed Jaynes questioning the need for a quantum theory of radiation at the 1966 conference on Coherence and Quantum Optics at Rochester, New York.

"Physics goes forward on the shoulders of doubters, not believers, and I doubt that QED is necessary," declared Jaynes. In his view, semiclassical theory—or "neoclassical" theory, with the addition of a radiation reaction field acting back on the atom – was sufficient to explain the Lamb shift, thought by most to be the best vindication yet of Dirac's field quantization and QED theory (see below). Another conference attendee, Peter Franken, challenged Jaynes to a bet. One of us (MOS) present at the conference recalls Franken's words: "You are a reasonably rich man. So am I, and I say put your money where your face is!" He wagered \$100 over whether the Lamb shift could or could not be calculated without QED. Jaynes took the bet that he could, and Willis Lamb agreed to be the judge.

In the 1960s and 70s, Jaynes and his collaborators reported partial success in predicting the Lamb shift using neoclassical theory.[17] They were able to make a qualitative connection between the shift and the physics of radiation reaction—in the absence of field quantization or vacuum fluctuations—but failed to produce an accurate numerical prediction which could be checked against experiment. For this reason, at the 1978 conference in Rochester, Lamb decided to yield the bet to Franken. An account of the arguments for and against this decision was summarized by Jaynes in his paper at the conference.[18] In the end, QED had survived the challenge of semiclassical theory, and vacuum fluctuations were indeed "very real things" to be reckoned with.

4.3 Seven Examples

Our first three examples below illustrate the reality of vacuum fluctuations in the electromagnetic field as manifested in the physics of the atom. The "photon" acquires a stochastic meaning in this context. One speaks of a classical electromagnetic field with fluctuations due to the vacuum. To be sure, one cannot "see" these fluctuations with a photodetector, but they make their presence felt, for example, in the way the atomic electrons are "jiggled" by these random vacuum forces.

4.3.1 Spontaneous Emission

In the phenomenon of spontaneous emission,[19] an atom in the excited state decays to the ground state and spontaneously emits a photon (see Figure 4.2). This "spontaneous" emission is in a sense stimulated emission, where the stimulating field is a vacuum fluctuation. If an atom is placed in the excited state and the field is classical, the atom will never develop a dipole moment and will never radiate. In this sense, semiclassical theory does not account for spontaneous emission. However, when vacuum fluctuations are included, we can think

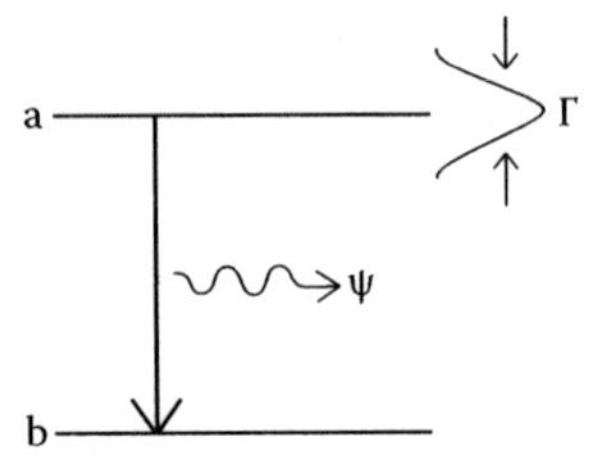

FIGURE 4.2
Spontaneous emission—Two-level atom, with upper-level linewidth Γ spontaneously emits a photon. Fluctuations in the vacuum field cause the electron in the excited state to decay to the ground state in a characteristic time Γ^{-1}.

conceptually of the atom as being stimulated to emit radiation by the fluctuating field, and the back action of the emitted light will drive the atom further to the ground state, yielding decay of the excited state. It is in this way that we understand spontaneous emission as being due to vacuum fluctuations.

4.3.2 Lamb Shift

Perhaps the greatest triumph of field quantization is the explanation of the Lamb shift[20] between, for example, the $2s_{1/2}$ and $2p_{1/2}$ levels in a hydrogenic atom. Relativistic quantum mechanics predicts that these levels should be degenerate, in contradiction to the experimentally observed frequency splitting of about 1 GHz. We can understand the shift intuitively[21] by picturing the electron forced to fluctuate about its first-quantized position in the atom due to random kicks from the surrounding, fluctuating vacuum field (see Figure 4.3). Its average displacement $\langle \Delta \mathbf{r} \rangle$ is zero, but the squared displacement $\langle \Delta \mathbf{r} \rangle^2$ is slightly nonzero, with the result that the electron "senses" a slightly different Coulomb pull from the positively charged nucleus than it normally would. The effect is more prominent nearer the nucleus where the

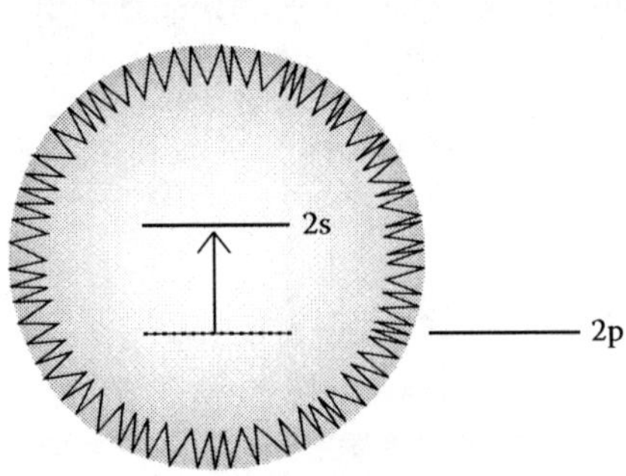

FIGURE 4.3
Lamb shift—Schematic illustration of the Lamb shift of the hydrogenic $2s_{1/2}$ state relative to the $2p_{1/2}$ state. Intuitive understanding of the shift as due to random jostling of the electron in the 2*s* orbital by zero-point fluctuations in the vacuum field.

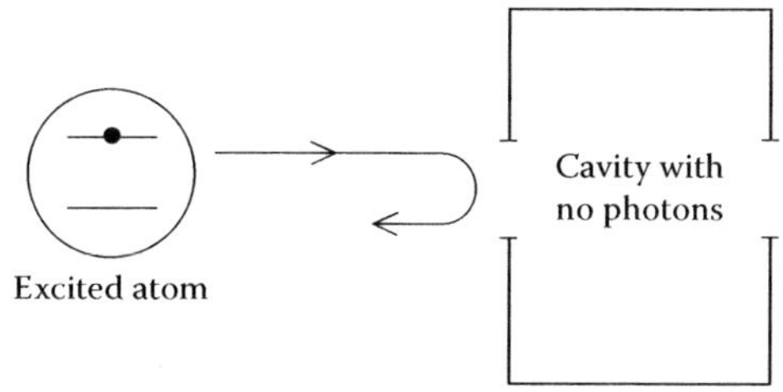

FIGURE 4.4
Scattering off the vacuum—An excited atom approaching an empty cavity can be reflected for slow enough velocities. The vacuum cavity field serves as an effective potential barrier for the center-of-mass wave function of the atom.

Coulomb potential falls off more steeply, thus the *s* orbital is affected more than the *p* orbital. This is manifested as the Lamb shift between the levels.

4.3.3 Micromaser—Scattering off the Vacuum

A micromaser consists of a single atom interacting with a single-mode quantized field in a high-Q cavity.[22] An interesting new perspective on vacuum fluctuations is given by the recent example of an excited atom scattering off an effective potential barrier created by a vacuum field in the cavity (see Figure 4.4).[23] When the atomic center-of-mass motion is quantized, and the atoms are travelling slow enough (their kinetic energy is smaller than the atom-field interaction energy), it is shown that they can undergo reflection from the cavity, even when it is initially empty, i.e., there are no photons. The reflection of the atom takes place due to the discontinuous change in the strength of the coupling with vacuum fluctuations at the input to the cavity. This kind of reflection off an edge discontinuity is common in wave mechanics. What is interesting in this instance is that the reflection is due to an abrupt change in coupling with the vacuum between the inside and the outside of the cavity. It is then fair to view this physics as another manifestation of the effect of vacuum fluctuations, this time affecting the center-of-mass dynamics of the atom.

Our next three examples involve the concept of multi-particle entanglement, which is a distinguishing feature of the quantized electromagnetic field. Historically, inter-particle correlations have played a key role in fundamental tests of quantum mechanics, such as the EPR paradox, Bell inequalities and quantum eraser. These examples illustrate the reality of quantum correlations in multi-photon physics. In recent years, entangled photons have been key to applications in quantum information and computing, giving rise to new technologies such as photon correlation microscopy (see below).

4.3.4 Quantum Beats

In general, beats arise whenever two or more frequencies of a wave are simultaneously present. When an atom in the excited state undergoes decay

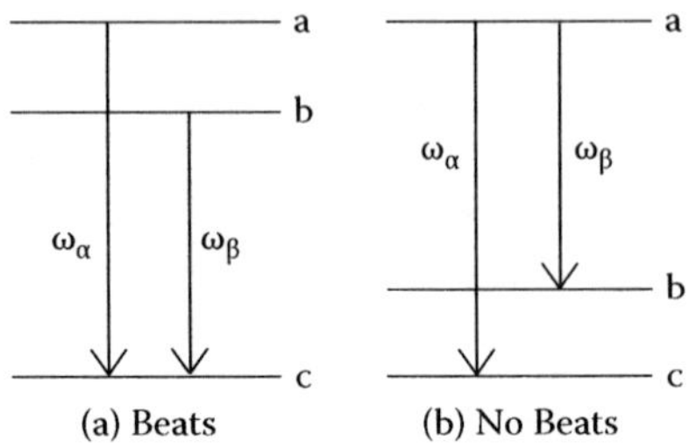

FIGURE 4.5
Quantum beats—a) When a single atom decays from either of two upper levels to a common lower level, the two transition frequencies produce a beat note $\omega_\alpha - \omega_\beta$ in the emitted photon. b) No beats are present when the lower levels are distinct, since the final state of the atom provides distinguishing information on the decay route taken by the photon.

along two transition pathways, the light produced in the process is expected to register a beat note at the difference frequency, $\omega_\alpha - \omega_\beta$, in addition to the individual transition frequencies ω_α and ω_β. However, when a single atom decays, beats are present only when the two final states of the atom are identical (see Figure 4.5). When the final states are distinct, quantum theory predicts an absence of beats.[24] This is so because the two decay channels end in different atomic states [$|b\rangle$ or $|c\rangle$ in Figure 4.5(b)]. We now have which-path information since we need only consult the atom to see which photon (α or β) was emitted—i.e., the entanglement between the atom and the quantized field destroys the interference. Classical electrodynamics, vis á vis semiclassical theory, cannot explain the "missing" beats.

4.3.5 Quantum Eraser and Complementarity

In the quantum eraser,[25] the which-path information about the interfering particle is *erased* by manipulating the second, entangled particle. Complementarity is enforced not by the uncertainty principle (through a measurement process), but by a quantum correlation between particles.[26] This notion can be realized in the context of two-photon interferometry.[27–29] Consider the setup shown in Figure 4.6, where one of two atoms $i = 1,2$ emits two photons ϕ_i and γ_i. Interference is observed in ϕ only when the spatial origin of γ cannot be discerned, i.e., when detector D_1 or D_2 clicks. Erasure occurs when the γ photon is reflected (rather than transmitted) at beamsplitter BS1 or BS2, which in the experiment occurs *after* the ϕ photon has been detected. Thus, quantum entanglement between the photons enables a realization of "delayed choice",[30] which cannot be simulated by classical optics.

4.3.6 Photon Correlation Microscopy

Novel interference phenomena arise from second-order correlations of entangled photons, such as arise from the spontaneous cascade decay of a

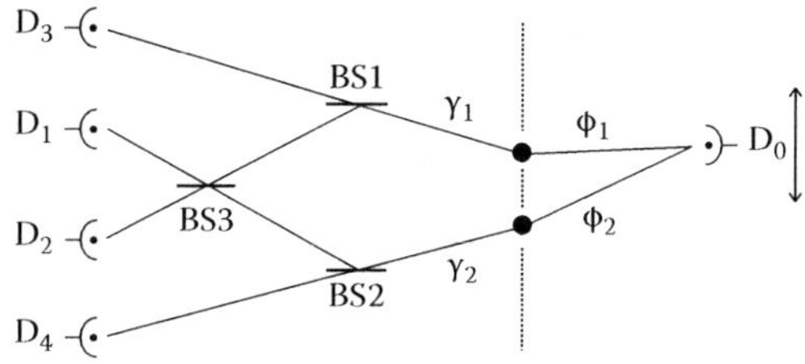

FIGURE 4.6
Quantum eraser—One of two atoms (solid circles) emits two photons ϕ_i and γ_i. Interference is observed in ϕ by scanning detector D_0. Beamsplitters BS1-BS3 direct γ to four detectors. A click in detectors *D*3 or *D*4 provides which-path information on γ, preventing interference in ϕ. A click in detectors D_1 or D_2 erases which-path information and restores interference in ϕ. Figure adapted from Ref. [29].

three-level atom (where the emitted photons are correlated in frequency and time of emission).[2] When two such atoms are spatially separated and one of them undergoes decay, a two-photon correlation measurement enables high-resolution *spectral* microscopy on the atomic level structure.[31] It can be shown that the resolution of the upper two levels *a* and *b* in each atom is limited only by the linewidth Γ_a, and not by Γ_a and Γ_b together (as is usually the case). This phenomenon relies on the path and frequency entanglement between the two photons arising from spatially separated cascade sources.

A further consequence of the two-atom geometry is the enhancement in *spatial* resolution that occurs because the photons are entangled in path—that is, the photon pair arises from one atom or the other, and their joint paths interfere. Coincident detection of the two photons (each of wavelength λ) shows a fringe resolution that is enhanced by a factor of two as compared to the classical Rayleigh limit, $\lambda/2$. This enables applications in high-resolution lithography.[32,33] The fringe doubling is due to the fact that the two photons propagate along the same path, and their sum frequency, 2ω, characterizes their joint detection probability. Path entanglement cannot be simulated by (co-propagating) classical light pulses.

4.3.7 Two-Site Downconversion Interferometry

In what follows, we consider a two-particle interferometry experiment that allows us to elucidate both facets of the photon considered above—vacuum fluctuations and quantum entanglement. The thought experiment we have in mind is based on an actual one that was carried out using parametric downconversion.[34] Consider the setup shown in Figure 4.7, where two atoms $i = 1, 2$ are fixed in position and one of them emits two photons, labeled ϕ_i and γ_i, giving rise to a two-photon state that is a superposition of emissions from each atom:

$$|\psi\rangle = \frac{1}{\sqrt{2}}(|\phi_1\rangle|\gamma_1\rangle + |\phi_2\rangle|\gamma_2\rangle). \tag{4.4}$$

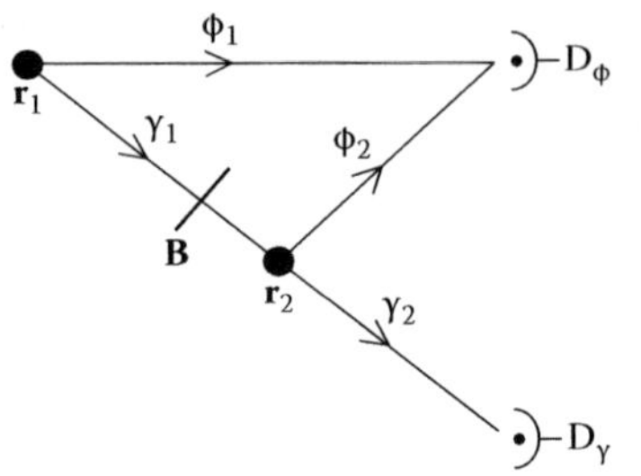

FIGURE 4.7
Two-site downconversion interferometry—Two atoms are located at $\mathbf{r}_1$ and $\mathbf{r}_2$, one of which emits two photons, labeled ϕ_i and γ_i. Detectors D_ϕ and D_γ measure the respective photons. Inserting the beamstop B in the path of γ_1 allows us to infer (potentially, by checking D_γ) which atom emitted the ϕ photon. This potential which-path information is sufficient to prevent the interference of ϕ_1 and ϕ_2 possibilities at D_ϕ. Setup models the experiment of Ref. [34].

This is an entangled state in the sense that an emission of ϕ_i is always accompanied by an emission of γ_i, for $i = 1$ or 2. Let us suppose that we are interested in interference of the ϕ photon only, as measured by varying the path lengths of ϕ_1 and ϕ_2 to detector D_ϕ. The γ photon serves as a marker that potentially records which atom emitted the ϕ photon. It is found that by inserting (or removing) a beamstop in the path of γ_1, the interference fringes can be made to vanish (or re-appear) at D_ϕ, even when D_γ is not actually observed.

It is interesting to explain this phenomenon using stochastic electrodynamics[35] (as was done with the Lamb shift). Let us replace the two photons ϕ and γ with *classical* light fields $E_i^\phi(\mathbf{r}, t)$ and $E_i^\gamma(\mathbf{r}, t)$, generated respectively by dipole transitions *a-b* and *b-c* in each atom *i*. If the atoms are initially in a superposition of states *a* and *c*, then zero-point fluctuations in the field mode γ will introduce population into level *b* (from *a*), with a random phase $\varphi_{\gamma,i}$. The first-order interference in the field mode ϕ will now depend on an ensemble average over the vacuum-induced two-atom phase difference: $\langle E_1^\phi E_2^\phi \rangle \propto \langle \exp[-i(\varphi_{\gamma,1} - \varphi_{\gamma,2})] \rangle$. This quantity goes to zero if the two phases are statistically independent, which is the case when the beamstop is in place between the two atoms. Thus, we have here a connection between vacuum fluctuation physics (which is responsible for spontaneous emission of photons), and two-particle correlation physics (which is the key to quantum erasure).

4.4 The Quantum Field Theory View

A quantum theory of radiation[2–5] is indispensable to understanding the novel properties of light mentioned above. Central to the theory is the idea of field quantization, which develops the formal analogy with the quantum mechanics of the harmonic oscillator. The position q and momentum p of an

UNIVERSITY OF BIRMINGHAM LIBRARY

oscillating particle satisfy the commutation relation $[\hat{q}, \hat{p}] = \hat{q}\hat{p} - \hat{p}\hat{q} = i\hbar$. In the case of the radiation field, q and p represent the electric (E) and magnetic (B) fields of the light in a given wave-vector and polarization mode k. Thus, the quantum electromagnetic field consists of an *infinite* product of such generalized harmonic oscillators, one for each mode of the field. A Heisenberg-type uncertainty relation applies to these quantized Maxwell fields:

$$\Delta E \Delta B \geq \hbar/2 \times \text{constant}. \tag{4.5}$$

Such field fluctuations are an intrinsic feature of the quantized theory. The uncertainty relation can *also* be formulated in terms of the in-phase ($\mathscr{E}_p$) and in-quadrature ($\mathscr{E}_q$) components of the electric field, where $E(t) = \mathscr{E}_p \cos \nu t + \mathscr{E}_q \sin \nu t$.

To introduce the notion of a photon, it is convenient to recast the above quantization of the field in terms of a Fourier decomposition, or in terms of the normal modes of a field in a cavity. These correspond to the positive frequency (going like $e^{-i\nu t}$) and negative frequency (going like $e^{i\nu t}$) parts of the electric field respectively (summed over all modes k):

$$\begin{aligned} E(\mathbf{r}, t) &= E^{+}(\mathbf{r}, t) + E^{-}(\mathbf{r}, t) \\ &= \sum_k [\alpha_k \mathscr{E}_k(\mathbf{r}) \exp(-i\nu_k t) + \alpha_k^* \mathscr{E}_k^*(\mathbf{r}) \exp(i\nu_k t)]. \end{aligned} \tag{4.6}$$

Here α_k is the amplitude of oscillation, and $\mathscr{E}_k(\mathbf{r})$ is a mode function like exp $(i\mathbf{k} \cdot \mathbf{r})$ for travelling waves in free space and sin $(\mathbf{k} \cdot \mathbf{r})$ for standing waves in a box. We consider the oscillator amplitudes α_k and α_k^*, corresponding to harmonic motion, to be quantized by replacing $\alpha_k \to \hat{a}_k$ and $\alpha_k^* \to \hat{a}_k^\dagger$. By analogy to the quantum mechanics of the harmonic oscillator, the application of $\hat{a}$ produces a field state with one *less* quantum of energy, and the application of $\hat{a}^\dagger$ produces a field state with one *more* quantum of energy. This naturally leads to discrete energies for the radiation field in each mode: $n_k = 0$, 1, 2, etc.

Both wave and particle perspectives are present in the quantum view—the former in the picture of a stochastic electromagnetic field, and the latter in the language of particle creation and annihilation. Combining these points of view, one can think of the "photon" as a discrete excitation of a set of modes $\{k\}$ of the electromagnetic field in some cavity, where the mode operators satisfy the boson commutation relation: $[\hat{a}_k, \hat{a}_k^\dagger] = 1$. Questions such as how to define the cavity, and what normal modes to use, cannot be answered once and for all, but depend on the particular physical setup in the laboratory (see quote by Willis Lamb at the beginning). Guided by this operational philosophy, we revisit the wave-particle debate on the nature of light in the guise of the following questions.

4.5 What Is a Photon and Where Is It?

In other words, in what manner (and to what extent) can we regard the photon as a true "particle" that is localized in space? When first introduced, the photon was conceived of as a particulate carrier of discrete light energy, $E = \hbar\upsilon$, a conception guided by considerations of the interaction between radiation and matter. From semiclassical arguments, we saw how this discreteness was related to finite energy spacings in the atom. Here, we pursue this line of reasoning further to inquire whether a fully quantized theory of matter-radiation interaction can lend a characteristic of *spatial discreteness* to the photon when it interacts with a finite-sized atom. This line of thinking derives from the quantum theory of photodetection[36] (which, incidentally, also relies on the photoelectric effect).

Closely related to the issue of photon localization is the (much debated) question of the existence of a photon wave function ψ (**r**, t),[2,7,8] analogous to that of an electron or neutrino (cf. Figure 4.8). The connection is that if such a wave function exists, then we can interpret $|\psi|^2\, dV$ as the probability of finding the photon in an infinitesimal volume element dV in space, and pursue the localization of the entire photon to an arbitrarily small volume constrained only by the uncertainty principle. Moreover, a "first-quantized theory" of the electromagnetic field would be interesting from the point of view of discussing various quantum effects that result from wave interference and entanglement. It would also allow us to treat the mechanics of the photon on par with that of massive particles, such as electrons and atoms, and enable a

	Photon	**Neutrino**
Eikonal physics	Ray optics (Fermat): $\delta\int nds = 0$	Classical mechanics (Hamilton): $\delta\int Ldt = 0$
"Wave" mechanics	Maxwell equations: $\dot{\Psi} = -\frac{i}{\hbar}\begin{bmatrix} 0 & -cs\cdot p \\ cs\cdot p & 0 \end{bmatrix}\Psi$	Dirac equations: $\dot{\Phi} = -\frac{i}{\hbar}\begin{bmatrix} 0 & -c\sigma\cdot p \\ c\sigma\cdot p & 0 \end{bmatrix}\Phi$
Quantum field theory	$\hat{E}^+(r,t) = \Sigma_k \hat{a}_k(t)\varepsilon_k(r)$ $|\dot{\psi}\rangle = -\frac{i}{\hbar}\mathcal{H}_\gamma|\psi\rangle$	$\hat{\phi}^+(r,t) = \Sigma_p \hat{c}_p(t)\phi_p(r)$ $|\dot{\phi}\rangle = -\frac{i}{\hbar}\mathcal{H}_\nu|\phi\rangle$

FIGURE 4.8
Comparison of physical theories of a photon and a neutrino. Eikonal physics describes both in particle terms, showing the parallel between Fermat's principle in optics and Hamilton's principle in classical mechanics (L is the Lagrangian). The Maxwell equations can be formulated in terms of photon wave functions, in the same form that the Dirac equations describe the relativistic wave mechanics of the neutrino. Here, ψ is a six-vector representing the wave functions associated with the electric and magnetic fields, $\mathbf{p} = (\hbar/i)\nabla$ as usual, and $\mathbf{s} = (s_x, s_y, s_z)$ are a set of 3×3 matrices that take the place of the Pauli matrices σ_x, σ_y and σ_z. See Ref [2] for details. Finally, quantum field theory gives a unified description of both the photon and the neutrino in terms of quantized field operators.

unified treatment of matter-radiation interaction that supersedes the semiclassical theory in rigor, but still avoids the language of field quantization.

Concerning the issue of "where" the photon is, one is reminded of an often asked question in introductory quantum mechanics: "How can a single particle go through both slits in a Young-type experiment?"

Richard Feynman answers this by saying "nobody knows, and its best if you try not to think about it." This is good advice if you have a picture of a single photon as a particle. On the other hand if you think of the photon as nothing more nor less than a single quantum excitation of the appropriate normal mode, then things are not so mysterious, and in some sense intuitively obvious.

What we have in mind (referring to Figure 4.1) is to consider a large box having simple normal modes and to put two holes in the box associated with the Young slits. If light is incident on the slits, we will have on the far wall of the box an interference pattern characteristic of classical wave interference, which we can describe as a superposition of normal modes. Now we quantize these normal modes and find that a photodetector on the far wall will indeed respond to the single quantum excitation of a set of normal modes which are localized at the peaks of the interference pattern, and will not respond when placed at the nodes. In this sense, the issue is a non sequitur. The photon is common to the box and has no independent identity in going through one hole or the other.

But to continue this discussion, let us ask what it is that the photodetector responds to. As we will clarify below, this is essentially what has come to be called the photon wave function.[2] Historical arguments have tended to disfavor the existence of such a quantity. For example, in his book on quantum mechanics,[37] Hendrik Kramers asks whether "it is possible to consider the Maxwell equations to be a kind of Schrödinger equation for light particles." His bias against this view is based on the disparity in mathematical form of the two types of equations (specifically, the number of time derivatives in each). The former admits *real* solutions ($\sin \nu t$ and $\cos \nu t$) for the electric and magnetic waves, while the latter is restricted to *complex* wave functions ($e^{i\nu t}$ or $e^{-i\nu t}$, but not both). Another argument is mentioned by David Bohm in his quantum theory book,[38] where he argues that there is no quantity for light equivalent to the electron probability density $P(x) = |\psi(x)|^2$:

There is, strictly speaking, no function that represents the probability of finding a light quantum at a given point. If we choose a region large compared with a wavelength, we obtain approximately

$$P(x) \cong \frac{\mathcal{E}^2(x) + \mathcal{H}^2(x)}{8\pi h\nu(x)},$$

but if this region is defined too well, $\nu(x)$ has no meaning.

Bohm goes on to argue that the continuity equation, which relates the probability density and current density of an electron, cannot be written for light. That is, a precise statement of the conservation of probability cannot be made for the photon. In what follows, we will see that we can partially overcome the objections raised by Kramers and Bohm.

Let us develop the analogy with the electron a bit further. Recall that the wave function of an electron in the coordinate representation is given by $\psi(\mathbf{r}, t) = \langle \mathbf{r} | \psi \rangle$, where $|\mathbf{r}\rangle$ is the position state corresponding to the exact localization of the electron at the point **r** in space. Now the question is, can we write something like this for the photon? The answer is, strictly speaking, "no," because there is no $|\mathbf{r}\rangle$ state for the photon, or more accurately, there is no particle creation operator that creates a photon at an exact point in space. Loosely speaking, even if there were, $\langle \mathbf{r}' | \mathbf{r} \rangle \neq \delta(\mathbf{r} - \mathbf{r}')$ on the scale of a photon wavelength. Nevertheless, we can still define the *detection* of a photon to a precision limited only by the size of the atom (or detector) absorbing it, which can in principle be much smaller than the wavelength. This gives precise, operational meaning to the notion of "localizing" a photon in space.

If we detect the photon by an absorption process, then the interaction coupling the field and the detector is described by the annihilation operator $\hat{E}^{+}(\mathbf{r}, t)$, defined in Eq. 4.6. According to Fermi's Golden Rule, the matrix element of this operator between the initial and final states of the field determines the transition probability. If there is only one photon initially in the state $|\psi\rangle$, then the relevant final state is the vacuum state $|0\rangle$. The probability density of detecting this photon at position **r** and time t is thus proportional to[2]

$$G_{\psi}^{(1)} = |\langle 0 | \hat{E}^{+}(\mathbf{r}, t) | \psi \rangle|^2 = \kappa | \psi_{\mathscr{E}}(\mathbf{r}, t) |^2. \tag{4.7}$$

Here, κ is a dimensional constant such that $|\psi_{\mathscr{E}}|^2$ has units of inverse volume. The quantity $\psi_{\mathscr{E}}(\mathbf{r}, t)$ may thus be regarded as a kind of 'electric-field wave function' for the photon, with $\{\langle 0 | \hat{E}^{+}(\mathbf{r}, t)\}^{\dagger} = \hat{E}^{-}(\mathbf{r}, t) | 0 \rangle$ playing the role of the position state $|\mathbf{r}\rangle$. That is, by summing over infinitely many wave vectors in Eq. 4.6, and appealing to Fourier's theorem, $\hat{E}^{-}(\mathbf{r}, t)$ can be interpreted as an operator that *creates* the photon at the position **r** out of the vacuum. Of course, we have to be careful not to take this interpretation too precisely.

It is interesting to calculate $\psi_{\mathscr{E}}(\mathbf{r}, t)$ for the photon spontaneously emitted by an atom when it decays. Consider a two-level atom located at $\mathbf{r}_0$, initially excited in level a and decaying at a rate Γ to level b, as shown in Figure 4.2. The emitted field state $|\psi\rangle$ is a superposition of one-photon states $|1_k\rangle$, summed over all modes k, written as

$$|\psi\rangle = \sum_k \frac{g_{ab,k} e^{-i\mathbf{k}\cdot\mathbf{r}_0}}{(\nu_k - \omega) + i\Gamma/2} |1_k\rangle, \tag{4.8}$$

where ω is the atomic frequency, and $g_{ab,k}$ is a coupling constant that depends on the dipole moment between levels a and b. The spectrum of the emitted field is approximately Lorentzian, which corresponds in the time domain to an exponential decay of the excited atom. Calculating $\psi_{\mathscr{E}}(\mathbf{r}, t)$ for this state, we obtain

$$\psi_{\mathscr{E}}(\mathbf{r}, t) = K\frac{\sin\eta}{r}\theta(t - r/c)\exp[-i(\omega + i\Gamma/2)(t - r/c)], \tag{4.9}$$

where K is a normalization constant, $r = |\mathbf{r} - \mathbf{r}_0|$ is the radial distance from the atom, and η is the azimuthal angle with respect to the atomic dipole moment. The step function $\theta\ (t - r/c)$ is an indication that nothing will be detected until the light from the atom reaches the detector, travelling at the speed c. Once the detector starts seeing the pulse, the probability of detection $|\psi_{\mathscr{E}}|^2$ decays exponentially in time at the rate Γ. The spatial profile of the pulse mimics the radiation pattern of a classical dipole.

To what extent can we interpret Eq. 4.9 as a kind of wave function for the emitted photon? It certainly has close parallels with the Maxwell theory, since it agrees with what we would write down for the electric field in the far zone of a damped, radiating dipole. We can go even further, and introduce *vector* wave functions $\Psi_{\mathscr{E}}$ and Ψ_{*} corresponding to the electric and magnetic field vectors **E** and **H** respectively, and show that these satisfy the Maxwell equations (see Figure 4.8). This formalism provides the so-called "missing link" between classical Maxwell electrodynamics and quantum field theory.[7] But we have to be careful in how far we carry the analogy with mechanics. For example, there is no real position operator $\hat{\mathbf{r}}$ for the photon in the wave-mechanical limit, as there is for a first-quantized electron. Nevertheless, the wave function $\Psi_{\mathscr{E}}(\mathbf{r}, t)$ does overcome the main objection of Kramers (since it is complex) and partially overcomes that of Bohm (photodetection events are indeed localized to distances smaller than a wavelength).

The real payoff of introducing a photon wave function comes when we generalize this quantity to two or more photons. A "two-photon wave function" $\Psi_{\mathscr{E}}(\mathbf{r}_1, t_1; \mathbf{r}_2, t_2)$ may be introduced along similar lines as above, and used to treat problems in second-order interferometry (see Ref [2], ch. 21). Entanglement between the two photons results in an inseparability of the wave function: $\Psi_{\mathscr{E}}(\mathbf{r}_1, t_1; \mathbf{r}_2, t_2) \neq \phi_{\mathscr{E}}\ (\mathbf{r}_1, t_1)\ \gamma_{\mathscr{E}}\ (\mathbf{r}_2, t_2)$, as in the example of the two-photon state in Eq. 4.4. The novel interference effects associated with such states may be explained in terms of this formalism.

Thus, the photon wave function concept is useful in comparing the interference of classical and quantum light, and allows us to home in on the key distinction between the two paradigms. In particular, through association with photodetection amplitudes, multi-photon wave functions incorporate the phenomenology of quantum-correlated measurement, which is key to explaining the physics of entangled light.

Conclusion

What is a photon? In this chapter, we have strived to address this concept in unambiguous terms, while remaining true to its wonderfully multifaceted nature. The story of our quest to understand the character of light is a long one indeed, and parallels much of the progress of physical theory. Dual conceptions of light, as wave and particle, have co-existed since antiquity. Quantum mechanics officially sanctions this duality, and puts both concepts on an equal footing (to wit, the quantum eraser). The quantum theory of light introduces vacuum fluctuations into the radiation field, and endows field states with quantum, many-particle correlations. Each of these developments provides us with fresh insight on the photon question, and allows us to hone our perspective on the waveparticle debate.

The particulate nature of the photon is evident in its tendency to be absorbed and emitted by matter in discrete units, leading to quantization of light energy. In the spatial domain, the localization of photons by a photodetector makes it possible to define a "wave function" for the photon, which affords a "first-quantized" view of the electromagnetic field by analogy to the quantum mechanics of material particles. Quantum interference and entanglement are exemplified by one-photon and two-photon wave functions, which facilitate comparisons to (and clarify departures from) classical wave optics. Moreover, this interpretive formalism provides a bridge between the two ancient, antithetical conceptions of light—its locality as a particle, and its functionality as a wave.

References

[1] M. O. Scully and M. Sargent III, *Physics Today* **25**, No. 3, March 1972.
[2] M. O. Scully and M. S. Zubairy, *Quantum Optics*, Cambridge Univ. Press, 1997, chs. 1 and 21.
[3] E. R. Pike and S. Sarkar, *Quantum Theory of Radiation*, Cambridge Univ. Press, 1995.
[4] R. Loudon, *The Quantum Theory of Light*, 2nd ed., Oxford Univ. Press, 1983.
[5] W. Schleich, *Quantum Optics in Phase Space*, Wiley-VCH, 2001.
[6] W. E. Lamb, Jr., *Appl. Phys.* **B66**, 77 (1995).
[7] I. Bialynicki-Birula, *Acta Phys. Polonica A*, **86**, 97 (1994).
[8] J. E. Sipe, *Phys. Rev. A* **52**, 1875 (1995).
[9] M. Planck, *Ann. d. Physik* **4**, 553, 564 (1901).
[10] A. Einstein, *Ann. d. Physik* **17**, 132 (1905).
[11] W. E. Lamb, Jr. and M. O. Scully in *Polarization, matter and radiation* (Jubilee volume in honor of Alfred Kastler), Presses Univ. de France, Paris, 1969.
[12] F. K. Richtmyer, E. H. Kennard and T. Lauritsen, *Introduction to Modern Physics*, 5th ed., McGraw Hill, New York (1955), p. 94.
[13] G. N. Lewis, *Nature* **118**, 874 (1926).

[14] M. Sargent III, M. O. Scully and W. E. Lamb, Jr., *Laser Physics*, Addison-Wesley, Reading, MA, 1974.
[15] I. I. Rabi, *Phys. Rev.* **51**, 652 (1937); F. Bloch, *Phys. Rev.* **70**, 460 (1946). See also L. Allen and J. H. Eberly, *Optical Resonance and Two-Level Atoms*, Wiley, New York, 1975.
[16] P. A. M. Dirac, *Proc. Roy. Soc. London A*, **114**, 243 (1927).
[17] M. D. Crisp and E. T. Jaynes, *Phys. Rev.* **179**, 1253 (1969); C. R. Stroud, Jr. and E. T. Jaynes, *Phys. Rev.* A **1**, 106 (1970).
[18] E. T. Jaynes in *Coherence and Quantum Optics IV*, ed. L. Mandel and E. Wolf, Plenum Press, New York, 1978, p. 495.
[19] V. Weisskopf and E. P. Wigner, Z. *Physik* **63**, 54 (1930).
[20] W. E. Lamb, Jr. and R. C. Retherford, *Phys. Rev.* **72**, 241 (1947).
[21] T. A. Welton, *Phys. Rev.* **74**, 1157 (1948).
[22] D. Meschede, H. Walther, and G. Müller, *Phys. Rev. Lett.* **54**, 551 (1985).
[23] M. O. Scully, G. M. Meyer and H. Walther, *Phys. Rev. Lett.* **76**, 4144 (1996).
[24] W. W. Chow, M. O. Scully and J. O. Stoner, Jr., *Phys. Rev. A* **11**, 1380 (1975); R. M. Herman, H. Grotch, R. Kornblith and J. H. Eberly, ibid. p. 1389.
[25] M. O. Scully and K. Drühl, *Phys. Rev.* A **25**, 2208 (1982).
[26] M. O. Scully, B.-G. Englert and H. Walther, *Nature* **351**, 111 (1991).
[27] P. G. Kwiat, A. M. Steinberg and R. Y. Chiao, *Phys. Rev. A* **45**, 7729 (1992).
[28] T. J. Herzog, P. G. Kwiat, H.Weinfurter and A. Zeilinger, *Phys. Rev. Lett.* **75**, 3034 (1995).
[29] Y.-H. Kim, R. Yu, S. P. Kulik, Y. Shih and M. O. Scully, *Phys. Rev. Lett.* 84, 1 (2000).
[30] J. A. Wheeler in *Quantum Theory and Measurement*, ed. J. A. Wheeler and W. H. Zurek, Princeton Univ. Press, Princeton, NJ, 1983.
[31] U. W. Rathe and M. O. Scully, *Lett. Math. Phys.* **34**, 297 (1995).
[32] M. D' Angelo, M. V. Chekhova and Y. Shih, *Phys. Rev. Lett.* **87**, 013602 (2001).
[33] A. N. Boto, P. Kok, D. S. Abrams, S. L. Braunstein, C. P. Williams and J. P. Dowling, *Phys. Rev. Lett.* **85**, 2733 (2000).
[34] X. Y. Zou, L. J. Wang and L. Mandel, *Phys. Rev. Lett.* **67**, 318 (1991).
[35] M. O. Scully and U. W. Rathe, *Opt. Commun.* **110**, 373 (1994).
[36] R. J. Glauber, *Phys. Rev.* **130**, 2529 (1963).
[37] H. A. Kramers, *Quantum Mechanics*, North-Holland, Amsterdam, 1958.
[38] D. Bohm, *Quantum Theory*, Constable, London, 1954, p. 98.

5

A Photon Viewed from Wigner Phase Space

Holger Mack and Wolfgang P. Schleich
Institut für Quantenphysik, Universität Ulm, Germany

CONTENTS

I don't know anything about photons, but I know one when I see one
Roy J. Glauber

We present a brief history of the photon and summarize the canonical procedure to quantize the radiation field. Our answer to the question "what is a photon?" springs from the Wigner representation of quantum mechanics as applied to a single photon number state.

© *2003 Optical Society of America*

5.1 Introduction

For centuries light in its various manifestations has been a pace maker for physics. We are reminded of the wave-particle controversy of classical light between Thomas Young and Isaac Newton. We also recall the decisive role of the null aether experiment of Albert A. Michelson in the birth of special relativity. Many more examples could be listed. However, three phenomena that opened the quantum era stand out most clearly. (i) Black-body radiation paved the way for quantum mechanics. (ii) The level shift in the fine structure of hydrogen, that is the Lamb shift marks the beginning of quantum electrodynamics, and (iii) the almost thirty years lasting debate[1] between Albert Einstein and Niels Bohr on the double-slit experiment could open the path way to quantum information processing in our still young millenium.

The photon as a continuous source of inspiration and its illusiveness has repeatedly been emphasized[2] by John A. Wheeler: Catchy phrases such as "the photon—a smoky dragon", "no elementary quantum phenomenon is a phenomenon until it is a recorded phenomenon", and "it from bit" were coined by him to express in a vivid way the seemingly acausal behavior of the photon in the delayed-choice experiment, the special role of the observer in quantum mechanics, and the concept of a participatory universe due to the measurement process, respectively.

The opposite view, one free from any mystery has been strongly advocated[3] by Willis E. Lamb. According to him the word "photon" should be striken from the dictionary since there is no need for it. The correct approach is: First define modes and then quantize them according to a harmonic oscillator. In the early days of the laser theory, that is the early sixties, Lamb handed out licences to physicists for the word "photon". Only those who were lucky enough to obtain such a license were allowed to use the word "photon". These days are long gone by. Today nobody applies for licenses anymore. We have again freedom of speech and photons appear everywhere even when there is no need for them. Often photons are used in a sloppy way like some people use phrases such as "You know what I mean" in conversations when even they themselves do not know what they mean. In these cases photons serve as a Charlie Brown security blanket.

Such a sloppy approach is not conducive to unravelling the deeper secrets of the photon that are still waiting to be discovered. We, therefore, welcome this opportunity to readdress the old question "what is a photon?" and argue in favor of the canonical approach to field quantization. At the same time we try to communicate the many fascinating facets of the photon. Needless to say, we do not claim to have understood all sides of the photon. Our position is probably best described by Roy J. Glauber's joke: "I don't know anything about photons, but I know one when I see one". This quote is a paraphrase of the well-known attempt of the American Supreme Court Justice Potter Stewart to define obscenity in the 1964 trial Jacobellis versus Ohio by stating "I know it when I see it". Glauber's application to our dilemma with the photon serves as the motto of our chapter. It is worth mentioning that Glauber after his lecture at the Les Houches summer school[4] 1963 was one of the very few people ever given a license for the photon and he had not even applied for one.

Our chapter is organized as follows. A brief historical summary of the quantum theory of radiation emphasizes the crucial roles of Max Born, Pascual

Jordan and Werner Heisenberg in introducing the quantum mechanics of the field.[5,6] This introductory section also alludes to the problem of a hermitian phase operator[7] that originated from Fritz London[8] and was ignored by Paul Adrian Maurice Dirac's seminal paper[9] on the quantum theory of the emission and absorption of radiation. We then outline the formalism[10] of the quantization of the field in a version well-suited for the description of recent experiments[11, 12] in cavity quantum electrodynamics. In this approach we expand the electromagnetic field into a complete set of mode functions. They are determined by the boundary conditions of the resonator containing the radiation. In this language a "photon" is the first excitation of a single mode. The Wigner phase space distribution[13, 14] allows us to visualize the quantum state of a system. We present the Wigner functions for a gallery of quantum states, including a single photon number state. Several proposals to measure the Wigner function have been made.[15] Recently experiments[11, 12, 16] have created and measured the phase space function of a single photon. We conclude by summarizing an approach pioneered by J. A. Wheeler in the context of geometrodynamics.[17] This formalism gives the probability amplitude for a given electric or magnetic field configuration in the vacuum state and does not make use of the notion of a mode function. A brief summary and outlook alludes to the question of a wave function of a photon,[18] addressed in more detail in the article by A. Muthukrishnan *et al.* in this issue.

5.2 History of Field Quantization

It was a desperate situation that Max Planck was facing at the turn of the 19th century. How to explain the energy distribution of black-body radiation measured in the experiments at the Physikalisch-Technische Reichsanstalt by Heinrich Rubens and coworkers with such an unprecedented accuracy? How to bridge the gap between the Rayleigh-Jeans law describing the data correctly for small frequencies and Wien's law valid in the large frequency domain? Planck's revolutionary step is well-known: The oscillators situated in the walls of the black-body resonator can only emit or absorb energy in discrete portions. The smallest energy unit of the oscillator with frequency Ω is $\hbar\Omega$, where in today's notation $\hbar$ is Planck's constant. It is interesting to note that Planck had initially called this new constant Boltzmann's constant—not to be confused with Boltzmann's constant k_B of thermodynamics.

Planck's discovery marks the beginning of quantum mechanics in its early version of Atommechanik à la Bohr-Sommerfeld and the matured wave or matrix mechanics of Erwin Schrödinger and W. Heisenberg. It also constitutes the beginning of the quantum theory of radiation. Although Planck got his pioneering result by quantizing the mechanical oscillators of the wall, it was soon realized that it is the light field whose energy appears in discrete portions. This discreteness suggested the notion of a particle which Einstein in 1905 called "light quantum". The concept of a particle was also supported by his insight that this light quantum enjoys a momentum $\hbar k$ where $k = 2\pi/\lambda$ is the wave number of the light of wave length λ. The name "photon" for the light quantum originated much later. It was the chemist[3] Gilbert N. Lewis at Stanford University who in 1926 coined the word "photon" when he suggested a model of chemical bonding. His model did not catch on, however the photon survived him. For more historical and philosophical details we refer to Ref. 3 and the chapter by A. Zajonc is this book.

The rigorous quantum theory of radiation starts in 1925 with the immediate reaction of Born and Jordan[5] on Heisenberg's deep insight[19] into the inner workings of the atom obtained during a lonely night on the island of Helgoland. Indeed, it is in this paper that Born and Jordan show that the non-commuting objects proposed by Heisenberg are matrices. This article[5] also contains the so-called Heisenberg equations of motion. Moreover, it applies for the first time matrix mechanics to electrodynamics. Born and Jordan recall that the electromagnetic field in a resonator is a collection of uncoupled harmonic oscillators and interpret the electromagnetic field as an operator, that is as a matrix. Each harmonic oscillator is then quantized according to matrix mechanics and the commutation relation $[\hat{q}, \hat{p}] = i\hbar$ between position and momentum operators $\hat{q}$ and $\hat{p}$, respectively. This work is pushed even further in the famous Drei-Männer-Arbeit[6] where also Heisenberg joined Born and Jordan. This paper elucidates many consequences of the quantum theory of radiation from the matrix mechanics point of view. In particular, it calculates from first principles the energy fluctuations of the black-body radiation. From today's demand for rapid publication in the eprint age, it is quite remarkable to recall the submission and publication dates of these three pioneering papers: July 26, 1925, September 27, 1925, November 16, 1925. All three papers were published in 1925.

A new chapter in the book of the quantized electromagnetic field was opened in 1927 when Dirac[9] considered the interaction of a quantized electromagnetic field with an atom which is also described by quantum theory. In this way he derived the Einstein A- and B-coefficients of spontaneous and induced emission. His paper defines the beginning of quantum electrodynamics leading eventually to the modern gauge theories.

Dirac's paper is also remarkable from a different point of view. He does not quantize the field in terms of non-commuting position and momentum operators but by decomposing the annihilation and creation operators $\hat{a}$ and $\hat{a}^\dagger$ into action $\hat{n}$ and angle $\hat{\phi}$ operators with $[\hat{n}, \hat{\phi}] = i\hbar$. However, such a decomposition is not well-defined, since $\hat{n}$ and $\hat{\phi}$ cannot be conjugate variables. Indeed, they have different type of spectra: The spectrum of $\hat{n}$ is discrete whereas the phase in continuous. The problems arising in the translation of classical action-angle variables which are at the heart of the Bohr-Sommerfeld Atommechanik to action-angle operators had already been pointed out by Fritz London[8] in 1926. He showed that there does not exist a hermitian phase operator $\hat{\phi}$. Since then this problem of finding the quantum mechanical analogue of the classical phase has resurfaced repeatedly whenever there was a substantial improvement in the technical tools of preparing quantum states of the radiation field. These periods are characterized by the development of the maser and laser, the generation of squeezed states, and the amazing one-atom maser. In particular, the generation of squeezed light in the mid-eighties has motivated Stephen Barnett and David Pegg[7] to propose a hermitian phase operator in a truncated Hilbert space.

Enrico Fermi independently developed his own approach[10] towards the quantum theory of radiation. In Ref. 10 Fermi applies the quantum theory of radiation to many physical situations. For example, he treats Lippmann fringes and shows that the radiation emitted by one atom and absorbed by another travels with the speed of light. Notwithstanding Fermi's analysis this problem was discussed later in many papers and it was shown that Fermi's model predicts instantaneous propagation.

5.3 Mode Functions

After this historical introduction we briefly summarize in the next two sections the essential ingredients of Fermi's approach towards quantizing the electromagnetic field. Here we concentrate on a domain of space that is free of charges and currents.

In the Coulomb gauge with $\vec{\nabla}\cdot\vec{A}=0$ we find from Maxwell's equations the wave equation

$$\left(\frac{1}{c^2}\frac{\partial^2}{\partial t^2}-\Delta\right)\vec{A}(t,\vec{r})=0 \tag{5.1}$$

for the vector potential $\vec{A}=\vec{A}(t,\vec{r})$ where Δ denotes the three-dimensional Laplace operator.

We shall expand $\vec{A}$ into a complete set of mode functions $\vec{u}_{\vec{k},\sigma}=\vec{u}_{\vec{k},\sigma}(\vec{r})$ defined by the Helmholtz equation

$$(\Delta+\vec{k}^2)\vec{u}_{\vec{k},\sigma}(\vec{r})=0 \tag{5.2}$$

and the boundary conditions set by the shape of the resonator.

For the example of a resonator shaped like a shoe box the mode functions are products of sine and cosine functions. In order to match the boundary conditions of vanishing transverse electric field on the metallic walls the components of the wave vector $\vec{k}$ have to be integer multiples of π/L_j where L_j denotes the length of the j-th side of the resonator. The vector character of the mode function $\vec{u}_{\vec{k},\sigma}$ is determined by the Coulomb gauge condition which for a rectangular resonator takes the form $\vec{k}\cdot\vec{u}_{\vec{k},\sigma}(\vec{r})=0$. Hence, the direction of $\vec{u}$ has to be orthogonal to the wave vector $\vec{k}$. The Coulomb gauge translates into a transverse vector potential which is the reason why this gauge is sometimes referred to as "transverse gauge". Since in general there are two perpendicular directions there are two polarization degrees indicated by the index σ.

At this point it is worthwhile emphasizing that the discreteness of the wave vector is unrelated to quantum mechanics. It is solely determined by the boundary conditions imposed on the Helmholtz equation. Indeed, the variable $\vec{r}$ indicating the position in space is a classical quantity and not a quantum mechanical operator.

For more sophisticated shapes of resonators the mode functions become more complicated. Nevertheless, their basic properties explained above for the elementary example of a box-shaped resonator still hold true. In particular, the mode functions $\vec{u}_\ell(\vec{r})$ are complete and enjoy the orthonormality relation

$$\frac{1}{\sqrt{V_\ell V_{\ell'}}}\int d^3r\, \vec{u}_\ell^*(\vec{r})\cdot \vec{u}_{\ell'}(\vec{r}) = \delta_{\ell,\ell'} \tag{5.3}$$

where V_ℓ denotes the effective volume of the ℓ-th mode. In order to simplify the notation we have combined the three components of the wave vector $\vec{k}$ and the polarization index σ to one index ℓ.

Due to the completeness of the eigenfunctions we can expand the vector potential

$$\vec{A}(t,\vec{r}) \equiv \sum_\ell \mathscr{A}_\ell q_\ell(t)\vec{u}_\ell(\vec{r}) \tag{5.4}$$

where $\mathscr{A}_\ell$ is a constant that we shall choose later in order to simplify the calculations. The time dependent amplitude q_ℓ of the ℓ-th mode follows from the differential equation

$$\ddot{q}_\ell(t) + \Omega_\ell^2 q_\ell(t) = 0 \tag{5.5}$$

of a harmonic oscillator of frequency $\Omega_\ell \equiv c|\vec{k}_\ell|$. Here a dot denotes differentiation with respect to time. This equation emerges when we substitute the expansion, Eq. 5.4, into the wave equation, Eq. 5.1, and make use of the Helmholtz equation, Eq. 5.2.

The notion of the field amplitudes in the modes as harmonic oscillators stands out most clearly when we calculate the energy

$$H \equiv \int d^3r \left(\frac{1}{2}\varepsilon_0 \vec{E}^2 + \frac{1}{2\mu_0}\vec{B}^2\right) \tag{5.6}$$

of the electromagnetic field in the resonator. Indeed, when we use the relations

$$\vec{E} = -\frac{\partial \vec{A}}{\partial t} = -\sum_\ell \mathscr{A}_\ell \dot{q}_\ell \vec{u}_\ell(\vec{r}) \tag{5.7}$$

and

$$\vec{B} = \vec{\nabla} \times \vec{A} = \sum_{\ell} \mathscr{A}_{\ell} q_{\ell} \vec{\nabla} \times \vec{u}_{\ell}(\vec{r}) \tag{5.8}$$

connecting in Coulomb gauge the electric and magnetic fields $\vec{E}$ and $\vec{B}$ with the vector potential $\vec{A}$ we find after a few lines of calculations[14]

$$H = \sum_{\ell} H_{\ell} = \sum_{\ell} \frac{1}{2} \dot{q}_{\ell}^2 + \frac{1}{2} \Omega_{\ell}^2 q_{\ell}^2. \tag{5.9}$$

Here we have used the orthonormality relation, Eq. 5.3, and have chosen the prefactor $\mathscr{A}_{\ell} \equiv (\varepsilon_0 V_{\ell})^{-1/2}$ in the expansion Eq. 5.4.

5.4 Field Operators

According to Eq. 5.9 the electromagnetic field is a collection of harmonic oscillators with conjugate variables q_ℓ and $p_\ell \equiv \dot{q}_\ell$. The natural method to quantize the field is therefore to replace the variables q_ℓ and p_ℓ by operators $\hat{q}_\ell$ and $\hat{p}_\ell$ satisfying the canonical commution relations $[\hat{q}_\ell, \hat{p}_{\ell'}] = i\hbar\delta_{\ell,\ell'}$. In this way we arrive at the operator

$$\hat{\vec{E}}(t,\vec{r}) = -\sum_{\ell} \mathscr{A}_{\ell} \hat{p}_{\ell}(t) \vec{u}_{\ell}(\vec{r}) \tag{5.10}$$

of the electric field and

$$\hat{\vec{B}}(t,\vec{r}) = \sum_{\ell} \mathscr{A}_{\ell} \hat{q}_{\ell}(t) \vec{\nabla} \times \vec{u}_{\ell}(\vec{r}) \tag{5.11}$$

of the magnetic field.

From the expressions Eqs. 5.10 and 5.11 we recognize that $\hat{\vec{E}}$ and $\hat{\vec{B}}$ must be conjugate variables since $\hat{\vec{B}}$ only contains generalized position operators $\hat{q}_\ell$ whereas $\hat{\vec{E}}$ only involves generalized momentum operators $\hat{p}_\ell$. Therefore, it is not surprising that in general it is not possible to measure the electric and magnetic field simultaneously with arbitrary accuracy. The limits put on the accuracy of field measurements has been the subject of two famous papers by N. Bohr and Leon Rosenfeld.[1]

We conclude by casting the quantum analogue

$$\hat{H} \equiv \sum_{\ell} \frac{1}{2} \hat{p}_{\ell}^2 + \frac{1}{2} \Omega_{\ell}^2 \hat{q}_{\ell}^2 \tag{5.12}$$

of the Hamiltonian Eq. 5.9 into a slightly different form. For this purpose it is useful to introduce the annihilation and creation operators $\hat{a}_\ell \equiv [\Omega_\ell/(2\hbar)]^{1/2}$ $(\hat{q}_\ell + i\hat{p}_\ell/\Omega_\ell)$ and $\hat{a}_\ell^\dagger \equiv [\Omega_\ell/(2\hbar)]^{1/2}$ $(\hat{q}_\ell - i\hat{p}_\ell/\Omega_\ell)$, respectively. The commutation relation $[\hat{a}_\ell, \hat{a}_{\ell'}^\dagger] = \delta_{\ell,\ell'}$ follows from the one of $\hat{q}_\ell$ and $\hat{p}_{\ell'}$. The Hamiltonian of the electromagnetic field then takes the form

$$\hat{H} \equiv \sum_\ell \hat{H}_\ell = \sum_\ell \hbar\Omega_\ell \left(\hat{n}_\ell + \frac{1}{2}\right) \tag{5.13}$$

where $\hat{n}_\ell \equiv \hat{a}_\ell^\dagger \hat{a}_\ell$ denotes the number operator.

The contribution 1/2 arises from the commutation relations and results in the familiar zero point energy. Since every mode contributes the energy $\hbar\Omega_\ell/2$ and there are infinitely many modes we arrive at an infinite zero point energy of the electromagnetic field. In general we drop this contribution since a constant shift in the energy, that is, in the Hamiltonian, does not influence the dynamics, even if it is infinite. Under certain circumstances this contribution becomes finite and gives rise to a physical effect. For example, we find an attractive force[20] between two neutral conducting metal surfaces. This Casimir force has also been observed experimentally.[14]

5.5 Quantum States

Operators are only one side of the coin of quantum mechanics. The other one is the description of the quantum system, that is, the electromagnetic field, by a quantum state. In general this state $|\psi\rangle$ is a multimode state, that is, it involves a quantum state $|\psi_\ell\rangle$ for each mode ℓ. In the most elementary situation the states of the individual modes are independent from each other and the state of the electromagnetic field is a product state

$$|\Psi\rangle \equiv \prod_\ell |\psi_\ell\rangle = \cdots |\psi_{-1}\rangle \otimes |\psi_0\rangle \otimes |\psi_1\rangle \cdots. \tag{5.14}$$

However, the most interesting states are the ones where two or more modes are correlated with each other. Schrödinger in his famous paper[1] *"On the current situation of quantum mechanics"* triggered by the Einstein-Podolsky-Rosen paper[1] asking the question *"Can quantum-mechanical description of physical reality be considered complete?"* coined the phrase "entangled states". In order to describe entangled states it is useful to first introduce the most elementary quantum states, namely photon number states $|n_\ell\rangle$.

The states $|n_\ell\rangle$ are eigenstates of the operator $\hat{n}_\ell$, defined by

$$\hat{n}_\ell |n_\ell\rangle = n_\ell |n_\ell\rangle \tag{5.15}$$

with integer eigenvalues. Since the states $|n_\ell\rangle$ are eigenstates of the Hamiltonian $\hat{H}_\ell$ of the ℓ-th mode the energy of the field in the state $|n_\ell\rangle$ is then (neglecting the zero-point energy) $n_\ell \hbar\Omega_\ell$, that is n_ℓ times the fundamental unit $\hbar\Omega_\ell$. This feature has led to the notion that n_ℓ quanta of energy $\hbar\Omega_\ell$ are in this mode. But we emphasize that this energy is distributed over the whole resonator. It cannot be localized at a specific position $\vec{r}$. Indeed, recall that we have found the Hamiltonian, Eq. 5.9, by integrating the energy density, Eq. 5.6, over the whole resonator. Due to the discreteness in the excitation of the mode in portions of units $\hbar\Omega_\ell$ the expression "photon" for this excitation is appropriate.

We are now in the position to discuss the notion of an entangled state. The state $|\psi\rangle$ of a given mode is in general a superposition of photon number states, that is

$$|\psi\rangle \equiv \sum_n \psi_n |n\rangle. \tag{5.16}$$

We emphasize that here the subscript n is not a mode index but counts the quanta in a single mode.

Two states $|\psi\rangle$ and $|\tilde{\psi}\rangle$ that are independent of each other are then described by a direct product, that is

$$|\Psi\rangle = |\psi\rangle \otimes |\tilde{\psi}\rangle = \sum_{m,n} \psi_m \tilde{\psi}_n |m\rangle |n\rangle. \tag{5.17}$$

In case the two states are entangled we find

$$|\Psi\rangle \equiv \sum_{m,n} \Psi_{m,n} |m\rangle |n\rangle. \tag{5.18}$$

where the expansion coefficients $\Psi_{m,n}$ do not factorize into a product of two contributions solely related to the two individual modes.

Entangled states are the essential ingredients of the newly emerging and rapidly moving field of quantum information processing.[21] They can be created by non-linear optical processes such as parametric down-conversion as discussed in the next section or by beam splitters as outlined in detail by R. Loudon and A. Zajonc in their chapters in the present volume.

5.6 Wigner Functions of Photons

In the following two sections we focus on states of a single mode of the radiation field and for the sake of simplicity suppress the mode index. We introduce the Wigner phase space distribution and discuss experiments measuring the Wigner function of a single photon.

A photon denoted by the quantum state $|1\rangle$ is an excitation of a mode of the electromagnetic field. But how to gain deeper insight into this state?

Here, the Wigner function offers itself as a useful tool to visualize the rather abstract object of a quantum state. It was introduced in 1932 by Eugene Paul Wigner in a paper[13] concerned with the corrections of quantum mechanics to classical statistical mechanics. It is remarkable that in a footnote Wigner shares the fame as the original proposer of this phase space distribution function. He states: "This expression was found by L. Szilard and the present author some years ago for another purpose".

However, no such paper by Leo Szilard and Wigner exists. Later in life Wigner explained that he had only added this footnote in order to assist Szilard in his search for a research position.[22] It is astonishing that Heisenberg[23] and Dirac,[24] who later was to become Wigner's brother in law, had already earlier introduced this phase space function. In particular, Dirac had also studied many of its properties and amazingly enough Wigner seemed to be unaware of Dirac's work.

We now turn to the definition of the Wigner phase space distribution. For this purpose it is useful to recall that the eigenstates $|E\rangle$ of the single-mode electric field operator $\hat{\vec{E}} = -\mathscr{A}_0\hat{p}\vec{u}(\vec{r})$ are proportional to the eigenstates $|p\rangle$ of the momentum operator $\hat{p}$. Likewise, the eigenstates $|B\rangle$ of the single mode magnetic field operator $\hat{\vec{B}} = \mathscr{A}_0\hat{q}\nabla\times\vec{u}(\vec{r})$ are proportional to the eigenstates $|q\rangle$ of the position operator $\hat{q}$.

The Wigner function $W = W(q, p)$ of a state $|\psi\rangle$ with wave function $\psi(q) \equiv \langle q|\psi\rangle$ is defined by

$$W(q,p) \equiv \frac{1}{2\pi\hbar}\int_{-\infty}^{\infty} d\xi\, e^{-ip\xi/\hbar}\, \psi^*\left(q-\frac{\xi}{2}\right)\psi\left(q+\frac{\xi}{2}\right) \tag{5.19}$$

where q and p are conjugate variables. For a massive particle they correspond to position and momentum whereas in the case of the electromagnetic field they represent the amplitude of the magnetic and electric field, respectively.

Hence, the problem of finding the Wigner function of a given wave function amounts to evaluating the integral Eq. 5.19. For the example of a photon number state $|n\rangle$ of a mode with frequency Ω the wave function $\varphi_n(q;\Omega) \equiv \langle q|n\rangle$ reads[14]

$$\varphi_n(q;\Omega) \equiv N_n(\Omega)H_n\left(\sqrt{\frac{\Omega}{\hbar}}q\right)\exp\left(-\frac{1}{2}\frac{\Omega}{\hbar}q^2\right) \tag{5.20}$$

where $N_n(\Omega) \equiv (\Omega/(\pi\hbar))^{1/4}\,(2^n n!)^{-1/2}$ and H_n denotes the n-th Hermite polynomial.

When we substitute this expression into the definition, Eq. 5.19, of the Wigner function and perform the integration we arrive at[14]

$$W_n(q,p) = \frac{(-1)^n}{\pi\hbar}L_n[2\eta(q,p)]\exp[-\eta(q,p)] \tag{5.21}$$

where $\eta(q,p) \equiv (p^2 + \Omega^2 q^2)/(\hbar\Omega)$ is the scaled phase space trajectory of a classical harmonic oscillator and L_n denotes the n-th Laguerre polynomial.

The two phase space variables q and p enter the Wigner function in a symmetric way. Moreover, the Wigner function is constant along the classical phase space trajectories, that is along circles. Its behavior along the radial direction is determined by the Laguerre polynomial. In order to study these features in more detail we now analyze and display in Fig. 5.1 the Wigner functions of the ground state, a one-photon and a six-photon state.

We start our discussion with the Wigner function of the ground state, that is $n = 0$ where according to Eq. 5.20 the wave function

$$\varphi_0(q;\Omega) = N_0(\Omega)\exp\left(-\frac{1}{2}\frac{\Omega}{\hbar}q^2\right) \tag{5.22}$$

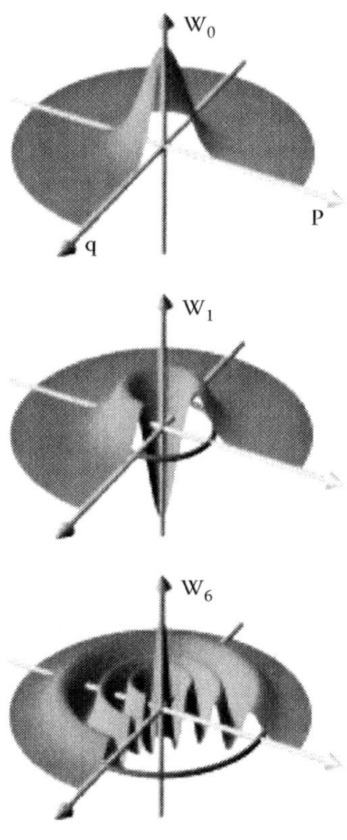

FIGURE 5.1
Gallery of Wigner functions of a single mode of the radiation field. The Wigner function of the vacuum (top) is always positive whereas the ones corresponding to a single photon (center) or six photons (bottom) contain significant domains where the phase space distribution assumes negative values. The circle visible in the quadrant of the foreground indicates where the phase space trajectory corresponding to the energy $\hbar\Omega(n + 1/2)$ runs. The scales on the axes are identical in all three cases.

is a Gaussian. The corresponding Wigner function

$$W_0(q,p) = \frac{1}{\pi\hbar}\exp\left[-\frac{1}{\hbar\Omega}(\Omega^2 q^2 + p^2)\right], \tag{5.23}$$

is then a Gaussian in the generalized position and momentum variables, that is in the electric and magnetic field amplitudes. Thus, the Wigner function of the ground state, that is a mode with no excitation, that is no photons, is everywhere positive.

We now turn to the Wigner function of a single photon, that is of the first excited state $|1\rangle$. Since the first Laguerre polynomial reads $L_1 = 1 - x$ the Wigner function, Eq. 5.21, takes the form

$$W_1(q,p) = \frac{(-1)}{\pi\hbar}(1-2\eta)e^{-\eta}. \tag{5.24}$$

Hence, at the origin of phase space the Wigner function assumes the negative value $W_1(0,0) = (-1)/(\pi\hbar)$. Figure 5.1 shows that the Wigner function is not only negative at the origin, but also in a substantial part of its neighborhood. It is the existence of negative parts that rules out a probability interpretation of the Wigner function. Nevertheless it can be used to develop a formalism of quantum mechanics in phase space,[14] that is equivalent to the one in Hilbert space.

The negative parts of the Wigner function are a consequence of the wave nature of quantum mechanics. This feature stands out most clearly when we consider the Wigner function of a photon number state with many photons in it. In Fig. 5.1 we show the Wigner function corresponding to the state $|6\rangle$. We recognize circular wave troughs that alternate with circular wave crests. The Wigner function repeatedly assumes negative values and contains $n = 6$ nodes. The last positive crest is located in the neighborhood of the classical phase space trajectory corresponding to the quantized energy $E = \hbar\Omega(n + \frac{1}{2})$ of this state. Hence, this positive-valued ring represents the classical part of the state $|n\rangle$. The fringes caught inside reflect the quantum nature of the state. In order to gain deeper insight into this separation of wave and particle nature, we recall that a photon number state is an energy eigenstate of a harmonic oscillator. In the limit of large n, that is many quanta of excitation, this state is the superposition of a right- and a left-going wave. Since the Wigner function, Eq. 5.19, is bilinear in the wave function the interference between these two waves manifests itself in the structures circumnavigated by the classical crest.

5.7 Measured Wigner Functions

Wigner functions of a single photon have recently been observed experimentally. Space does not allow us to present these experiments in every detail, nor can we provide a complete theoretical description. Here we only

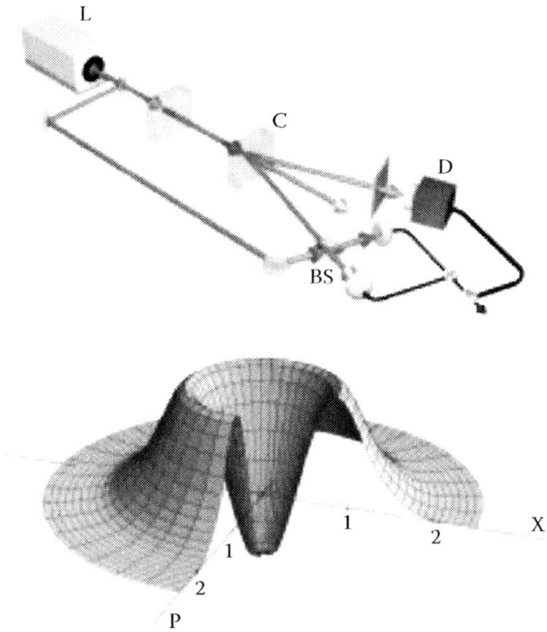

FIGURE 5.2
Quantum state tomography of a single photon. Generation of entangled photons and triggered homodyne detection (top) leads to the reconstruction of the Wigner function (bottom). A laser L creates through a non-linear interaction in a crystal C a pair of photons in two modes. The photon in the upper mode triggers a detector D and the photon in the lower mode gets mixed on a beam splitter BS with a portion of the original laser field which serves as a local oscillator. The difference in the two mixed photo-currents (homodyne detector) is correlated with the detection of the photon in the upper mode. The current distributions for various phases of the laser field together with a mathematical algorithm—the Radon transform—yield the Wigner function of a single photon. After Lvovsky et al., *Phys. Rev. Lett.* **87**, 050402 (2001).

try to give the flavor of these experiments and refer to the literature[11, 12] for more details.

There are essentially two types of experiments. The first approach shown in Fig. 5.2 uses the method of quantum state tomography to reconstruct the Wigner function, whereas the second technique summarized in Fig. 5.3 obtains the Wigner function from the output of a Ramsey set-up.

In the tomography approach the quantized light field to be investigated is mixed on a beam splitter with a classical field of rather well-defined phase. The currents emerging from two photodetectors are subtracted. In contrast to many other experiments which only measure the average of the current for the reconstruction of the Wigner function we need the full statistics of the current fluctuations, that is the probability distribution of the current. These measurements are repeated for many different phases of the classical field. A mathematical algorithm, the so-called Radon transform,[14] enables us to obtain from this set of data the Wigner function of the underlying state.

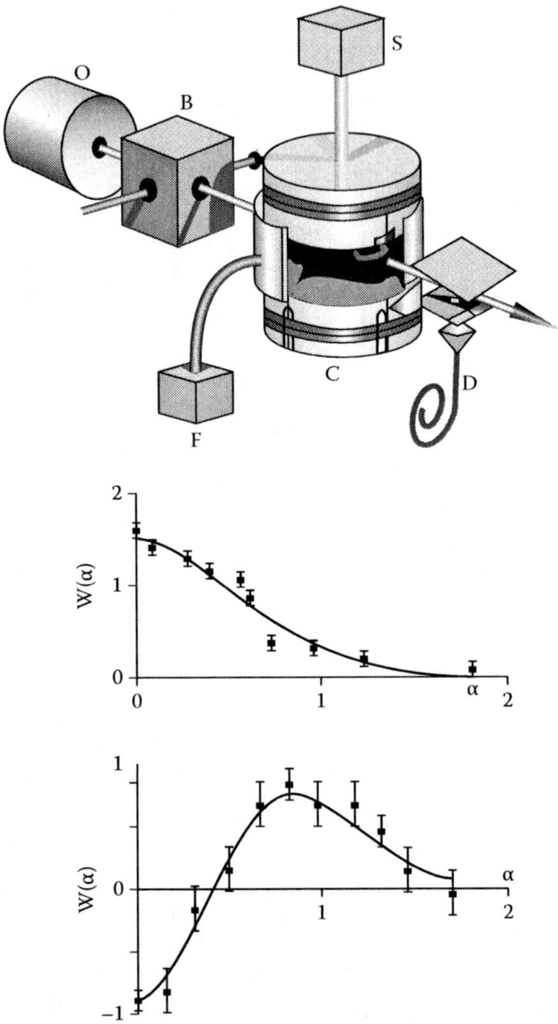

FIGURE 5.3
Ramsey interferometry (top) to reconstruct the Wigner functions (bottom) of the vacuum (bottom upper) and a single photon (bottom lower) in an ideal cavity. An atomic beam emerging from an oven O and prepared in B in a Rydberg state probes the field in a cavity C. For this purpose two classical light fields F first prepare and then probe two internal levels of the atom: The first field prepares a dipole whereas the second field reads out the change of the dipole due to the interaction with the cavity field. A detector D measures the populations in the two levels as a function of the phase difference between the two classical fields. These Ramsey fringes are recorded for various displacements of a classical field S injected into the cavity. The contrast of the fringes for a given displacement α determines the Wigner function at the phase space point α. After P. Bertet *et al.*, *Phys. Rev. Lett.* **89**, 200402 (2002).

Figure 5.2 shows the so-reconstructed Wigner function[11] of a single photon state created by a parametric process in a crystal. We recognize the negative parts around the origin.

The second experiment[12] is from the realm of cavity QED. Here an atom probes the quantum state of the field inside a resonator. This field has been prepared earlier by one or more atoms. In this method of state reconstruction the information about the state is stored in the internal states of the atom. In order to be sensitive to interference in the field the atoms enter and are probed in a coherent superposition of their internal states. For the sake of simplicity we have assumed here only two internal states. A detector at the exit of the device measures the populations in the two states as a function of the amplitude of a classical field injected into the resonator. The contrast of the interference structures determines the value of the Wigner function.

In Fig. 5.3 we show the radial cut of the so-obtained Wigner function of the vacuum and a single photon. Whereas the vacuum enjoys a Gaussian Wigner function, Eq. 5.23, that is positive everywhere the one corresponding to a single photon, Eq. 5.24, displays clearly substantial negative parts around the origin.

5.8 Wave Functional of Vacuum

Find the mode functions appropriate for the problem at hand and quantize every mode oscillator according to the canonical prescription—that is the one-sentence summary of the quantum theory of radiation. The excitations of these modes are the photons. The situation when all mode oscillators are in their ground states defines the vacuum of the electromagnetic field.

This approach relies heavily on the concept of a mode function. We now briefly review a treatment[17] that does not involve mode functions but refers to the complete electromagnetic field given by all modes. This formulation provides us with a probability amplitude $\Psi = \Psi[\vec{B}(\vec{r})]$ for a given magnetic field configuration $\vec{B} = \vec{B}(\vec{r})$ being in the ground state.

In order to motivate this expression we first consider a single mode of frequency Ω_ℓ characterized by the mode index ℓ. We assume that the field in this mode is in the ground state. According to Eq. 5.22 the corresponding probability amplitude $\varphi_0(q_\ell;\Omega_\ell)$ to find the value q_ℓ determining the magnetic field via Eq. 5.8 is then the Gaussian distribution

$$\psi_\ell(q_\ell) = \mathcal{N}_\ell \exp\left(-\frac{1}{2}\frac{\Omega_\ell}{\hbar}q_\ell^2\right) \tag{5.25}$$

where $\mathcal{N}_\ell \equiv N_0(\Omega_\ell)$ denotes the normalization constant.

The probability amplitude Ψ for the vacuum of the complete electromagnetic field, that is all modes in the ground state, with the scaled magnetic field q_{-1} in the mode −1, and the field q_0 in the mode 0, the amplitude q_1 in mode 1 and … is the product

$$\Psi = \cdots \psi_{-1}(q_{-1}) \cdot \psi_0(q_0) \cdot \psi_1(q_1) \cdots = \prod_\ell \psi_\ell(q_\ell) \tag{5.26}$$

of the ground state wave functions ψ_ℓ of these modes. This product in wave function space is an example for a multimode state $|\Psi\rangle$ expressed in Eq. 5.14 in terms of state vectors.

When we recall the Gaussian wave function, Eq. 5.25 and make use of the property $e^A \cdot e^B = e^{A+B}$ of the exponential function we arrive at

$$\Psi = N \exp\left(-\frac{1}{2\hbar} \sum_\ell \Omega_\ell q_\ell^2 \right). \tag{5.27}$$

Here, we have introduced the normalization constant $N \equiv \Pi_\ell \mathcal{N}_\ell$.

In the derivation of the Hamiltonian Eq. 5.9 we have used the relation

$$\frac{1}{\mu_0} \int d^3 r \vec{B}^2(\vec{r}) = \sum_\ell \Omega_\ell^2 q_\ell^2. \tag{5.28}$$

The integral of the square of the magnetic field translates into a sum of the squares of the mode amplitudes. Hence, we should be able to express the sum in the ground state wave function Ψ, Eq. 5.27, in terms of a bilinear product of magnetic fields. However, in contrast to Eq. 5.28 Ψ involves Ω_ℓ only in a linear way. Hence, the connection between the sum in Eq. 5.27 and the magnetic field must be more complicated. Indeed, Wheeler showed[17] that such a connection exists which finally yields

$$\Psi[\vec{B}(\vec{r})] = N \exp\left[-\frac{1}{16\pi^3 \hbar c} \int d^3 r_1 \int d^3 r_2 \frac{\vec{B}(\vec{r}_1) \cdot \vec{B}(\vec{r}_2)}{|\vec{r}_1 - \vec{r}_2|^2} \right]. \tag{5.29}$$

The quantity Ψ is the ground state functional. It is not an ordinary function but a functional since it depends not on a point but a whole function $\vec{B} = \vec{B}(\vec{r})$. Indeed, it is the probability amplitude to find the magnetic field distribution $\vec{B} = \vec{B}(\vec{r})$ in the vacuum state. In this approach no explicit mentioning of a mode function is made.

Conclusions

The photon has come a long way. From Planck's minimal portion of energy triggering the quantum revolution at the end of the 19th century, via the quantum of excitation of the electromagnetic field dominating the physics of the 20th century, to entangled photons as resources of quantum cryptography and teleportation. In this version photons will surely be central to the quantum technology of the 21st century. At last we have achieved a complete understanding of the photon, we might think.

Is our situation not reminiscent of 1874 when professor of physics Phillip von Jolly at the University of Munich tried to discourage the young Planck from studying theoretical physics with the words: "Theoretical physics is an alright field ... but I doubt that you can achieve anything fundamentally new in it" (German original: "Theoretische Physik, das ist ja ein ganz schönes Fach ... aber grundsätzlich Neues werden sie darin kaum mehr leisten können")? In hindsight we know how wrong von Jolly was in his judgement.

Today there exist many hints that the photon might again be ready for suprises. For example, we do not have a generally accepted wave function of the photon. Many candidates[18] offer themselves: Should we use the classical Maxwell field, the energy density, or the Glauber coherence functions?[4] The pros and cons of the various approaches have been nicely argued in the paper by A. Muthukrishnan *et al.* in this volume. But could it be that there is no such wave function at all? Would this exception not point into a new direction?

Closely related to the problem of the proper photon wave function is the question of the position operator of a photon.[25] Might there be a completely new aspect of the photon lurking behind these questions?

D. Finkelstein's article in this volume is even arguing that there is still too much commutativity in quantum mechanics—restricting it further might lead to an even richer land of quantum phenomena.

Make no mistake, we have learned a lot since Einstein's famous admission about his lack of deeper insight into the photon. Nevertheless, we have only started to scratch the surface. Many more exciting discoveries can be expected to appear in the next hundred years of a photon's life.

Acknowledgments

For many years we have enjoyed and profited from numerous fruitful discussions with A. Barut, I. Bialynicki-Birula, R. Glauber, W. E. Lamb, M. O. Scully, H. Walther, and J. A. Wheeler on the question "what is a photon?" We are grateful to them for sharing their insights with us. We thank A. I. Lvovsky and J. M. Raimond for providing us with the experimental figures of Refs. 11 and 12. Moreover, we appreciate the kind support provided by the Alexander von Humboldt-Stiftung, the European Community and Deutsche Forschungsgemeinschaft.

For more information, e-mail W. P. Schleich at wolfgang.schleich@uni–ulm.de or go to http://www.physik.uniulm. de/quan/.

References

1. These seminal articles are reprinted in and commented on in J. A. Wheeler and W. Zurek, *Quantum Theory and Measurement* (Princeton University Press, Princeton, 1983).
2. J. A. Wheeler, Frontiers of time, in: *Problems in the foundations of physics, International School of Physics "Enrico Fermi"*, course LXXII (North-Holland, Amsterdam, 1979).
3. See for example, W. E. Lamb Jr., Anti-photon, *Appl. Phys.* B **60**, 77–84 (1995).
4. R. J. Glauber, in: *Quantum Optics and Electronics*, Les Houches, edited by C. DeWitt, A. Blandin and C. Cohen-Tannoudji (Gordon and Breach, New York, 1965), p. 331–381.
5. M. Born and P. Jordan, Zur Quantenmechanik, *Z. Physik* **34**, 858–889 (1925).
6. M. Born, W. Heisenberg, and P. Jordan, Zur Quantenmechanik, *II, Z. Phys.* **35**, 557–615 (1925).
7. See for example the special issue on Quantum phase and phase dependent measurements edited by W. P. Schleich and S. M. Barnett, *Physica Scripta* **T48** (1993).
8. F. London, Über die Jacobischen Transformationen der Quantenmechanik, *Z. Phys.* **37**, 915–925 (1926); Winkelvariable und kanonische Transformationen in der Undulationsmechanik, *Z. Phys.* **40**, 193–210 (1927).
9. P. A. M. Dirac, The quantum theory of the emission and absorption of radiation, *Proc. Roy. Soc.* A **114**, 243–265 (1927).
10. E. Fermi, *Quantum theory of radiation*, Rev. Mod. Phys. **4**, 87–132 (1932).
11. A. I. Lvovsky, H. Hansen, T. Aichele, O. Benson, J. Mlynek, and S. Schiller, Quantum state reconstruction of the single-photon Fock state, *Phys. Rev. Lett.* **87**, 050402 (2001).
12. P. Bertet, A. Auffeves, P. Maioli, S. Osnaghi, T. Meunier, M. Brune, J. M. Raimond, and S. Haroche, Direct measurement of the Wigner function of a one-photon Fock state in a cavity, *Phys. Rev. Lett.* **89**, 200402 (2002).
13. E. P. Wigner, On the quantum correction for thermodynamic equilibrium, *Phys. Rev.* **40**, 749–759 (1932).
14. See for example W. P. Schleich, *Quantum Optics in Phase Space* (Wiley-VCH, Weinheim, 2001).
15. See for example P. Lougovski, E. Solano, Z. M. Zhang, H. Walther, H. Mack and W. P. Schleich, Fresnel representation of the Wigner function: An operational approach, *Phys. Rev. Lett.* **91**, 010401 (2003) and references therein.
16. B. T. H. Varcoe, S. Brattke, M. Weidinger and H. Walther, Preparing pure photon number states of the radiation field, *Nature* **403**, 743–746 (2000).

17. J. A. Wheeler, *Geometrodynamics* (Academic Press, New York, 1962), Z. Bialynicka-Birula and I. Bialynicki-Birula, Space-time description of squeezing, *J. Opt. Soc. Am.* B **4**, 1621–1626 (1987).
18. M. O. Scully and M. S. Zubairy, *Quantum Optics* (Cambridge University Press, Cambridge, 1997); a nice review and novel approach can be found in I. Bialynicki-Birula, *Photon wave function*, Progress in Optics, Volume XXXVI (North Holland, Amsterdam, 1996), p. 245–294 or On the wave function of the photon, *Acta Physica Polonica* A **86**, 97 (1994).
19. W. Heisenberg, Über quantentheoretische Umdeutung kinematischer und mechanischer Beziehungen, *Z. Phys.* **33**, 879–893 (1925).
20. See for example, E. Elizalde and A. Romeo, Essentials of the Casimir effect and its computation, *Am. J. Phys.* **59**, 711–719 (1991); there is an interesting "classical" Casimir effect. Two ships that lie parallel to each other in a harbour attract each other as discussed in S. L. Boersma, A maritime analogy of the Casimir effect, *Am J. Phys.* **46**, 539–541 (1996).
21. *The Physics of Quantum Information*, edited by D. Bouwmeester, A. Ekert and A. Zeilinger (Springer, Berlin, 2000), *Quantum Information*, edited by G. Alber *et al.*, *Springer Tracts in Modern Physics,* Vol. 173 (Springer, New York, 2001).
22. R. F. O'Connell (private communication).
23. W. Heisenberg, Über die inkohärente Streuung von Röntgenstrahlen, *Physik. Zeitschr.* **32**, 737–740 (1931).
24. P. A. M. Dirac, Note on exchange phenomena in the Thomas atom, *Proc. Camb. Phil. Soc.* **26**, 376–395 (1930).
25. The question of the position operator of a photon has been raised by T. D. Newton and E. P. Wigner, Localized states for elementary systems, *Rev. Mod. Phys.* **21**, 400–406 (1949); see also E. R. Pike and S. Sarkar, Spatial dependence of weakly localized single-photon wave packets, *Phys. Rev. A* **35**, 926–928 (1987).

Section 2

Epistemological Origin of Logical Contradiction

6

Inevitable Incompleteness of All Theories: An Epistemology to Continuously Refine Human Logics Towards Cosmic Logics

Chandrasekhar Roychoudhuri

Physics Department, University of Connecticut, Storrs, Connecticut

CONTENTS

Abstract

This article proposes a methodology of thinking (epistemology) to assist scientific exploration of real physical processes in nature (ontology). Our first assumption is that whatever we sense (experimentally or observationally), always represents real interactions between physical entities in nature. Our second assumption is that nature evolves through causal (logical) interactions between different entities, which are validated by the very successes of our logical mathematical theories. So, our objective is to understand and visualize all the processes taking place in nature, which are at the root of cosmic and biospheric evolution. Unfortunately, we do not know any of the natural entities completely. Further, the transformations (changes) that we measure or observe do not provide us with the complete information regarding neither all the forces that the interactants are experiencing, nor can they relay to us through our measuring device(s) all the information regarding any particular transformation they experience in any experiment. Thus we are forever challenged to create a causal theory about nature without inventing (imaginary) human logics to fill in the gap of incomplete information to construct a theory that hopefully will map the cosmic logics behind the interactions we are studying. To overcome this "incomplete information paradigm", we need a scientific epistemology to iteratively keep on refining our human logics in all theories and move them closer and closer to our goal of mapping the cosmic logics. This "incomplete information paradigm" underscores the inevitability of paradoxes, contradictions and confusions in our conceptual interpretations of any theory. In this article, we explore these paradoxes regarding wave-particle duality of photons and suggest possible resolutions of such paradoxes.

6.1 Introduction

The purpose of this chapter is to generate sufficient doubt in the minds of the readers regarding the current definition of photon by proposing a new paradigm of thinking for doing science. Hopefully, this will entice the readers to explore the out-of-box proposals regarding what photons are presented in Section III.

But, do we really need another paradigm change in thinking for doing science? We think so because some of the leading thinkers like Smolin [1] Laughlin [2] and Penrose [3] are expressing doubt about the direction of physics research. Conferences for out-of-the-box thinkers are being organized [4,5], although these are miniscule in size compared to main stream conferences. And this book itself is an attempt to inspire thinking about photons beyond the currently accepted definition—a monochromatic Fourier mode of the vacuum. We want to underscore that our approach is that of reverse engineers by accepting nature as a creative system engineer. Everything in the micro and macro domains of nature, single cells or galaxies, are all very complex systems constantly undergoing orderly and creative transformations through assembly, dis-assembly and re-assembly.

Today we have over half a dozen or more "solved puzzles" or theories that are logically congruent and self consistent in mapping the behavior of different domains of nature: (i) classical theory, (ii) special relativity, (iii) general relativity, (iv) quantum mechanics, (v) quantum field theories, (vi) cosmology, (vii) string theory, etc. But we have been failing to merge these separate "solved puzzle" pieces into one harmonious bigger puzzle even though the number of operating forces behind all possible transformations are only four, so far. It is important to appreciate that mathematics being pure logic, an equation "working" in modeling nature represents causally connected terms (states of interactants) by appropriate symbols (interacting force between the interactants and outcome). Physical meaning, the reality, or visualization of the interaction processes behind the equation, is a matter of human interpretation, and not a mathematically derivable set of statements. Hence, interpretations of any equation should not be considered as either unique or final. Thus, we must maintain serious scientific doubts on the imposition of interpretations like non-causality on causal mathematical relations and the underlying interactions as non-local when they represent interactions between physical interactants through forces, which are always of finite range. Therefore, our interpretation process requires a well structured methodology of thinking, or an epistemology to sort out the difference and connectivity between different *human logics* (epistemology) that have organized the theories and the *cosmic logics* (ontology) that run all the real *interaction processes* in our universe. If we treat all the "working" theories as inviolable, we will never succeed advancing science very much further. Almost thirty years of failure to find anything fundamentally new in physics clearly tells us that we need to reassess all the hypotheses that are behind all these different "successful" theories [1-3] and revisit the purpose of physics. We believe that the motto of classical physics, understanding and visualizing the physical processes undergoing in nature, should be our key guidance.

It is generally acknowledged that framing a question determines the answer we create by developing a theory around various observations. The frame of our enquiring mind, or the model of our thinking, which is varied and quite complex, determines how we frame our questions. This makes debating different interpretations of the same theory sometimes confusing, the best example being the unresolved [6] "Bohr-Einstein debate" over *reality* about quantum mechanics [7]. Another good example is our insistence on the same questions like, "what are light quanta?" [8], which has yielded very little new information about the deeper nature of light for over a century. Semi classical analysis yields most of the light-matter interactions [9]. The formalism of quantum mechanics (QM) "works" very well and Schrödinger's equation has opened up a flood gate of accurate predictions about the quantum world of micro universe.

Obviously, QM must have captured a good amount of fundamental *realities* regarding interaction *processes* behind atoms, molecules and their interactions. Instead of accepting conceptual problems of QM as a guide to discover better or newer theories [10], we are mystifying nature to be non-causal whenever our attempt to visualize the micro world becomes unsuccessful. Logically it is more self consistent to accept emergence of a chaotic and apparent non-causal macro system out of constituent entities interacting causally but randomly. But, it is difficult to accept the emergence of our causally evolving macro universe to be built out of fundamentally non-causal micro interactions between elementary particles. Culturally we have become so accustomed to accept "nobody understands quantum mechanics" that we do not question the current interpretations and accept that QM is "complete". We are still engaged in creating wide ranges of non-causal, non-local interpretations leading to accept teleportation, delayed superposition, etc., to accommodate Dirac's statement, "photon interferes only with itself", which perhaps appeared logical in 1930.

We ought to urge students with proactive encouragements that there must be something seriously wrong with the current interpretations of QM and initiate efforts towards finding better interpretations and eventually frame a better theory to supersede QM, just as QM superseded classical mechanics. A broadly accepted simple and rational epistemology could facilitate our understanding how we have become more inclined to invent many mathematical realities for nature rather than staying focused on discovering actual realities in action. These realities, however elusive they may be to visualize, are manifest through incessant interaction *processes* between diverse entities, both in the macro and micro domains of the entire universe.

We need to develop a better methodology of thinking, debating and scrutinizing information gathered from new and old experiments and theories and learn to re-phrase our exploratory questions and re-evaluate the current state of understanding. In this article we propose an epistemology that will encourage the next generation to carry out such re-evaluation to advance physics [11]. We must also acknowledge at the outset that the proposed epistemology itself being a product of *human logics*, it must be scrutinized, modified, changed as we progress farther towards mapping *cosmic logics* with increasing accuracy.

6.2 Classical Physics Nurtured the Emergence of Quantum Physics by Seeking Reality in Nature

Maxwell presented his comprehensive equations on electromagnetism in 1864 by synthesizing the already discovered rules of electricity and magnetism developed by Coulomb, Ampere, Gauss, and Faraday, all of whom contributed during the period 1736 and 1867. Lorentz utilized this knowledge to correctly attribute the generation of light by atoms as due to dipole like undulations of electrons in atoms validated by observation of Zeeman effect in 1896 in which magnetic field splits the spectral lines. This dipole model with multiple absorption lines led to the development of a quite accurate model of dispersion theory with distinct "oscillator strengths" for the different absorption lines, which was corroborated many decades later after quantum theory was fully developed.

Before the end of the 19th century, the Rydberg-Ritz formula,

$$\nu_{mn} = cR_y\left(\frac{1}{n^2} - \frac{1}{m^2}\right) \tag{6.1}$$

was correctly mapping the discrete spectroscopic frequencies found from gas discharge lamps, where R_y is the Rydberg constant and n & m are integers that turned out to be the "principal quantum numbers" by both Bohr's early heuristic quantum theory and later formal Quantum Mechanics. By 1900 Planck also captured another very important quantum nature of light regarding its emission and absorption through his heuristic representation of the classical experimental energy density curve for "blackbody" radiation as:

$$u(\nu, T) = \frac{8\pi h\nu^3}{c^3}\frac{1}{\exp(h\nu/kT) - 1} \tag{6.2}$$

Some 25 years later, quantum theory did find that all light-matter interactions do correspond to quantized energy exchange of $\Delta E_{mn} = h\nu_{mn}$, establishing also the logical congruence between the Eqns. 6.1 and 6.2. Noteworthy also was the derivation of "A and B coefficients" by Einstein for stimulated absorption and emission from atoms, which gave birth to lasers much later during 1960's. In view of Jaynes' [9] successes in showing that most light-matter interactions can be analyzed by semi classical approach, Dirac's a, a^+ do not appear to help any better understanding of the realities than Einstein's "A and B coefficients" regarding light-matter interactions. After all, photon wave packets are always "created" and "annihilated" by atoms and molecules, not by the "vacuum" that only sustains their propagations. It is important to note that the classical motto of visualizing the physical entities was at the root of Einstein's 1905 hypothesis of photon as a quantum and Bose's derivation of Planck's black body relation in 1922 using statistics of indistinguishable particles, which became the quantum mechanical foundation of Bose-Einstein statistics for spin integral particles. Several recent Nobel prizes went to people in recent years demonstrating applications of BE statistics.

Our point in summarizing these elementary classical achievements of various observed phenomena is to underscore that the platform for the birth of Quantum Mechanics (QM) and the necessary structure for formulating it were already embedded in classical physics. *Classical physics, by staying focused on how to figure out the actual processes behind various interactions in nature, succeeded in nurturing the minds of the scientists for the next revolutionary changes in our theories.* In contrast, QM, based on its rapid successes beyond expectations in computing the observable results with extreme accuracy, marginalized (and even opposed) the concept of seeking *reality* in the micro world. It taught us not to waste our energy in imagining and visualizing the actual processes going on in nature. Even after more than 80 years of maturity, QM has failed in its leadership role to facilitate the next revolution in constructing new concepts to map processes of the micro world with further depth. We believe that this is due to the belief system established by some of the key founders and developers of QM. For generations, we have been systematically pushed to believe that: (i) QM is a *complete* theory of the micro world; (ii) visualizing the actual processes in the micro world is beyond QM and hence beyond human capability of imaginations; (iii) the "lack of knowledge" of humans as to which way light or particle beams travel to the detector is essential to the emergence of interference patterns, etc. Heisenberg's indeterminacy relation for measurements [6,12,13] is essentially a corollary of the Fourier theorem, which itself is not a principle of nature [14]. But it has been re-interpreted as incessant violation of causality in the micro world. Do we really need to, or do we measure more than one physical parameter of the same entity in any one experiment? Is the progress of physics really fundamentally limited by our lack of simultaneous measurement of two related parameters of a single entity, whether they commute or not? It is generally agreed upon in the scientific culture that all organized bodies of knowledge in use today are necessarily provisional and incomplete because they have been constructed based on the incomplete knowledge of the universe. Yet, our enquiring mind has been trained to ask only those questions which are congruent within the logical bounds of the accepted "working" theories and their interpretations, effectively ensuring that we will never find our way out beyond the current framework of QM.

All the startup classical physics rules ("laws") were firmly rooted on seeking reality, or the deeper cosmic logics in operation in nature. The mathematical relationships were such that all the symbols represented some dynamic and/or static parameter of the state of a physical entity and the operating symbols implied some actual interaction (force law) or evolving process constrained by some conservation rule. Unfortunately, rapidly accumulating successes of the mathematical QM formalism and the concomitant exuberance diverted us from keeping ourselves anchored to repeated refinement of our starting human logics towards actual cosmic logics. We misplaced our objective of doing science as figuring out and visualizing the actual processes behind all the magnificent cosmic evolutionary events to become mere data gatherers and data correlators. We have become equation-crunchers as computers are our number crunchers. By demeaning our visualization and imagination faculties, we have made our enquiring mind subservient to a belief system that elegance, esthetic beauty and symmetry of mathematical relationships give us the power over nature and tell her how she ought to behave in carrying out physical processes.

6.3 Accepting a Higher Order Challenge to Seek Cosmic Realities

The *purpose* of science needs to be redefined as incrementally becoming wiser and wiser towards understanding the *purpose* of the orderly evolving universe, which will then help us define our *purpose* as humans in this universe. Irrespective of our divergent belief systems as to whether there is a pre-ordained *purpose*, we will evolve to define one for ourselves simply because of our innate desire to keep on evolving. Our sciences, so far, have wisely stayed focused on understanding and/or predicting the outcomes of interaction *processes* going on in the material universe, which are the causation behind the cosmic and biospheric evolution and appears objective (reproducible). Here we are addressing the issue of refining the methodology of studying physical *processes* in nature and leave the subjective issue of defining our "*purpose*" for social scientists.

Nonetheless, we believe that a deeper understanding of the inter-related cosmic *processes*, when sufficiently well organized by human logics (*working rules*) and refined towards cosmic logics (*laws of the universe*), our wisdom will be capable of defining and slowly refining our *purpose* hypothesis congruent with our desire for sustainable evolution and the laws of cosmic evolution. As of now, none of our organized set of human knowledge system, however successful they are, can claim to have reached the level of refinement as to have become identical with the pure cosmic logics. Accordingly, any attempt to define our "*purpose*" in the cosmic universe is bound to produce many different subjective interpretations developed by human logics belonging to different epistemological groups. Let us leave the reader with the following question. Is it possible to enlighten ourselves in understanding the *cosmic purpose* by understanding the *cosmic processes* [15], *the domain of scientific studies*?

6.4 "Incomplete Information Paradigm" or Fundamental Limits in Information Accessible Through Observations

We have created impediments towards our scientific progress by ignoring the roots of unavoidable limitations in gathering information about nature (interaction processes) from even the best organized experimental apparatus. We can "see" (or sense, or measure) incidents in the universe only indirectly through the "eyes" of the various sensors (detectors or interactants). First, none of these interactants are completely known to us. We still do not know what an electron is. Second, all interactants have inherently limited capabilities to "see" (or, respond to) all the input signals (forces or potential gradients) around it and generate discernable and measurable transformations (change) in a particular experiment. Third, all interactants have limited capabilities to relay all that it experiences through the various parts of any practical detecting system, which constitute, at a minimum, a "classical" device as the final measuring meter. We may characterize the situation this way. All sensors (interactants) "see" through vision-limiting "goggles" and "speak" to us through band-limited "channels" that are characteristically unique for each of them and not quite known to us.

We need to appreciate the deep consequences of this "incomplete information paradigm" thrust upon us by nature. We are forced to develop our logically complete "working" equation by using incomplete experimental information by inserting innovative human logics (hypotheses) to fill the information gap, which may not be exactly mapping the cosmic logics (cosmic laws) that we are seeking to map. Thus all theories are necessarily provisional and incomplete since they are predicting only correctly measured but limited reports gathered about the interactants. Such a theory automatically limits our progress in integrating new behaviors of nature that are not logically congruent with those limited set of human logics that has already constructed the "working" equation! New parameters may not be "plugged" in arbitrarily. The fact that decades of attempts of introducing "hidden variables" to aid the visualization of the invisible micro world phenomena could not be accommodated within the framework of QM implies that QM, inspite of its successes, is logically closed to logics behind "hidden variables". They are logically incongruent. Instead of declaring that nature is not visualizable, we should be building a new theory that can accommodate causality and locality within its framework.

A working equation needs to be almost logically "complete" (hence "closed") for it to be successful. Such an equation (theory) to work for a small segment of the undivided universe, by necessity, it must have ignored many other potential interactions due to other forces and/or under logically very different contexts. Thus, the only way to integrate multiple successful theories, akin to partially solved jigsaw-puzzles of the universe, is to break them apart and try to re-assemble them as one bigger jigsaw-puzzle by selectively rejecting and/or modifying some of the human logics towards mapping infallible cosmic logics. Therefore, we should be careful not to jump to conclusions with any working theory that we have correctly captured all the necessary cosmic logics behind the set of interactions represented by the theory.

6.5 Identifying Logical Process Steps Behind All Observations as SEMT

6.5.1 SEMT and Locality of Interactions Defined

SEMT stands for *Superposition Effects as Measured Transformations.* We are implying that all scientific measurements, classical or quantum, arise out of interaction between our chosen interactants. Since all of our validate-able information about any phenomenon comes through experimental observations and the gathered information is always incomplete, it is necessary for us to identify all the logical process steps behind all measurements. This would help the process of applying human logics to construct the best possible mathematical equation to map the observations under consideration while filling the missing gaps of information that cannot be provided by the experiments. When a working theory is already well matured, we can re-assess the human logics behind its construction by re-visiting the related experimental process steps while being cognizant that there was missing information that is essential to refine our theory towards mapping cosmic logics more accurately.

- (i) We can scientifically measure only re-producible quantitative *transformations* (changes in states) that are experienced by our interactants (or detector-detectee, or sensor-sensee interaction).
- (ii) Any transformation in a measurable physical parameter requires *energy exchange* between the interactants.
- (iii) The energy exchange must be guided by at least one *force of interaction* between the interactants and it must be strong enough to facilitate the exchange of energy, which are usually constrained by unique characteristics of each interactant.
- (iv) All force rules being range (distance) dependent, energy exchange between the interactants requires that they must experience each other as *local* or *physically superposed* entities (experience each other within their sphere of influence).

In summary, the interactants in an experiment must be physically *superposed* (present) within the range of the *interacting force* that will allow for some *energy exchange* followed by some *transformations* that is measurable for us through some classical meter. Superposition effect is thus an *active causal and local process*, and not a passive mathematical principle only! Interpretations of successful mathematical formulation must recognize this *reality*. Operationally, real physical superposition, as implied by our dissection of all interaction processes, is a concept of high physical significance both in classical and quantum mechanics because it implies *locality* for all interaction processes. This understanding also provides a path to reduce the epistemological gap between the classical mechanics and quantum mechanics. The purpose of physics is to map, visualize and articulate the physical interaction *processes* that facilitate the energy exchange leading to change and evolution.

6.5.2 Generalized Validity of SEMT Reality

We have claimed *locality* for all physical interactions, classical or quantum mechanical. In view of the dominant role of currently accepted interpretation of QM, we feel that following explanations will be useful to accept our broad proposition behind SEMT.

(i). Gravitational force (GF): GF is weak; its range is long. Our planets within our solar system constitute, of course, a strongly bound *local* and superposed system. Air molecules in our lower atmosphere are tethered by Earth's gravity, but cannot effectively display the influence of the significantly weaker Sun's gravity. Yet, all cosmic entities, from galaxies, stars, planets, atoms and elementary particles, the entire observable material universe is effectively superposed on each other or *local* as far as GF

is concerned; however, the degree of influence on each other is dictated by their mass and distance.

(ii). Electromagnetic force (EMF): EMF is relatively stronger than GF, but the range is generally shorter. Atoms within a molecule are superposed and *local* to each other by EMF. Stability of atoms, molecules and their all possible transformations, including their interactions with electromagnetic waves are all dictated by this force. The dominant part of the biospheric evolution is driven by this force. The superposition effects due to the EMF from the molecules within a biological cell may or may not be effective depending upon the type of molecule and their physical separations.

(iii). Weak Nuclear force (WNF): Radioactivity and related isotopic nuclear transmutations are a by product of this force. The range of WNF is of the order of the size of the atomic nuclei. The superposition effects due to two radioactive atomic nuclei within the same bound molecule are negligible within the first order analysis.

(iv). Strong nuclear force (SNF): Our slow physical evolution relies on the stability of an array of nuclei held together by this SNF, built into stable atoms and molecules by the EMF and held on to the surface of the Earth under the atmosphere by the Earth's GF. Different atoms within the same molecule are superposed as far as electromagnetic force is concerned, but their nuclei are not superposed as far as SNF is concerned within the first order analysis.

Thus, *locality* as we have defined in the context of SEMT is unique and force dependent. Even though the physical range varies from the size of a nucleon to almost "infinity" (for galaxies under mutual gravitational influence), it is logically self consistent for any interaction process to generate the measurable transformation. Physical entanglement (measurable influence) between different entities can be operative only within the range of the operating force. Interaction free energy exchange or measurable transformation is not allowed by our SEMT platform. We understand that our reality epistemology is a stronger demand than EPR [7], but it is in the spirit of the very first sentence of this controversial, but highly stimulating paper: "In a complete theory there is an element corresponding to each element of reality". By demanding such a *process driven interpretation* we will be able to check and re-check our assumptions behind all theories as our knowledge evolves and expands.

6.6 Proposed Epistemology for Refining Human Logics Toward Unknown Cosmic Logics

6.6.1 Defining CC-LC-$(ER)_{1,2}$ Epistemology

We believe that the "trouble" is not with physics [1], but lies with the lack of application of a well articulated epistemology. All organized human bodies of knowledge in general and physics in particular has evolved by applying the CC-LC epistemology. We seek out *Conceptual Continuity* (CC) among a group of diverse but related set of observations. We iteratively and creatively impose *Logical Congruence* (LC) among the entire set to find a higher level of organization leading to a coherent map or a theory. Human *belief* in this CC-LC epistemology and intuitive *faith* in one continuous and logically functioning universe have been paying off enormously. Our cumulative successes in physics indicate that nature's evolutionary processes do consist of logical patterns & organizations. Otherwise, our mathematical theories based on pure logic, would not be so successful. Thus far, the CC-LC-epistemology has helped us "solve" several separate little pieces out of the giant cosmic jigsaw-puzzle. But we are having trouble in integrating them into one coherent puzzle.

We should also recognize that mathematics is a secondary by-product of our rational thinking and imaginations. Mathematics must be subordinate to our thinking and imaginations, not the other way around. Newton invented differential calculus because he needed a tool that has the built-in capability of enforcing *logical congruence* (LC) among apparently very different kinds of observations (those of Brahe and Keppler; Galileo's "stone and feather" falling, his own "apple falling", acceleration of objects, etc.) under one conceptually continuous (CC) or a harmonious model of nature.

As articulated earlier, all of our "successful" theories are constructed based on limited information gatherable from experiments. But however limited, the very success implies that the theory has captured some cosmic truth in some form. Accordingly, it is time for further attempts in *Extracting and Extrapolating Reality* $(ER)_1$ from the working theory. There are two great benefits. First, *extraction* of reality aids visualization of some correctly predicted phenomenon that was not originally anticipated. Second, *extrapolation* of potential reality either to visualize some processes deeper than before or an attempt to integrate a different phenomenon within this theory will help us understand the limits and "bottle necks" of the theory. This step of reality epistemology will help refine a theory and may also help find the limits of its validity in accommodating new observation, which will then pave the way for a new logical frame work to construct a higher level theory.

The state of classical physics went through this $(ER)_1$ epistemology phase during the last quarter of 19th century and the first quarter of the 20th century, which paved the way for the discovery of quantum theory. However, we have been neglecting the power of this $(ER)_1$ epistemic process by not applying them on the quantum theory, which could have paved the way for discovering next generation of higher level theories.

Current physics has been developed based essentially on reductionism—matter into elementary particles and radiations into photons. We have neglected to develop a formal methodology of thinking that would help appreciate the emergence of new complex properties and rules when a complex system is formed out of very many simpler elements or sub-systems. We now need to add another iterative feed back loop of $(ER)_2$—*E*mergentism and *R*eductionism on to CC-LC-$(ER)_1$ and create a higher level of methodology, CC-LC-$(ER)_{1,2}$ epistemology. We need to understand the real physical processes behind the emergence of both the irreducibly stable elementary particles as well as the most complex systems out of these elementary particles.

6.6.1.1 $(ER)_2$ Example, Rainbow as an Emergent Phenomenon

It may be worth examining a classical example of $(ER)_2$ to appreciate that we are not proposing anything fundamentally new. Consider how we see a rainbow. Classical physics has *reduced* the physical principles (refraction, reflection and dispersion of EM waves by water droplets in clouds) behind the generation of a physical rainbow. But the real rainbow never exists physically! Photons are not colored; the water droplets are not colored; but we see vivid colors. Even its orientation varies with the position of the observer. A rainbow is an emergent phenomenon. It is not in the cloud even through it is the cloud that helps it become manifest with the help of the sunlight. The rainbow is "visible" only to an observer (eye or camera) having a color sensitive registration material along with an optical focusing system and oriented with the sun behind. No rainbow will be observable if we enter inside the cloud. Similarly, there could be other phenomena that become emergent only because of the restricted behavior of the sensors to a superposed set of other entities, but no mutual interactions (transformations) in the absence of the right kind of sensors.

6.6.1.2 $(ER)_2$ Example, Interference as an Emergent Phenomenon

In fact, optical "interference" is an emergent phenomenon that we have been neglecting to recognize with the consequent erroneous interpretations of superposition effects due to light beams under various circumstances. The superposition effects can become manifest only when detecting dipoles with the right QM property are inserted within the volume of superposed beams [10]. There is no physical *interference* between light beams. Two chapters in this book [16; Chapters 25 and 26] elaborate these points. Like the rainbow, *interference* is what the detectors "see", not what the light beams or the photons do beyond just the simultaneous stimulations and energy they provide to the detecting dipoles. The dipoles then sum up the simultaneous stimulations. This is the physical process behind the "+" sign we use for superposition in equations. The rate of energy absorption (QM transitions) is proportional to the square modulus of these joint dipole amplitude stimulations. Slow countable rate of "clicks" at very low flux level of light become un-countable fast rate of "clicks" at high flux of light. These discrete "clicks" are due to all photo detectors being quantum mechanical [10]. These detected "clicks", being quantum property of the detector, cannot conclusively prove that light beams consist of discrete indivisible quanta. Low light level experiments only re-validate that the atomic and molecular world is definitely quantum mechanical. If self-interference of indivisible single photon were the general behavior in nature, the universe would have been in a constant chaotic state, instead of being always in a state of change that is very orderly. Validation of light as discrete quanta will require carrying out very careful experiments with isolated single atom emitter and single atom detector [17]. Careful experiments with extremely reduced intensity from a laser do demonstrate that expected diffraction pattern rings cannot be recorded simply by increasing the recording time [18]. Recognition that *interference* is as an emergent phenomenon as detectors' behavior has enormous consequences both in the classical and quantum optics that we have been neglecting at the cost of progress in physics! After centuries of unresolved struggles with wave-particle (or corpuscular) duality of light, if we keep on framing our enquiring question as "what are light quanta?", we cannot get any better answer than already given by Copenhagen Interpretation of quantum mechanics. However, the readers are advised to consult the articles summarizing the dominant main stream views [8,19,20].

6.6.2 The Purpose of CC-LC-$(ER)_{1,2}$ Epistemology

This reality seeking epistemology will help us iteratively refine, reject and re-define some of the founding human logics behind our current "successful" puzzle pieces (theories) and let them evolve closer and closer to the actual operating cosmic logics. Thereby, make the various theories more congruent (amenable) to each other towards possible unification, through CC-LC epistemology but at a higher level. As we have underscored earlier, logically closed equations, mapping successfully different subsets of cosmic phenomena based on incomplete knowledge of the universe, will necessarily require modifications on their original fundamental premises (hypotheses) before they can accommodate, or amalgamate into one coherent model. We do not have any other options but to start with human logics, organize related observations into small solved puzzles and then reorganize and/or break them to create a bigger puzzle, and so on, to move closer towards solving the cosmic puzzle. Application of such iterative feedback loop is akin to successful biological evolutionary intelligence.

Four molecules GACT (Guanine, Adenine, Cytosine and Thiamine) in all possible permutations in the DNA-helix, starting with the simple combinations of GC and AT, have been gathering and processing feedback information from the real world into intelligence and wisdom allowing our sustainable evolution. CC-LC-$(ER)_{1,2}$ epistemology explicitly calls for utilization of all possible feed back loops within and between theories to refine, enhance and integrate them to higher level theories while facilitating the visualization of the real physical processes behind all interactions that we are modeling [Fig. 6.1]. The key goal of real genes (or their genetic algorithm) is sustainable evolution of all biological specie collectively. Accordingly, if CC-LC-$(ER)_{1,2}$ epistemology succeeds in understanding and emulating real genetic algorithm, it will be applicable not only in science, but also in developing and advancing all organized bodies of human knowledge, which are deeply connected to our sustainable evolution. After all, from biospheric processes to human thinking, they are all physical processes bound by the same set of

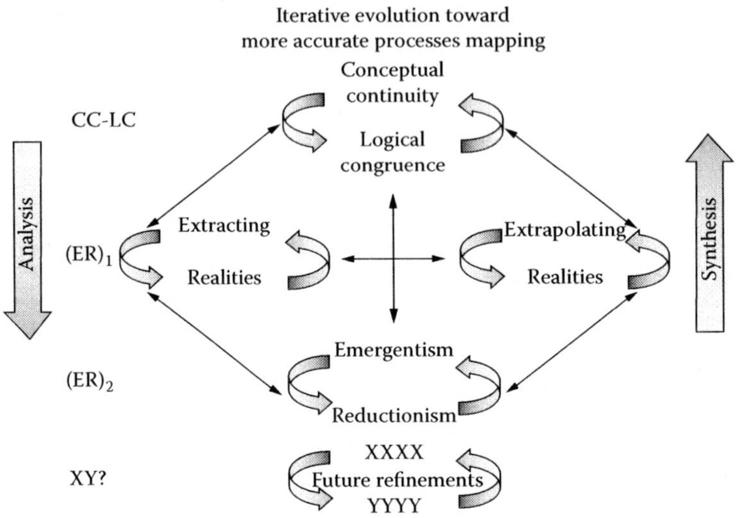

FIGURE 6.1
Logical flow-diagram for CC-LC-$(ER)_{1,2}$ epistemology. The lower (XY) segment is undefined to underscore that all of our epistemology must remain flexible and open to radical revision as our understanding of the universe advances. Our proposal attempts to emulate our biological genetic algorithm, which is at the root of intelligence, derived from the desire to assimilate all possible feed back information applied proactively towards a specific goal of sustainable evolution even though the actual path of the evolution is not known. Similarly, we do not know the path that will take us from the starting platform of human logics to the final goal of understanding the cosmic logics, yet we must attempt to create one. We still do not fully understand how our genetic system creates the intelligence. Our fundamental premise is that we are starting from ignorance; we are deprived from gathering complete information about anything and hence we must remain open to ever refining epistemology for advancing our science. Yet, the proposed epistemology, attempting to emulate biological genetic algorithm, is very generic. Accordingly, it is applicable to developing and advancing all organized bodies of human knowledge that are deeply connected to our sustainable evolution.

basic laws of nature we are trying to understand. Understanding nature's evolutionary processes has a deeper pragmatic value for us. Our successful and sustainable evolution clearly demands better and better technologies over uncontrollable natural calamities by developing newer technologies to protect ourselves. That is the teaching of our DNA.

6.6.2.1 *Ancient Example of* CC-LC-$(ER)_{1,2}$ *Epistemology*

Some 2500 years ago Gautama Buddha of India gave the best allegorical story on how to apply CC-LC-$(ER)_{1,2}$ to visualize and understand the subtle and elusive "material" universe. How would a group of people, blind from birth, describe and visualize an elephant? It applies equally well to us today as we are trying to describe and visualize the cosmic elephant. First we need to recognize that as far as scientific vision is concerned, we are literally blind.

We never see or sense the world directly. Even our human vision is essentially a set of interpretations created by our brains that is convenient for our evolution, not what the actual image is on the retina. We see vivid colors in bright light and we recognize the same colors even in faint light, even though the photons do not have any colors at all. The images we "see" are erect, even though the actual image on the retina is always inverted! We do not see anything! We only interpret the patterns registered by the rods and cones, congruent with our biological evolutionary needs, dictated collectively by the molecules GACT, which are behind the intelligence of our DNA!

In seeking reality about the elephant, the blind people have to search for conceptual continuity (CC) among all their individual sensory inputs by iteratively applying all possible logical congruence (LC) among them. [Diversity of input is critically important.] Even then they will only get the outer shape of the elephant. A deeper level of understanding about how such a shape can be a conscious living being requires the blind people to iteratively refine the model of elephant by first applying $(ER)_1$—extracting and extrapolating their perceived realities to become commensurate with models of other living species they are aware of. Then they need to apply $(ER)_2$ – *emergentism and reductionism*, to delve deeper into understanding the emergence of elephant's living behavior out of many parts and organs. Today, we "scientifically" understand the *emergence* (E) of any living being out of molecules and DNA's that are highly *reduced* (R) constituents, but we still do not fully understand the emergence of consciousness. Buddha's story also underscores that the existence of the elephant is real, irrespective of whether the blind people sensed it at all or understood its existence in the strict sense. So, philosophies giving serious credence to questions like, "did the tree fall if nobody heard of it?" is a useless diversion if we want to seriously explore the realities of the emergent cosmic universe. The bacteria in the woods are fully cognizant of the availability of lots of food from the fallen tree! Human philosophy cannot hinder their evolutionary physical drives.

We can learn to visualize the invisible interaction processes in the domains of atoms and elementary particles only when we gather the wisdom to acknowledge that we are literally blind. We do not see anything; we only interpret using incomplete information!

6.6.3 Why Elegant Mathematics and Visualizations Are not Enough

Although it is obvious from the prolonged stagnant state of physics that elegance and symmetry of mathematics is not complete guide to explore nature, we present two simple examples to underscore the necessity of constantly applying CC-LC-$(ER)_{1,2}$ epistemology. First, let us revisit why we have rejected Ptolemy's geocentric model. It required several free parameters to allow Ptolemy to construct "epicycles" for each planet separately to accommodate relative "wiggle" motion relative to our Earth. Kepler, based on Copernicus' suggestion, showed that Helio-centric model fits the observations more coherently and logically without many free parameters except a central force of attraction by the sun. Over the following centuries, Newton formalized the "central force" as gravity, Einstein generalized it to "curvature of space" and we are still encounter dilemmas as to how to accommodate the measured velocity discrepancy of the stars in the outer periphery of the galaxies. The point is we need continuous refinements in our modeling based on discovering actual realities rather than inventing mathematically elegant ones. But, if we take the example of today's "successes" of various String Theories using many dozens of free parameters, Ptolemy's geocentric model can be revived with many fewer free parameters than the String Theories require.

Let us look at another example with elementary mathematics. Pythagoras' relation can be replaced by a pair of relations that I discovered in my 7th grade school from a particular example of a right angled triangle with sides 5, 4 and 3, as many other students must have:

$$[c^2 = a^2 + b^2] \quad \text{vs.} \quad [c = 2a - b \text{ where } (b/a) = (3/4)] \tag{6.3}$$

Even though Pythagoras' quadratic relation can be derived from the pair of linear relations suggested above, which makes the two relations mathematically equivalent, my teacher favored the visualizing power of the geometric construct proposed by Pythagoras. Because, one can literally construct the unit squares on each side of a right angled triangle and see for himself why Pythagoras' relation makes sense, which is not so obvious from the other approach based on a particular geometric ratio of the sides. Advanced physics is replete with many such examples like (i) the equivalency of Heisenberg's matrix formulation vs. Schrödinger's "wave" equation and (ii) equivalency of Feynman's "path-integral" vs. Tomonaga-Schwinger's "variational method". (iii) Sudarshan showed that Wolf's classical coherence formulation is equivalent to Glauber's QM representation. Can one of the mathematical constructs guide us better than the other in seeking and visualizing the actual interaction processes in nature? This is a relevant question from the stand point of the epistemology we are proposing. The key point is to recognize that not all "working" human logic has a unique one-to-one relation to the cosmic logic. Thus we must develop a methodology of rational iteration process that can help us keep on refining our working human logic towards the "nirvana", the cosmic logic. Continuous debate and rational doubt over even the most successful theory is at the core of doing science. No human organized theory is ever complete!

6.6.4 Fourier Theorem in Optics and Interference as an Emergent Phenomenon

The Fourier theorem that effectively represents superposition principle in mathematical form is quite enigmatic [14]. It has never been declared as a principle of nature but it plays a principal role in all sciences, especially in physics. Its pervasive success in physics and optics derives from its foundation. It represents linear superposition of harmonic functions. Physics deals with fields and particles that are all based on different kinds of harmonic undulations (may or may not be waves). Because of its diverse successes, we have started pretending that it is equivalent to the superposition principle of nature, creating epistemological problems of enormous magnitude. This section will demonstrate that inspite of mathematical correctness of the Fourier transform (FT), we have been using it incorrectly in a number of places. This recognition will strengthen our view point that quantization of EM field as a Fourier monochromatic mode of the vacuum may not be sound physics.

6.6.4.1 *Space-Space Transform; Optical Signal Processing*

This is the only FT-formalism that is in a sound platform because the Huygens-Fresnel Integral, a proposed principle of nature, morphs into a FT integral under the far field condition because the quadratic curvatures of the secondary wavelets drop off. Optical signal processing is a highly matured field based on this FT-formalism. However, one should be aware of pitfalls of modeling higher order diffracted intensity distribution due to an ultra short light pulse; it is not a serious problem for imaging applications since the relative delays in the image plane is essentially zero [21–24].

6.6.4.2 *Delay-Frequency Transform; Fourier Transform Spectroscopy (FTS)*

FTS is on a sound platform as long as one does not use (i) fast detectors and (ii) the maximum interferometer delay is smaller than the pulse width. Otherwise, differential amplitude induced visibility reduction would artificially broaden the recovered spectrum [25, 26]. One should be aware of the built-in contradiction behind FTS. The key assumption is that different optical frequencies are *incoherent* to each other. This is a wrong assumption but correct observation as long as the photo detector has a long time constant for integrating photo electric current. During the days of slow retinal observations followed by photographic recordings, this signal integration requirement was built-in. But, after the discovery of fast photo detectors [27] we have developed heterodyne or light beating spectroscopy, which is quite common these days. Light beams of different optical frequencies are really not *incoherent* to each other.

6.6.4.3 *Time-Frequency Transform, Classical Spectrometry*

Classical spectrometry also gives numerically correct results but only for light pulses that are definitely longer than the instrument's characteristic time constant, $\tau_0 = R\lambda/c$, R being the classical resolving power. For some unknown reasons, this time constant is not explicitly recognized in classical spectrometry [28]. We have shown that the true spectrometer impulse response must be derived by time domain propagation of a pulse, which converges to the classical CW formulation for signal duration longer than this time constant. *Time integrated fringe broadening* due to a pulse do correspond to the convolution of the CW intensity impulse response with the Fourier intensity spectrum by virtue of conservation of energy (Parseval's theorem) [29]. Recognition of this subtlety has two important consequences.

The first consequence relates to classical spectrometry. It tells us that the traditionally accepted time-frequency bandwidth limit $\delta t \delta \nu \geq 1$ is observationally correct because $\delta \nu$ represents the *time integrated* physical fringe broadening, but not the physical generation of a new set of frequencies by a linear diffraction grating (or a pair of Fabry-Perot beam splitters). New frequency generation generally requires nonlinear, Raman or n-photon stimulations of a material medium by the incident field. In other words, $\delta t \delta \nu \geq 1$ does not represent physical presence of new frequencies. This opens up the door to designing algorithms and instruments to achieve spectral super resolution. The summary of the necessary derivation and some experimental results can be found in these references [29, 30].

The second consequence relates to the demand of QED for a photon to be a Fourier monochromatic mode of the vacuum [31–33], as if required by the combination of QM requirement that the frequency of the spontaneously emitted "photon" has to be uniquely defined through the relation $\Delta E = h\nu$ while classical observation $\delta t \delta \nu \geq 1$ apparently claims that it cannot be a space and time finite wave packet, which is a conceptual mistake perpetuated by classical physics and co-opted by QM. Accordingly, we have proposed [34] that a spontaneously emitted photon is a "mode of the vacuum" but as a space and time finite wave packet with a unique carrier frequency ν as demanded by $\Delta E = h\nu$. The envelope function is dominantly an exponential function with a very sharp rise time to accommodate the observation that the time integrated line width of spontaneous emission is approximately Lorentzian.

6.6.4.4 Time-Frequency Transform, Coherence Theory

First let us appreciate that all light signals must necessarily be time and space finite pulse dictated by the principle of conservation of energy. Even a CW laser has to be turned on and off in the real world. The physical spectrum of a pulse is its actual carrier frequencies (undulations of the E & B field vectors) contained in it, and not the FT of the amplitude envelope. This position is validated by the observations made in the last section. Measurable fringe visibility (modulus of autocorrelation function) can be degraded (i) by unequal amplitudes of same frequency light pulse, (ii) by displaced fringe locations due to variable phase delays produced by the same path delay, but, due to multiple frequencies, $\nu_x \tau = m_x$ (order of interference), or (iii) due to presence of light with nonparallel states of polarizations. Today, we do not distinguish between temporal coherence (due to a time finite pulse with a single carrier frequency) and the spectral coherence (due to CW light containing multiple carrier frequencies). Pitfalls of traditional Wiener-Khintchine theorem can be compared from these references [25, 33]. These understandings will provide the platform for better characterization of ultra short light pulses whose spectral content (distribution of E-vector undulation frequencies) may be different even for the same intensity envelopes. We mentioned earlier that FTS works using slow detectors under the assumption that beams of light containing different optical frequencies are *incoherent*. In reality, different optical frequencies are coherent and they do produce oscillatory beat or heterodyne currents in fast detectors. The concept of coherence needs to be revisited through the "eyes" of photo detectors.

This last point can be appreciated further by exploring why orthogonally polarized light beams produced from the same single mode laser do not produce superposition fringes. Obviously, the two beams from the same laser mode cannot suddenly become *incoherent* (phase random or multi frequency) by inserting orthogonal polarizers. Our proposed hypothesis is that it is the quantum property of the detecting dipoles (see Eqs. 6.5–6.9) embedded in the susceptibility property χ that dictates the observed results. The energy absorption is not modulated by the "cross term" when the two beams are orthogonally polarized. Orthogonality of the inducing dipolar stimulation makes this term zero [see Eqs. 6.6–6.11 below]. In effect, the complex amplitudes due to simultaneous but orthogonal stimulations cannot be summed by detecting dipoles. This limiting quantum property of detecting dipoles should not be assigned to orthogonal EM fields as being *incoherent*. EM fields are never *incoherent*. Integration time and the dipolar properties of detectors determine the degree of fringe visibility, mathematically equivalent to the modulus of the autocorrelation of the superposed fields. Any wave group by definition consists of a collective *coherent* set of undulations.

If indivisible single photon really "interferes only with itself", all thermal sources could be converted into *coherent* sources simply by putting a narrow band spectral filter followed by an absorber to allow only single photons to emerge!

6.6.4.5 Time-Frequency Transform, Laser Mode Locking

It is standard practice to express mode locked laser pulses as the summation of periodic longitudinal modes of a laser cavity, irrespective of whether the characteristics of the laser gain media are homogeneously or inhomogeneously broadened. But we know from discussions in the previous sections that light beams by themselves do not re-group their energy. We also know from the key requirements for designing actual mode locked lasers that it is the insertion of devices like a saturable absorber (or, its equivalent, a nonlinear Kerr medium) that really generates the short pulses by behaving as a temporal on-off switch. Interactions between the cavity fields and the dipoles of the devices jointly create the temporal on and off durations of these mode locking devices. So, the ultra short pulse generation community has correctly kept their engineering focus more on the material properties of the gain media, saturable absorber, Kerr medium, etc., rather than on just the phases of the longitudinal modes. Besides, we doubt that simple intra-cavity insertion of a mode locking device can make a homogeneous gain medium to oscillate in multiple longitudinal modes. We believe that truly transform limited pulses contain a single carrier frequency [35–37].

6.6.4.6 *Time-Frequency Transform, Pulse Dispersion*

Based on the correctness of the diffraction theory and our success in reformulating classical spectroscopy [29-30], we believe that "pulse dispersion" is actually pulse stretching [38] due to time diffraction. This is the counter part of diffractive spatial spreading of a beam when it is cut off by a small aperture. When people use FDTD (finite difference time domain) method of computation to propagate short pulses using directly Maxwell's wave equation, they are computing time diffraction [39]. Molecules in media usually respond in the femto second domain to the local amplitude and carrier frequency (-ies) at the moment of their exposure. They do not have memory and they cannot wait to determine the Fourier frequencies due to pulses of long durations and shapes. Thus, as in classical spectrometry, propagating Fourier transformed frequencies may give "correct" *time integrated* pulse broadening in limiting cases, but that is not the correct physical modeling. Counter examples to establish our point can be found in these references [37, 40].

6.7 Bell's Theorem and Interference as an Emergent Phenomenon

Our proposed reality epistemology, CC-LC-$(ER)_{1,2}$, requires imposing real physical meaning to the symbols and mathematical operators of key working equations. Even in pure mathematics, for equations to be correct, the meaning and operation of all the symbols and the connecting operators must be clearly defined. This is an essential component of the reality epistemology we are promoting [10]. Superposition effects emerge as measurable transitions in photo detectors. Thus the detector's first-order susceptibility ${}^{(1)}\chi$ to polarization induced by the superposed E-vectors is an important physical parameter that is not normally taken into account when writing equations for interferometry when the basic superposition process is linear, but we need to:

$$\begin{gathered}\text{Field}: E(t) = a(t)e^{i2\pi\nu t}; \quad \text{Stimulation:}\psi(t) = {}^{(1)}\chi a(t)e^{i2\pi\nu t}; \\ \text{Transformation: } D(t) = \langle \psi^*\psi \rangle\end{gathered} \tag{6.4}$$

Ensemble averaged photo current $D(t)$ is the measurable transformation due to real physical superposition of the EM field on the detecting molecules. QM prescription to compute has two built in steps, taking square modulus of the dipole stimulation and the ensemble average. The susceptibility to polarization of the dipole ${}^{(1)}\chi$ contains all the classical and quantum response properties of the detecting molecules. Note that while normally we use only the linear (first order susceptibility), in reality all EM fields induce all possible linear and non-linear susceptibilities all the time. We normally neglect these higher order effects until we encounter molecules with strong nonlinear polarizability that is becoming more and more common with time. In reality, the total dipole stimulation due to an EM field should be written as:

$$\text{Stimulation:} \Psi(t) = \sum_n {}^{(n)}\chi E^n(t); \text{ Transformation: } D(t) = < \Psi^*(t)\Psi(t) > \quad (6.5)$$

While Eq. 6.5 already looks complex for general situations, it is even more complex in reality, because both the susceptibility and the EM field should be treated as vectors to accommodate the angle between them in anisotropic media as is done by the specialists in nonlinear optics. Consider the simple case of a two beam Mach-Zehnder interferometer containing two rotate-able linear polarizers in the two arms and illuminated by a linearly polarized single mode laser beam. Neglecting the possible phase and polarization changes that can be introduced by the beam splitters and mirrors, the output beams can be represented as $\vec{a}_1 \exp(i2\pi\nu t - t_1)$ & $\vec{a}_2 \exp[i2\pi\nu(t - t_2)]$ where $\tau = (t_1 - t_2)$ is the propagation induced relative time delay between the two beams. When these two superposed output beams are received by a detector, the sum of the induced dipolar undulation amplitudes experienced by the detector is:

$$\begin{aligned}\Psi(t) = \psi_1(t) + \psi_2(t) = {}^{(1)}\chi\vec{a}_1 e^{i2\pi\nu(t-t_1)} + {}^{(1)}\chi\vec{a}_2 e^{i2\pi\nu(t-t_2)} \equiv {}^{(1)}\hat{\chi}_1 \, {}^{(1)}\chi a_1 e^{i2\pi\nu(t-t_1)} \\ + {}^{(1)}\hat{\chi}_2 \, {}^{(1)}\chi a_2 e^{i2\pi\nu(t-t_2)}\end{aligned} \quad (6.6)$$

The unit vectors ${}^{(1)}\hat{\chi}_{1,2}$ in Eq. 6.6 represent the two physical directions of undulations induced on the detecting molecule (or cluster). The detectable transition d can be written as, assuming θ is the angle between the induced dipole stimulations:

$$\begin{aligned}d = \Psi^*\Psi &= \left|{}^{(1)}\hat{\chi}_1 \, {}^{(1)}\chi a_1 e^{i\phi_1} + {}^{(1)}\hat{\chi}_2 \, {}^{(1)}\chi a_2 e^{i\phi_2}\right|^2 = |\vec{\psi}_1 + \vec{\psi}_2|^2 = \vec{\psi}_1^* \cdot \vec{\psi}_1 + \vec{\psi}_2^* \cdot \vec{\psi}_2 \\ &+ \vec{\psi}_1^* \cdot \vec{\psi}_2 + \vec{\psi}_2^* \cdot \vec{\psi}_1 = {}^{(1)}\chi^2\left[a_1^2 + a_2^2 + 2a_1a_2({}^{(1)}\hat{\chi}_1 \cdot {}^{(1)}\hat{\chi}_2)\cos 2\pi\nu\tau\right] \\ &= {}^{(1)}\chi^2\left[a_1^2 + a_2^2 + 2a_1a_2 \cos\theta \cos 2\pi\nu\tau\right]\end{aligned} \quad (6.7)$$

When the polarizers within the interferometers are lined up with the incident vertically polarized beam, $\theta = 0$, and we can recover from Eq. 6.7 the traditional intensity pattern multiplied by a constant ${}^{(1)}\chi^2$ that we routinely neglect and yet contains most of the details behind the real physical processes:

$$d = {}^{(1)}\chi^2\left[a_1^2 + a_2^2 + 2a_1a_2 \cos 2\pi\nu\tau\right] = a_0[1 + \gamma \cos 2\pi\nu\tau] \quad (6.8)$$

Here $\gamma \equiv 2a_1a_2/(a_1^2 + a_2^2)$ represents the fringe visibility quotient and $a_0 \equiv {}^{(1)}\chi^2 \times (a_1^2 + a_2^2)$. When the two beams within the MZ are deliberately made orthogonally polarized, $\theta = 90^0$, then the detectable transition becomes simply proportional to the sum of the two intensities multiplied by ${}^{(1)}\chi^2$; the interference term drops out:

$$d = {}^{(1)}\chi^2\left[a_1^2 + a_2^2\right] \tag{6.9}$$

Photodetecting molecular complexes cannot respond to the different phase information brought by the EM fields if they are orthogonally polarized, ${}^{(1)}\hat{\chi}_1 \cdot {}^{(1)}\hat{\chi}_2 = 0$. Since EM fields do not interfere with each other by themselves, we should not attribute the absence of fringes because "orthogonally polarized light beams do not interfere". Again, we must recognize that we "see" light through the "eyes" of dipoles. Further, any time light passes through any material and/or is reflected or scattered by some material surface, some of its intrinsic physical properties (frequency, phase, amplitude, and polarization) very likely will change. This is built into Maxwell's wave equation when one applies the "boundary conditions". Thus, if we think in terms of propagating photons, most of the time the "re-directed" photon is no longer the same photon that originally impinged on the surface of the medium.

Accordingly, the Bell's theorem [41] to be relevant at all for superposition (*interference*) experiments, it has to be re-derived for each interferometer in terms of physical dipole undulations of not only the detector molecules but also of those of dielectric or metal coating boundary molecules of beam splitters and mirrors that introduce differential phase shifts for "internal" vs. "external" reflections and the states of polarizations [25], etc. Our point should be obvious from Eqs. 6.5–6.9 even though they consider the very simple case where no relative phase or polarization changes are introduced by the two separate arms of the MZ mirrors and beam splitters. We have demonstrated the consequences in the fringe intensity and location changes produced by an MZ illuminated by a beam containing two orthogonally polarized lights having an asymmetric case of gold and a dielectric mirror [see Chapter 26 of this book]. Simple sum of the EM fields with two different phases, as represented by the Bell's theorem, is not what we measure or what emerges as transformations in detectors.

Equations 6.5 through 6.9 essentially represent classical relation for energy absorption. Let us now apply $(ER)_1$-epistemology on the Eq. 6.8 and take a deeper look at the significance behind the QM prescription of taking an ensemble average of $\Psi^*\Psi$. The expression for the fringes represented by Eq. 6.8 is re-written below with the reminder that all photo detectors are quantized and that each individual transition (photo counting "clicks") needs to absorb a unique "quantum cup" of energy given by $(\Delta E)_{m-n} = h\nu_{m-n}$, where the suffix "m-n" refers to quantum transition between levels (or bands) m and n.

$$d \equiv (\Delta E)_{m-n} = h\nu_{m-n} \overset{?}{=} a_0[1 + \gamma\cos 2\pi\nu\tau] \tag{6.10}$$

If d represents a single quantum transition event in a detector that always requires the absorption of a fixed quantity of energy $(\Delta E)_{m-n}$ to be delivered by a radiation of well defined frequency ν_{m-n}, then can it be equated to a quantity that varies sinusoidally with the delay τ by an interferometer? Obviously, the absorbed energy cannot vary for any individual transition even when we vary τ (as long as the frequency remains fixed). An individual count at any value of τ cannot provide very useful information regarding the superposition effect we are studying. The right hand side of Eq. 6.10 must now be re-interpreted as the rate of discrete transitions in the photo detector; *it is no longer a simple energy balance equation*. We just wanted to underscore the conceptual shift from "discrete photons" to discrete detector transition. The energy equation has become a quantum statistical rate equation determined by the flux of the propagating light energy, which is classical. Accordingly the founders of QM have wisely developed the necessity of ensemble average that completes the picture:

$$D = \langle \Psi^* \Psi \rangle = \langle a_0 [1 + \gamma \cos 2\pi\nu\tau] \rangle \tag{6.11}$$

However, Ψ to us, is not an abstract "probability amplitude". It represents the strength of the resultant physical amplitude of the dipole undulation induced by all the simultaneously present EM fields provided their frequencies and polarizations conform to the QM allowed stimulation rules. Superposition principle naturally allows a quantum detector to collect the necessary quantum of energy $(\Delta E)_{m-n}$ for any single transition by gathering energy from multiple fields as long as they are congruent with the QM rules. We do not need to hypothesize that only an "indivisible single photon" can trigger a detector transition. We should not unnecessarily assign the quantum behavior of detectors to the EM fields. Further, if Ψ represents actual dipole amplitude induced by the EM field, then it can be characterized as a joint "quantum compatibility dance" jointly carried out by the field and the detector before the dipole can undergo an allowed transition. There is no arbitrary "collapse of wave function"; a finite number of dipole undulation goes on before the allowed transition takes place. Quantum processes are visualizable.

From the perspective of communication theory, the relative phase delay $\tau = (t_1 - t_2)$ is derived from two pieces of separate information that has to evolve as propagational delays experienced by the two separate light beams in the two arms of the MZ, which must be jointly delivered on to the detectors for taking action. This is part of the same causality in nature that we are underscoring. We agree that information is "physical", as is now claimed in literature [42], but it does not have separate existence outside of physical entities that we can detect and manipulate. In general, physical information is manifest as changes in values of some dynamic physical parameters of some naturally manifest entity that are accessible to control by other physical means.

6.8 Applying CC-LC-$(ER)_{1,2}$ to Model a Photon

We believe that an attempt to re-define the photon is called for inspite of the current state of very broad acceptance of photons as indivisible quanta propagating as various Fourier modes of the vacuum, which "interferes only with itself" [43], perhaps, because they are Bosons. A summary of the mainstream views and related references can be found from these review articles [8, 19-20], which accepts non-causality, non-locality and the consequent teleportation, etc. Our position is that the interactions between elementary particles are causal, albeit probabilistic. So we should try to model a causal "photon" to bring back *reality* in physics.

A Fourier monochromatic mode of the vacuum is not a starting causal model for a photon since Fourier modes are physically non-causal, existing over all time that violates conservation of energy. So, the "CC-LC" component of our epistemology demands a causal model for the photon and when we press to also apply $(ER)_1$ (*extract and extrapolate reality*) out of various classical and quantum optics theories and observations, we find the following model. Our proposed photon is a mode of the vacuum as QED claims, but with two caveats. First, it is a space and time finite packet of EM wave evolving and propagating out following Maxwell's classical wave equation from the moment the emitting molecule releases the quantum of energy $(\Delta E)_{m-n} = h\nu_{m-n}$ and undulating the "vacuum" with a carrier frequency ν_{m-n}. This "perturbation" then evolves (diffracts) out, following Maxwell's equation, under the space and time finite 3D exponential-like amplitude envelope [34 or Ch. 27]. This far, our model is congruent with the correct demand of QM, $(\Delta E)_{m-n} = h\nu_{m-n}$. The next issue is to reconcile with the measured natural line width of spontaneous emission to be a Lorentzian. Classical physics (Lorentz) has solved the problem by proposing the emission envelope to be exponential whose Fourier transform is Lorentzian. We have analytically shown that the *time integrated* fringe broadening observed in classical spectrometers due to time-finite pulses does mathematically appear to be equivalent to the presence of a broad spectrum given by the Fourier spectral intensity of the amplitude envelope [29–30].

Let us now apply again $(ER)_1$ along with $(ER)_2$ (*emergentism and reductionism*). The HF diffraction model, also supported by Maxwell's wave equation, is holding out as a remarkably accurate model for light propagation from all the macro to nano photonic devices. So, it must have captured some cosmic logic in it. Its key proposition is that every single point on the wave front behaves as a new source point. We are proposing to accept this point to be literally true. This implies that the cosmic "vacuum" holds a stationary and uniform electromagnetic tension field (EMTF) everywhere in a state of equilibrium [44-45]. The light wave (photon) is simply a propagating wave group that is an undulation of the EMTF induced by the released energy ΔE by an excited molecule while undulating as a dipole at a frequency ν. The photon wave packet is an emergent phenomenon out of the stationary EMTF.

The model is quite congruent with all classical material-based undulations that inherently propagate out with diffraction. The wave on the water surface is simply an undulation of the surface against the surface tension when displaced by an external energy source out of its state of equilibrium. Same is true for sound waves where the tension in equilibrium is the air pressure due to Earth's gravitational attraction on the air molecules. The similarity between the Maxwell's wave equation and the material based wave equation is remarkable. The displaced point out of the state of equilibrium, whether EMTF or water surface under tension, wants to come back to its original state of equilibrium and delivers its "displacement energy" to the next domain making it the next ("secondary") source of wave while generating propagating wave and also validating Huygens' hypothesis over Newton's "corpuscular" model, although a space and time finite wave packet (energy conservation) do imply the "corpuscular" existence of light! Propagating wave is an emergent and collective phenomenon. By applying CC-LC on all the material based wave phenomena (water wave, sound, string and percussion instrument vibrations, etc.) we find that the root of the generation and propagation of the waves lay with the respective "tension field" in equilibrium held by the material media over extended domain.

"Do photons have mass?" may be the wrong question to ask. In reality, all wave propagation is effectively a perpetual motion of some "form", not of matter, which is energetically supported by the tension energy of the medium that wants to stay in its state of equilibrium! Mass-less energy transfer from one point to another through the manifestation of propagating waves is obvious in classical physics in any medium under uniform tension. In classical medium, the wave energy propagates out leveraging local kinetic movement but without transfer of any mass to the distant places where the wave arrives. Considering the similarity in the structure of various wave equations, it is logical to extend the EMTF-like tension concept on the cosmic medium. After all, Maxwell's wave equation does find that the velocity of light, $c = 1/\sqrt{\varepsilon_0 \mu_0}$, which is actually a manifestation of the properties of the vacuum, ε_0 (dielectric constant) and μ_0 (magnetic permeability). If EMTF-hypothesis is correct, then the cosmic space holds an enormous amount of un-manifest potential energy. Only a tiny fraction of this EMTF energy is manifest as propagating photon wave packets whirling in every direction of the universe carrying the messages from one set of atoms and molecules to another distant set. Could possibly this EMTF energy be the "Dark Energy" the astrophysicists have been looking for? No cosmic or local communication waves would have been possible without the existence of such an EMTF in a state of quiet equilibrium! This concept is very different from "luminiferous ether" of the nineteenth century because such a field cannot possess traditional matter like properties. The point is obvious from the considerations that light of wave length 500nm can be easily transported or collected by 10nm guides and a mega watt laser beam can be focused and passed through a pinhole of diameter no bigger than two wavelengths without any distortion in any of its fundamental properties. The energy is transported locally by a very steep gradient of the field, EMTF. The important question may be: What holds, or generates, this cosmic EMTF?

This model raises another question, how can one construct the stable particles out of this tension field? Maybe they are some form of vortex [46-47]. Or, more likely as a self-looped wave train propagating forever in resonance with itself, leveraging EMTF and giving rise to the key properties of *matter* like rigidity and inertial opposition (*mass*) to any of the 3D lateral translation. Schrödinger's "wave equation" already contains the time varying internal harmonic undulation factor, $\exp(i2\pi Et/h)$ [44-45]. We already know from $E = mc^2$ that mass is definitely not an immutable property of nature; Relativity validates that mass is some form of inertia. After all, the key premise of Huygens-Fresnel principle, that wave energy at every point becomes the source of wave energy for the next point, is possible only when the wave is manifest as an undulation of a uniform tension filed existing in a state of equilibrium.

Summary and Discussions

All theories have to start with human logics that help organize a selected set of measurements into a logically congruent group with the implied dream of refining the theory to eventually map the actual cosmic logics behind the physical processes making the measured transformation happen.

The core contribution of this chapter is to underscore that we are forever challenged in gathering complete information about any phenomenon through experiments alone because the measurable transformations relayed by our instruments are rarely all that they have experienced. Thus working theories (equations) have to be made logically closed as an equation and self consistent by filling in the information gaps with imagined (*invented*) human logics some of which may not be correctly mapping the cosmic logics (*realities*), which we are trying to *discover.* Nature being fundamentally logical and causal, as evidenced by the very successes of our logical mathematical theories, we should be able to develop a rational epistemology to move towards the reality ontology that lies behind the evolving universe. However, based on several centuries of successes demonstrated by our mathematical modeling, we have developed the tendency to *invent* realities and impose that on nature whenever our elegant theories are falling short of making a causal and visualizable model of the very *processes* we are trying to model. This chapter is an attempt to overcome this troubling trap [1].

We have accepted that all experimental observations, classical and quantum mechanical, as *causal and "local"* superposition effects as measured transformation (SEMT). Dissection of SEMT informs us of the eternal information gap or the "incomplete information paradigm" of all experimental observations that we are forced to accept. This awareness creates the opportunity for us to appreciate that our *human logics* behind "working" theories need continuous refinements to move them closer to the *cosmic logics* that are driving the cosmic evolutionary *processes*, which are undeniably real irrespective of whether humans had evolved to observe them or not.

Then we have proposed a model methodology of thinking, CC-LC-$(ER)_{1,2}$ epistemology [Fig. 6.1]. The utility and power of this epistemology has been demonstrated by summarizing the successes and hidden failures in the field of optics that uses ubiquitous Fourier theorem. We have used our epistemology to argue that superposition effects are necessarily local and that "photons" may be space and time finite undulation of a hitherto undiscovered electromagnetic tension field (EMTF) filling the entire volume of cosmic space. We have also presented our view that superposition effects being local interactions with detecting molecules, Bell's theorem is not the right guide to overthrow causality in nature. It's ineffectiveness may also lie with the faulty derivation of the joint probability distribution, as has been claimed by some [48].

The proposed epistemology can guide us to continuously refine our human logics towards correctly mapping cosmic logics. The model attempts to emulate our biological genetic algorithm, the stuff out of which we are built. From the very early stages of evolution the GACT's (Guanine, Adenine, Cytosine, and Thiamin) moved to create the DNA molecules, and then the viruses and the living cells. They all function as little creative engineers, effectively following the interaction processes allowed by nature's limited set of laws. As very complex systems, as conscious humans, we will be better off by being humble and honest creative reverse engineers. This is not a philosophy. This is emulating successful evolutionary engineering of nature for our own sustainability. However, neither the path to sustainability is defined for us, nor can we acquire complete information from any observation or experiment. Perhaps, this is a deliberate design to keep our mind challenged towards choosing a better evolutionary direction!

By virtue of "incomplete information paradigm", our proposed CC-LC-$(ER)_{1,2}$ epistemology must remain as a "work in progress" for ever. Then only can we assure ourselves that one dominant epistemology cannot slow down the progress of scientific investigations and thinking [1–3]. All "correct" scientific theories must be superseded and/or invalidated by new theories! Therefore, the younger generation should be constantly asking: How can we stay focused on *discovering* actual *realities* in nature driven by *cosmic logics* rather than stay limited to *inventing* realities that are esthetically pleasing to our *human logics*?

We hope this chapter will inspire our readers to give serious attention to the various out-of-the-box proposals for photons presented in the next section.

References

1. L. Smolin, *Trouble with Physics*, Houghton Mifflin (2006).
2. R. Laughlin, *A Different Universe: Reinventing Physics from the Bottom Down*, (2006), Basic Books.
3. R. Penrose, *Road to Reality*, Alfred Knopf (2005).

4. Quantum Theory—Reconsideration of Foundations; QTRF-4 Conference, 2007, Vexjo University, Sweden. http://www.vxu.se/msi/icmm/qtrf4/
5. SPIE biannual conference series on "The Nature of Light: What Are Photons?" Contents of 2007 conference: http://spie.org/x648.xml?product_id=721469&Search_Results_URL=http://spie.org/x1636.xml&search_text=6664&category=All&go=submit
6. D. Home and A. Whitaker, *Einstein's Struggle With Quantum Theory*, Springer, 2007.
7. A. Einstein, B. Podolsky and N. Rosen, *Phys.Rev.* **47**, 777 (1935); "Can quantum mechanical description of physical reality be considered complete?"
8. C. Roychoudhuri and R. Roy, Guest Editors, *The Nature of Light: What is a Photon?*; special issue of Optics and Photonics News, October 2003. http://www.osa-opn.org/abstract.cfm?URI=OPN-14-10-49.
9. E. T. Jaynes, "Is QED Necessary?" in Proceedings of the Second Rochester Conference on Coherence and Quantum Optics, L. Mandel and E. Wolf (eds.), Plenum, New York, 1966, p. 21. See also: Jaynes, E. T., and F. W. Cummings, *Proc. IEEE.* **51,** 89 (1063), "Comparison of Quantum and Semiclassical Radiation Theory with Application to the Beam Maser". http://bayes.wustl.edu/etj/node1.html#quantum.beats.
10. C. Roychoudhuri, *Phys. Essays* **19** (3), September 2006; "Locality of superposition principle is dictated by detection processes".
11. C. Roychoudhuri, Conf. Proc. QTRF-4, 2007 at Vaxjo U., Sweden, to be published by AIP; "Shall we climb on the shoulders of the giants to extend the *reality* horizon of Physics?"
12. W. Heisenberg, *The Physical Principles of the Quantum Theory*, Dover publications, 1930.
13. C. Roychoudhuri, *Found. of Physics* **8** (11/12), 845 (1978); "Heisenberg's Microscope—A Misleading Illustration".
14. C. Roychoudhuri, SPIE *Conf. Proc.* Vol. **6667**, paper #18 (2007); Invited paper; "Bi-centenary of successes of Fourier theorem! Its power and limitations in optical system designs".
15. E. J. Chaisson, *Cosmic evolution: the rise of complexity in nature*, Harvard University Press, 2001.
16. See the two articles by this author in section three, III-19 and III-20.
17. K. O. Greulich, *SPIE Proc.* Vol. **6664** 0B (2007), "Single photons cannot be extracted from the light of multi-atom light sources".
18. E. Panarella, "Nonlinear behavior of light at very low intensities: the photon clump model", p.105 in *Quantum Uncertainties—recent and future experiments and interpretations*, Eds. W. M. Honig, D. W. Kraft & E. Panarella, Plenum Press (1987). For a summary, see pp.218-228 of Ref. C. Roychoudhuri, K. Creath and A. Kracklauer, eds., *The Nature of Light: What Is a Photon*, SPIE Proceeding, Vol. **5866** (2005).
19. (i) A. Zeilinger, et al., *Nature* **433**, pp.230–238 (2005), "Happy centenary, Photon". (ii) Simon Groblacher, et. al., Nature Vol. **446** (2007), doi:10.1038/nature05677: "An experimental test of non-local realism".
20. Alain Aspect, "To be or not to be local", *Nature* Vol. **446**(19), p.866, April 2007.
21. J. W. Goodman, *Introduction to Fourier Optics*, McGraw-Hill, 1988.
22. C. Roychoudhuri, *Boletin. Inst. Tonantzintla,* **2**(3), 165 (1977); "Causality and Classical Interference and Diffraction Phenomena".

23. C. Roychoudhuri, N. Prasad and Q. Peng, *Proc. SPIE* Vol.6664-24 (2007); "Can the hypothesis 'photon interferes only with itself' be reconciled with superposition of light from multiple beams or sources?
24. J. D. Gaskill, *Linear Systems, Fourier Transforms and Optics*, John Wiley (1978).
25. C. Roychoudhuri, *Proc. SPIE* Vol. **6108**-50(2006); "Reality of superposition principle and autocorrelation function for short pulses".
26. C. Roychoudhuri; *Bol. Inst. Tonantzintla* **2**(2), 101 (1976); "Is Fourier Decomposition Interpretation Applicable to Interference Spectroscopy?"
27. A. T. Forrester, R. A. Gudmundsen and P. O. Johnson, *Phys. Rev.* **99**, 1691(1955); "Photoelectric mixing of incoherent light".
28. C. Roychoudhuri; *J. Opt. Soc. Am.;* **65**(12), 1418 (1976); "Response of Fabry-Perot Interferometers to Light Pulses of Very Short Duration". (The analysis of this paper is followed and cited in two books: a. *Fabry-Perot Interferometers;* G. Hernandez, Cambridge U., 1986 and b. "The Fabry-Perot Interferometer"; J. M. Vaughan; Adam Hilger, 1989.)
29. C. Roychoudhuri, D. Lee, Y. Jiang, S. Kittaka, M. Nara, V. Serikov and M. Oikawa, *Proc. SPIE* Vol. **5246**, 333-344, (2003) **Invited**; "Limits of DWDM with gratings and Fabry-Perots and alternate solutions".
30. C. Roychoudhuri and M. Tayahi, *Intern. J. of Microwave and Optics Tech.*, July 2006; "Spectral Super-Resolution by Understanding Superposition Principle & Detection Processes", manuscript ID# IJMOT-2006-5-46: http://www.ijmot.com/papers/papermain.asp.
31. R. Loudon, *The quantum theory of light*, Oxford University Press (2000).
32. M. Scully and M. S. Zubairy, *Quantum optics*, Cambridge University Press (1997).
33. L. Mandel and E. Wolf, *Optical coherence and quantum optics*, Cambridge University Press (1995).
34. C. Roychoudhuri and N. Tirfessa, *Proc. SPIE* Vol. **6372,** paper-29 (2006), "Do we count indivisible photons or discrete quantum events experienced by detectors?"
35. C. Roychoudhuri, D. Lee and P. Poulos, *Proc. SPIE* Vol. **6290**-02 (2006); "If EM fields do not operate on each other, how do we generate and manipulate laser pulses?"
36. C. Roychoudhuri and N. Prasad, Invited Talk; "Various ambiguities in reconstructing laser pulse parameters", proceedings of the IEEE-LEOS Annual Conference, October, 2006.
37. C. Roychoudhuri, N. Tirfessa, C. Kelley & R. Crudo, *SPIE Proceedings*, Vol. **6468**, paper #53 (2007); "If EM fields do not operate on each other, why do we need many modes and large gain bandwidth to generate short pulses?".
38. N. H. Schiller, *Opt. Comm.*, **35**, pp.451–454 (1980); "Picosecond characteristics of a spectrograph measured by a streak camera/video readout system".
39. R. J. P. Engelen et al., *Nature Phy.* 2007, doi:10.1038/nphys576; "Ultrafast evolution of photonic eigenstates in *k*-space".
40. C. Roychoudhuri, *Proc. SPIE* Vol. **5531**, 450-461(2004); "Propagating Fourier frequencies vs. carrier frequency of a pulse through spectrometers and other media".
41. J. S. Bell, *Speakable and unspeakable in quantum mechanics*, Cambridge U. Press, (1997).
42. S. Lloyd, *Programming the universe*, Alfred A. Knopf, (2007).

43. P. A. M. Dirac, *The Principles of Quantum Mechanics*, Oxford University Press (1974).
44. C. Roychoudhuri and C. V. Seaver, *Proc. SPIE* Vol. **6285**-01, **Invited**, (2006); "Are dark fringe locations devoid of energy of superposed fields?"
45. C. Roychoudhuri, *Proc. SPIE* Vol. **6664**-2 (2007); "Can a deeper understanding of the measured behavior of light remove wave-particle duality?"
46. F. L. Walker, *Physics Essays* 15 (2), pp. 138–155 (2002); "The Fluid Space vortex: Universal Prime Mover".
47. C. Rangacharyulu, "No point particles, definitely no waves", see p329 in D. Aerts, S. Aerts, B. Coecke, B. D'Hooghe, T. Dart and F. Valckenborgh, in *New Developments on Fundamental Problems in Quantum Mechanics*, Eds. M. Ferrero & A. van der Merwe (Kluwer Academic, 1997).
48. A. F. Kracklauer, *Optics & Spectroscopy, 103* (3) 457–460 (2007); "Nonlocality, Bell's ansatz and probability."

7

"Single Photons" Have not Been Detected: The Alternative "Photon Clump" Model*

Emilio Panarella

Physics Essays, 2012 Woodglen Crescent, Ottawa, Ontario K1J 6G4, Canada

CONTENTS

Abstract

There continues to be a common belief that the registration of single photographic grains or emission of single photo electrons at a time validates the assertion that the interference and diffraction patterns are built through the contribution of individual photons (hν). A careful analysis of the past literature indicates that these experiments actually were not able to ascertain that one photon at a time interacted with the photo detector. This chapter reviews a series of experiments carried out during the early eighties, which suggest that the simultaneous presence of multiple photons (multiple units of hν) makes possible the registration of a single photographic blackening spot or the emission of a single photoelectron. The congruency with the paradigm of "wave-particle duality" is now better maintained by assuming that the photons, after they are emitted and then propagate from the source, develop the "bunching" property, which we proposed as a "photon clump" in 1985 and explained with a plausible extension of the Heisenberg's Uncertainty Principle.

7.1 Introduction

Einstein's path breaking photo electric paper inspired many "single photon" experiments [1]. With the advent of formal Quantum Mechanics and then field quantization by Dirac, many more "single photon" experiments were carried out [2–6]. We carefully analyzed these papers regarding the certainty of the presence of a single photon in the beam, in contrast to single detection event at a time [12]. A firm corroboration was lacking, because the quantum efficiency of detection is never 100%. Then we carried out a series of carefully designed diffraction experiments in the early eighties with attention to the number of photons per second in the experimental beam. We found that both for photographic and for photoelectric detectors, simultaneous presence of multiple photons (multiple units of hν) was more likely required to record any single successful event. Naturally, this posed a conflict as to whether the Dirac's famous assertion, "photon interferes only with itself" is still valid [7].

The lack of a direct experimental demonstration of the wave-particle duality for assured single photons led us to consider the hypothesis that perhaps isolated photons do not exist. We put forward a model of light in which a photon is invariably accompanied by other photons, all clumped together. If the individual photons in a clump are arranged on a distribution with maxima and minima of number density (i.e., a wave distribution), one is able to retrieve from this model not only an explanation for our experimental results, but also for those of Hanbury-Brown and Twiss [8], of Pfleegor and Mandel [9], of Clauser [10], and of Grangier, Roger, and Aspect [11]. Moreover, in the light of this model, Dirac's dictum that a photon interferes only with itself [7] must be reinterpreted as meaning that a clump or cluster of photons has already imprinted in it all the characteristics of interference or diffraction. Consequently, an interferometer must be viewed now as an instrument that does not do anything to the photons to let them interfere (because they have already interfered and positioned themselves on a wave geometrical arrangement with maxima and minima of distribution, even before entering the interferometer) but, by changing slightly the direction of motion of two outgoing clumps or conglomerates of photons originating from a single clump, makes them change the initial geometrical arrangement into an arrangement which can be clearly seen as a wave pattern. In short, an interferometer acts as an amplifier of the fringe separation or as a microscope to see more easily the interference or diffraction pattern already existing in the clumps of photons.

Considering the overwhelming success of the paradigm of "wave-particle duality", we developed a "photon clump" model to bring consistency with our observations. These experiments and related discussions and the details of the "photon clump" model had been presented at a NATO Advanced Research Workshop [12]. Due to limited presentation time, we will only summarize the experiments in this chapter. As to the existence and the origin of clumps of photons, this matter has been already dealt with classically to some extent by Dicke [13], who pointed out that individual atoms in a source of thermal light cannot emit photons independently of each other, because they are constantly interacting with a common radiation field. Therefore, incoherent photons are not emitted as random isolated particles, but have certain characteristic bunching properties [14]. However, we developed our "photon clump" model from a novel point of view, namely from an analysis of the Heisenberg Uncertainty Principle for photons. An interaction law for photons was derived from this analysis, which led naturally to the general form of Kirchhoff's equation and functionally to the bunching or clumping effect.

Our diffraction experiments carried out with extremely low level light using both a photographic plate and a photoelectric detector indicated that registration of a single unit of blackening or the emission of a single electron is more likely due to the simultaneous presence of multiple units of photons (hν) on the target.

7.2 Recording Very Low Intensity Diffraction Pattern by a Photographic Plate

We have chosen the simplest possible diffraction aperture, a small pinhole. However, we arranged the experiment very carefully to be able to quantify the total power (number of photons per second) received by the detector. Fig. 7.1 shows the arrangement. Two identical small pinholes (50.8μ) were

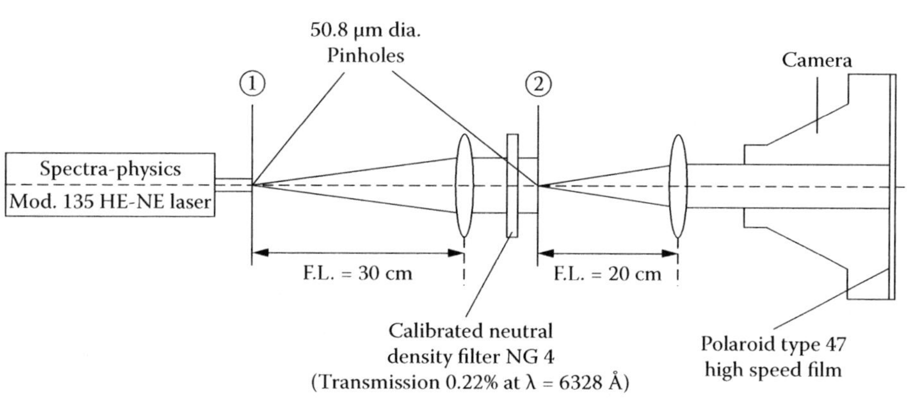

FIGURE 7.1
Experimental apparatus used to reveal the effect of the degree of statistical independence on the photon distribution on a photographic film. Without the neutral density filter along the light path, a clear diffraction pattern can be recorded on the film. With the neutral density filter inserted, the diffraction pattern does not appear as clearly as before, even when the total number of photons impinging on the film is more than two orders of magnitude larger than before.

used. The first one was used as a "spatial filter" to derive a clean Gaussian beam centered on the second pinhole, the actual diffracting aperture. Placing a well calibrated neutral density filter before the pinhole allowed us to control the arrival of the number of photons on the photographic film. A lens following the diffracting aperture was arranged to record the diffraction fringes corresponding to the Fraunhofer (far field) pattern which is the well known Airy diffraction pattern (or $J_1(r)/r$ function).

A 5 mW CW TEM_{00} mode Spectra-Physics Model 135 He-Ne laser was the source of light. The laser emitted a Gaussian beam of radius $a = 0.35$ mm at $1/e^2$ points. The peak light intensity in the central part of the beam was:

$$I_p = (2P_0/\pi a^2) = 2.59\ W.cm^{-2};\ [P_0 = 5.10^{-3}\ W]$$

The light intensity profile was smoothed out by means of a pinhole of diameter $d = 5.08 \times 10^{-3}$ cm positioned at the center of the beam, at the point of maximum light intensity. The resultant emerging bright central disc of the Airy pattern was collimated by means of a simple double-convex lens located at a distance from the pinhole equal to the lens focal length $f = 30$ cm. The intensity of light at the center of Airy pattern resulted in [15].

$$I_0 = (AP_1/\lambda^2 f^2) = 2.95 \times 10^{-4}\ W.cm^{-2}$$

where $A = \pi d^2/4$ is the pinhole area and $P_1 = I_p A$. The diffracting aperture, also of diameter $d = 5.08 \times 10^{-3}$ cm (drilled in aluminum foil 1.27×10^{-3} cm thick) was positioned at the center of the Airy disc. Since the light intensity across this pinhole was essentially constant, the photon flux entering the pinhole was 1.90×10^{10} photons sec^{-1}.

The diffracted light out of this second pinhole was then re-collimated by means of a simple double-convex lens located at a distance from the pinhole equal to the lens focal length $f = 20$ cm and the diffraction pattern was recorded by means of a camera equipped with Polaroid type 47 high speed film. The resulting intensity of light at the center of the second Airy disc was 7.57×10^{-8} $W.cm^{-2}$. The reasonable assumption was then made that such intensity was constant over a small circular area of radius equal to the diameter of the pinhole. Hence, the photon flux resulted in being 1.95×10^{7} photons sec^{-1}.

The objective of the experiment was the following. Irrespective of the photon flux reaching a detector, the diffraction pattern is considered to be the result of the superposition of the patterns created by each individual photon, which diffracts only with itself. On the other hand, if one photon were sufficient to activate a photographic grain, an identical number of photons reaching the film should provide identical diffraction patterns. Fig. 7.2 shows the experimental results. Fig. 7.2a was obtained with the apparatus just described. The photograph was exposed for 20 sec and 3.91×10^{8} photons produced the clearly defined diffraction pattern shown in the figure. We then reduced the intensity of light by inserting a calibrated neutral density

(a) (b) (c)

FIGURE 7.2
(a) Regular diffraction pattern obtained with a total of 3.91×10^8 statistically independent photons reaching the photographic film (20 sec exposure time); (b) Picture obtained when a total of 2.27×10^9 photons reach the photographic film (17h36m exposure time); c) Picture obtained when a total of 5.19×10^{10} photons reach the photographic film (336h20 m exposure time). The (b) and (c) pictures show that the diffraction pattern is missing, although the number of photons impinging on the film is ~1 order of magnitude, or even 2 orders of magnitude, respectively, larger than that which was capable of producing a clear diffraction pattern in (a).

filter (type NG4-homogeneous filter-transmission 0.22% at λ = 632.8 nm) along the light path (see Fig. 7.1). The intensity of light crossing the second pinhole was reduced in this way by a factor of 454 and only 4.29×10^4 photons reached the film per second. In order to have the same diffraction pattern as in Fig 7.2a, it was calculated that an exposure time of 2h32 m was required. The first experiments with such exposure time failed to provide the expected result in that the film did not record any light at all. Only when the exposure was increased to 17h36 m, or when 2.72×10^9 photons reached the plate (i.e., a number of photons almost an order of magnitude larger than before) were we able to obtain a meaningful photograph (Fig 7.2b.). Finally, when the exposure time was pushed up to over 2 weeks (more exactly, 336h20m, or 5.19×10^{10} photons on the plate) the resultant photograph was better defined, although the expected diffraction pattern did not appear, as Fig 7.2c shows.

These experimental results bring therefore new evidence that a diffraction pattern on a photographic plate is not preserved when the intensity of light is extremely low, even when the total number of photons reaching the film is larger than that which is capable of producing a clear diffraction pattern. In other words, a diffraction pattern does not build up linearly with light intensity, as the wave-particle duality requires.

The foregoing experimental results can be explained if one refers to the theory of photographic grain developability, as put forward by Rosenblum [16] and experimentally verified by Polovtseva et al. [17] The details have been worked out in the reference [12]. Both the experiment and the theory point out that packets of at least four photons are required for diffraction effects to be revealed by a photographic plate. Single photons are not recorded and their dual nature cannot be demonstrated with the photographic technique.

7.3 Recording Very Low Intensity Diffraction Patterns by a Photoelectric Detector

The experimental apparatus used for the photoelectric detection of the photons was essentially the one previously described. Only the camera has been replaced by a high gain photomultiplier mounted on a motor-driven translation unit (Fig. 7.3). For good fringe resolution, the photomultiplier is provided with a small orifice of 5.08×10^{-2} cm diameter drilled on its front cover. The fringe pattern is then vertically scanned and recorded on an oscilloscope.

The detection system consisted of a fourteen-stage, flat-faceplate RCA photomultiplier type 7265 having a multialkali photodiode ([Cs]Na_2KSb) with S-20 response. The photomultiplier current amplification was 2×10^7. The tube was normally operated at 2000 V, i.e., below the maximum permissible voltage of 2400 V, in order to reduce the dark current from thermionic emission and to increase the signal to noise ratio.[19] However, when maximum amplification was required, the tube was operated at 2400 V. In order to further reduce the dark current, the photomultiplier was cooled with a blanket of dry ice to –15°C. Light uniformity over the photocathode area was achieved by inserting a diffuser within the photomultiplier case, right

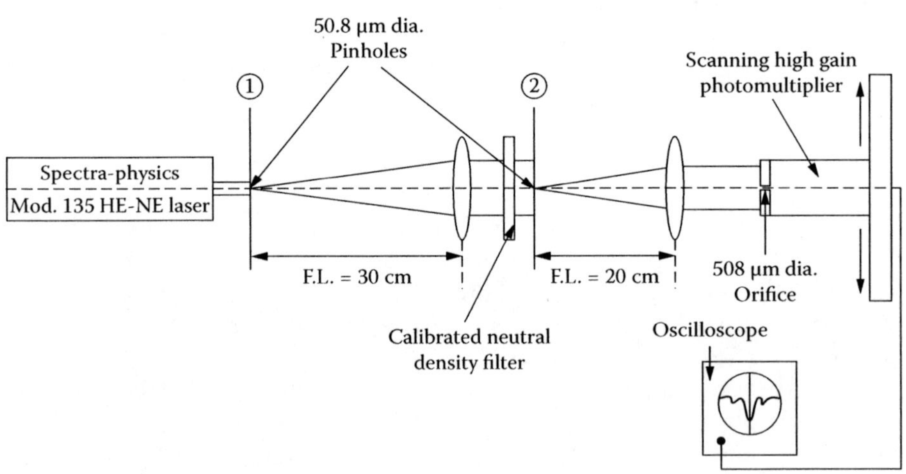

FIGURE 7.3
Experimental apparatus used to reveal the effect of the degree of statistical independence on the photon distribution on a diffraction pattern. With a photon flux $\Lambda = 1.90 \times 10^{10}$ photons/sec within the interferometer (i.e. between pinholes 1 and 2), a clear diffraction pattern is recorded on the oscilloscope. When a neutral density filter of transmission T = 2.09% is inserted in the light path, so that the light flux is reduced by a factor of 48 to $\Lambda = 4 \times 10^8$ photons/sec, the same clear diffraction pattern as before does not appear, despite an overall increase of amplification of the detection system by a factor of 441.

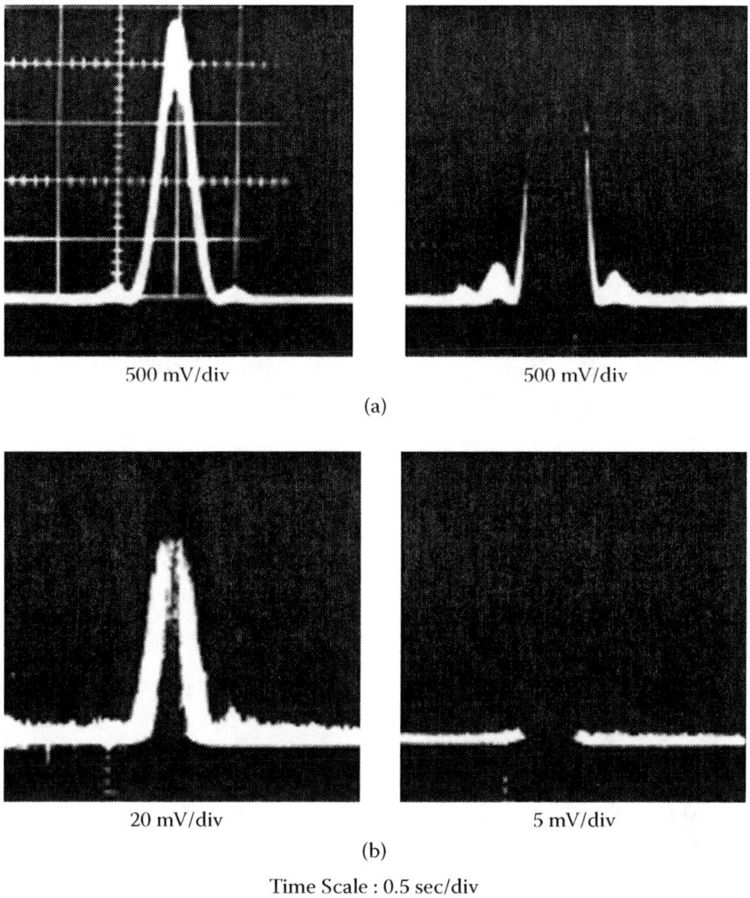

FIGURE 7.4
(a) Regular diffraction pattern obtained with a photon flux $\Lambda = 1.90 \times 10^{10}$ photons/see within the interferometer. (b) The diffraction pattern is affected and the lateral fringes do not appear when the light flux is reduced to $\Lambda = 4 \times 10^{8}$ photons/sec, despite the fact that the amplification of the detection system has been increased 441-fold while the light flux went down only 48-fold from (a) and (b).

behind the entrance orifice. Finally, the signal from the photomultiplier was sent to a Tektronik type 555 oscilloscope where it was recorded by means of a type D high-gain preamplifier unit. Fig. 7.4 reports the experimental results. As in the case when the diffraction pattern was recorded with the photographic film, Fig. 7.4a (left) shows that, with the beam unimpeded by any filter and photon flux $\lambda = 1.90 \times 10^{10}$ photons.sec^{-1} within the interferometer, the diffraction pattern (Airy pattern) is clearly defined and is composed of a central peak surrounded by two subsidiary maxima. The latter can be seen more clearly in Fig 7.4a (right), where the central peak has been amplified by a factor 2.55 to ~10.2 divisions (by "division" we mean, of course, the separation between two consecutive solid horizontal lines on the photographic grid) from the original ~4 divisions, by increasing the photomultiplier voltage from 2000 V to 2200 V. According to classical optics [18] the first subsidiary maximum on the diffraction pattern should have amplitude equal to 0.0175 times the central peak amplitude, that is

$$0.0175 \times 10.2 = 0.178 \text{ division}$$

As Fig. 7.4a shows, this is indeed so and the first subsidiary maximum is clearly seen. Actually, even the second subsidiary maximum is revealed, whose amplitude is [18]:

$$0.0042 \times 10.2 = 0.042 \text{ division}$$

The fringes in this case are justified by considering that, with high probability, more than one photon is present within the interferometer at any one time. In fact, the probability analysis carried out in Sec. II.3.2.1 of Ref. 12 shows that, for the case of Fig 7.4a:

$$\Lambda\tau = 1.90 \times 10^{10} \times 1.4 \times 10^{-9} = 26.6 >> 0.69$$

where Λ is the flux of photons per unit time crossing the interferometer and τ is defined differently by different researchers, either the coherence time [16], or the characteristic time of the emulsion materials [17]. This means that the probability of interference with two or more photons is by far greater than the probability of interference with one photon and the fringes might then be created by packets of photons rather than single photons.

Now, if the same experiment is repeated with reduced light intensity, the fringes (or subsidiary maxima) should be seen again provided the amplification is sufficiently high to yield a fringe amplitude of 0.178 division or higher. We inserted therefore in the light path (Fig. 7.3) a calibrated neutral density filter of transmission $T = 2.09\%$ at the laser wavelength, thus reducing the light intensity within the interferometer by a factor of -48 to -4×10^8 photons.sec^{-1} ($\Lambda\tau$ of the probability analysis is now $0.56 < 0.69$) and obtained the picture of Fig. 7.4b (left) which shows only the central peak of amplitude ~2.5 divisions. No sign of fringes or subsidiary maxima is present in this picture. In an attempt to retrieve the fringes, the amplification of the photomultiplier was increased by a factor of 4.41 to its maximum value, by allowing the maximum permissible photomultiplier voltage (2400 V). Also, the oscilloscope amplification was increased by a factor of 4 from 20 mV/division to 5 mV/division in going from Fig. 7.4b (left) to 7.4b (right) (this means that the oscilloscope amplification was increased by 100 from the initial 500 mV/division—Fig. 7.4a – to 7.5. mV/division—Fig. 7.4b (right). Despite an overall amplification of 441 in going from Fig. 7.4a (left) to Fig. 7.4b (right), the subsidiary maxima did not appear. This is surprising because the amplitude of the first subsidiary maximum on Fig. 7.4b (right) should have been:

$$0.0175 \times 2.5 \times 4.41 \times 4 \text{ divisions} = 0.77 \text{ divisions}$$

i.e., larger than in Fig. 7.4a (right), where it was detected. Consequently it seems that the expected fringes did not exist.

In order to analyze more in depth these unexpected results and to discover if a valid reason exists for the absence of the fringes, we have reproduced in Fig. 7.5 the photographs of Fig. 7.4a and 7.4b (right) and added another oscilloscope record (Fig. 7.5c) obtained with a much lower light flux of 7×10^7 photons.sec^{-1} within the interferometer, i.e., a photon flux lower by a factor of ~273 than the initial one. Moreover, beside each oscilloscope record, we show the same diffraction pattern drawn with a thin line passing in the middle of the baseline or in the middle of the broadened trace.

Now, if one looks at Fig 7.5a, and more specifically at the figure on the right, one observes that the first subsidiary maximum has amplitude of ~250 mV. If one reduces the light intensity by a factor of ~48, the amplitude of such subsidiary maximum should be reduced accordingly to ~5 mV. This signal is of sufficient amplitude and a clear upward displacement of the baseline in Fig. 7.5b at the position of the fringes should have occurred. [The absence of the fringes cannot be justified on the ground that, since the photomultiplier gives very short pulses, they do not overlap at the position of the fringes when the light intensity is weak, thus precluding the trace to elevate from the baseline. In fact, at the position B on the central peak at the same height as point A of the fringe (see Figs. 7.5a and 7.5b) the light intensity is just as weak. Although in B the trace does not elevate from the baseline by as much as it should, namely ~5 mV, still an upward displacement takes place by ~1 mV, whereas nothing of this happens on the fringes (A. Gozzini, C.W. McCutchen, E.S. Hanff, private communication).]

On the other hand, the experimental results reported in Fig. 7.5 seem to indicate a departure from the predictions of wave optics and an apparent approach to the predictions of geometrical optics. Such indication is provided in particular by Figs. 7.5b to 7.5c which show a sudden discontinuity of light intensity at points which are closer to the optical axis of the system as the light intensity goes down. In other words, although the central fringe seems to be maintaining always, at the light intensities we have investigated, a width of 13.5 mm, the light intensity distribution presents a sudden discontinuity (very similar to a shadow effect) which is closer to the geometrical axis of the system as the light intensity goes down.

To conclude this section, it seems that the absence of fringes is due to nonlinearity of detection at very low light intensity, and this assumption will receive a confirmation from the experiment to be reported in the next section.

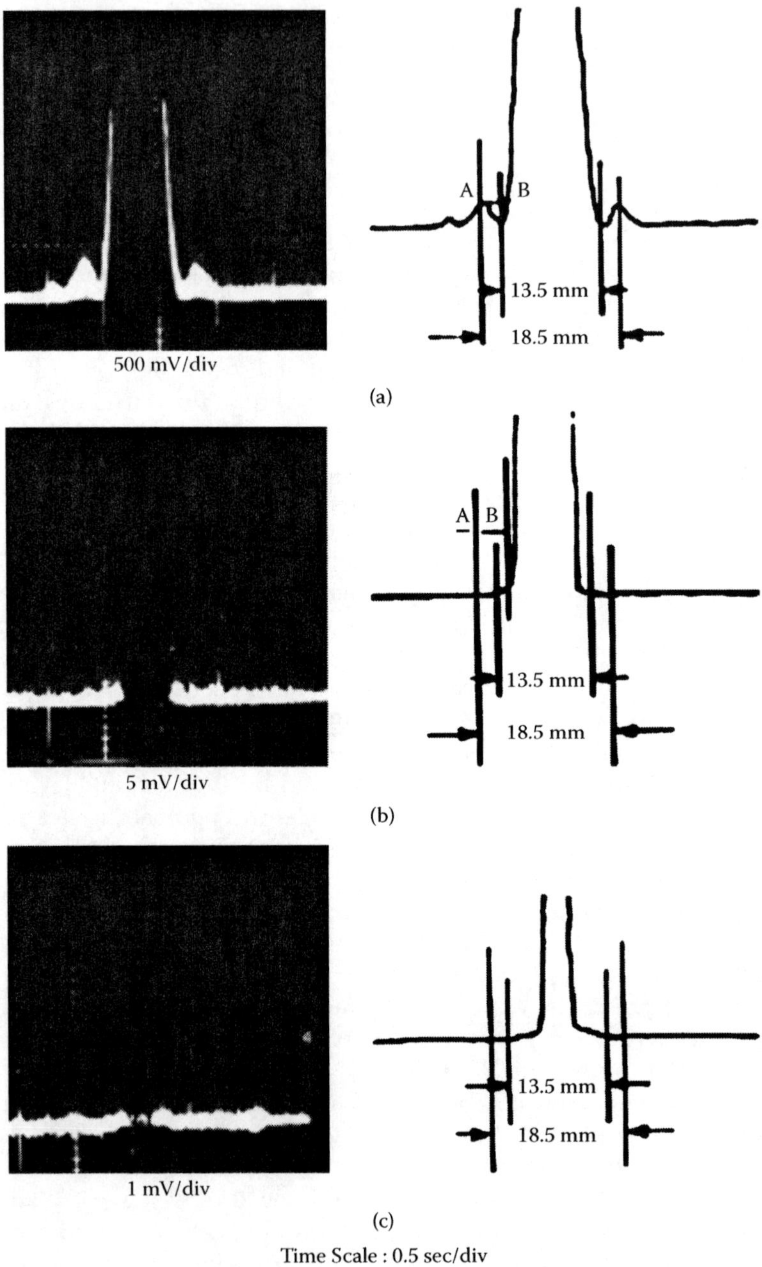

FIGURE 7.5
The regular diffraction pattern obtained in (a) with a photon flux $\Lambda = 1.90 \times 10^{10}$ photons/sec within the interferometer is not preserved and the lateral fringes do not appear when the light flux is reduced to $\Lambda = 4 \times 10^{8}$ photons/sec (b) and to $\Lambda = 7 \times 10^{7}$ photons/sec (c), despite the fact that the amplification of the detection system has been increased 441-fold from (a) to (b), and 2200-fold from (a) to (c), while the light intensity went down only 48-fold and 273-fold, respectively.

7.4 Photoelectric Detection and Photon Counting Along a Diameter of the Diffraction Pattern

Our latest experiment, in which we counted the photons along a diameter of the diffraction pattern, was done with basically the same experimental apparatus as previously described (see Fig. 7.3), but the oscilloscope is replaced by the combination of an amplifier and a pulse counter. The photomultiplier was now operated at a constant voltage of 2050 V. In order to eliminate any stray light entering the photomultiplier, the entire apparatus containing the laser and related optics was enclosed within a black box, so that only a small opening was available for the laser beam to get out of pinhole No. 2. As to the residual light from the laser discharge tube going through the pinhole, it was cut almost completely out by placing in front of the photomultiplier a high-pass filter having transmission 84% at the laser wavelength $\lambda = 632.8$ nm and rapidly falling down to 0.03% at $\lambda = 554.0$ nm. Finally, the entire experiment was carried out in a small windowless dark room completely shielded from any external light.

The experiment consisted in moving the photomultiplier by equidistant steps of 5/1000 of an inch ($= 1.27 \times 10^{-2}$ cm) and arresting it at each step just for the time required for pulse counting. The counting was done with a Tennelec 546P Scaler and 541A Timer, the signal from the photomultiplier having been amplified by a factor of 10 through an amplifier having input resistance 1000Ω.

The counting time was chosen rather short, 2×10^{-3} sec and 2 sec for the two experiments that we ran, respectively, because this offered some distinct advantages over long counting times. For one thing, one avoids in this way problems of photomultiplier fatigue and decrease of sensitivity [19]. For another, the dark count can be greatly reduced with an appropriate choice of short counting time.

The experimental results are reported on Fig. 7.6. The solid circles represent the counts obtained when the photon flux within the interferometer (i.e., between pinholes 1 and 2) was 1.90×10^{10} photons.sec^{-1} (the average photon separation is 1.57 cm, much less than the length of the interferometer 42 cm) and the counting time 2×10^{-3} sec. The open circles are the counts obtained when the photon flux was decreased 769-fold to 2.47×10^{7} photons.sec^{-1} by the insertion of a calibrated neutral density filter along the light path (the average photon separation is now 1214 cm, much greater than the interferometer length) and the counting time increased 1000-fold to 2 sec. One can see that the two diffraction patterns do not overlap (actually, the second pattern should be 30% greater than the first because of the factor $1000/769 = 1.3$). On the other hand, a well defined diffraction pattern appears in the first instance—the high light intensity case – with a clear fringe or subsidiary maximum on the left side of the central peak (the other on the right is absent because we did not scan the full diffraction pattern). Also, the fringe amplitude is what one would expect [18], namely 0.0175 times the central peak amplitude:

$$0.0175 \times 335 = 6 \text{ counts}$$

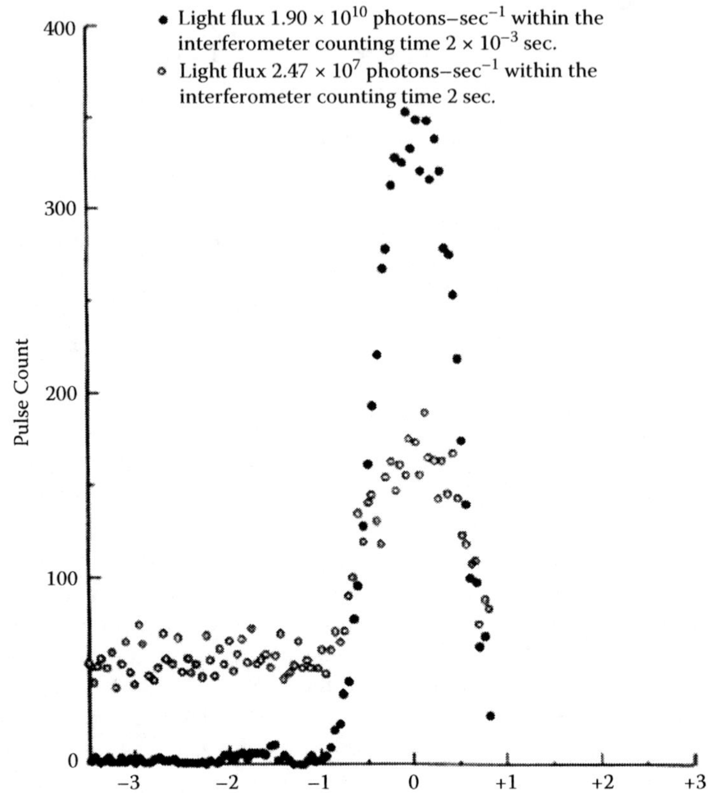

FIGURE 7.6
Solid circles: regular diffraction pattern obtained with a photon flux $\lambda = 1.90 \times 10^{10}$ photons/sec within the interferometer and counting time 2×10^{-3} sec. Open circles: the diffraction pattern does not have the same amplitude as before when the light flux is reduced 769-fold, despite having increased the counting time 1000-fold to 2 sec.

In the second instance, the low light intensity case, the diffraction pattern, besides lacking the lateral fringe, which can be justified because its amplitude is below the noise level, does not have the expected central peak amplitude of

$$1000/769 \times 335 + 52 \text{ (average noise)} = 487 \text{ counts}$$

but only an amplitude of 163 counts.

In order to have a measure of the detection nonlinearity, we subtracted the noise-free signal amplitude of the low light intensity case from the expected noise-free signal, and divided the difference by the former amplitude:

$$\frac{435 \text{ (expected)}2111 \text{ (found)}}{111 \text{ (found)}} = 2.91 = 291\%$$

This is quite a large nonlinearity.

In conclusion, these experimental results confirm the nonlinearity of photoelectric detection of the previous section. Moreover, they indicate that such nonlinearity, at very low light intensities, is no different, as far as the effects are concerned, from the nonlinearity of the photographic detection and that both constitute an obstacle for proving that we are dealing with a single particle phenomenon.

Discussion

The wave-particle duality for single photons can be demonstrated only if a wave phenomenon, such as an interference or diffraction pattern, is unequivocally associated with a single particle phenomenon, for which linearity of photon detection with light intensity is required. All three experiments reported above have shown that, at very low light intensities, the phenomenon is nonlinear. Moreover, they indicate that the flux for which the nonlinearities start to appear is of the order of 10^4 photons.sec^{-1} at the detector. It is interesting to find that, apparently, never before the linearity of photomultipliers response at such low light fluxes has been carefully investigated [19]. Fig. 7.7 reports the linearity characteristics of typical RCA photomultipliers [19]. It is to be noticed that these instruments are linear within a large range of photon fluxes (10^5–10^{13} sec^{-1}), but their linearity characteristics have not been tested right where they should be for our purposes of verifying

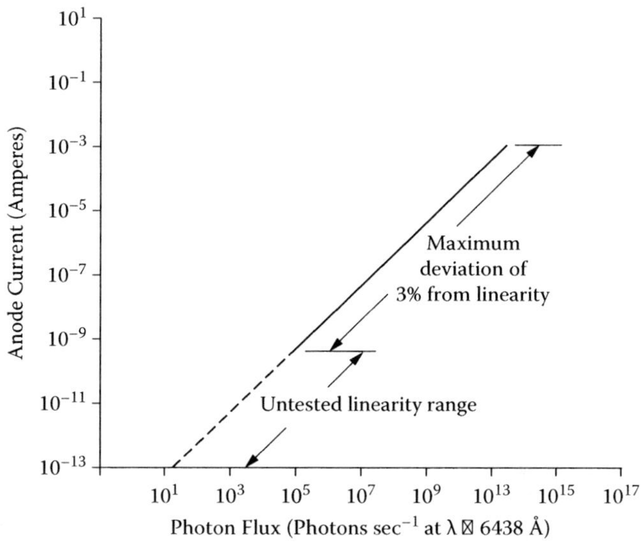

FIGURE 7.7
Linearity characteristics of RCA photomultipliers.

the wave-particle duality hypothesis, namely below 10^4 photons/sec. In summary, because of the nonlinearities found, the wave-particle concept for single photons remains at the "status quo ante", namely as that of a theoretical hypothesis or postulate.

One could explain, of course, the photoelectric results reported here in the same manner as it was done with the photographic results, in terms of some possible cause for the detection nonlinearity. One of these possible causes, for instance, is that the higher the light flux, the higher the noise generated within the photomultiplier. The problem with this approach is that it does not serve its purpose. In fact, the justification of the nonlinearity in this way will require the assumption that the wave-particle duality hypothesis is correct and that linearity of photoelectric detection with light intensity is to be expected. But then, any justification of the departure from such linearity cannot be used to prove the original hypothesis. To put it more clearly, a hypothesis (the wave-particle duality) cannot be proven by starting with the assumption that the wave-particle duality hypothesis is correct. What is required, in other words, is a direct and clear demonstration of linearity of photon detection with light intensity (at very low light intensities) in order to prove that we are dealing with a single particle phenomenon.

In the case of several photons within the interferometer, or in what we would call the regular intensity case, the wave-particle duality is proven: clear interference fringes appear and the phenomenon is linear. It is unfortunate, however, that we cannot unequivocally ascribe the wave phenomenon to single particles because there are many of them within the interferometer, which could collectively act to create the fringes.

To summarize our results, Fig. 7.8 reports in graphical form, for comparative purposes, the two diffraction patterns obtained in the latest of our experiment with the photon counting technique. We observe that, at regular photon flux (= 1.95×10^7 sec^{-1}), such that the total number of photons reaching the detector is 39000, we obtain diffraction pattern B which peaks at ~350 counts. When we lower the photon flux to 2.53×10^4 sec^{-1} and let a larger number of photons (= 50600) reach the detector, we obtain diffraction pattern A of smaller amplitude (= 130 counts). The nonlinearity is clearly present. Such nonlinearity is not unique to our experiments. Reynolds, Spartalian and Scarl published [20] diffraction patterns recorded with two photon fluxes of 200 sec^{-1} and 30 sec^{-1} such that the total number of photons were 72000 and 14400, respectively, the ratio between these two numbers being 5. We have measured the ratio of the densities of the two photographs and found it to be 8.125. Thus, the nonlinearity present is 62.5%, a quite appreciable nonlinearity.

In conclusion, the series of experiments reported here on the detection of diffraction patterns from a laser source at different low light intensities confirms the wave nature of collections of photons but tends to dispute it for single photons. In other words, our experiments underscore the absence of unambiguous proof of "single photon" interference.

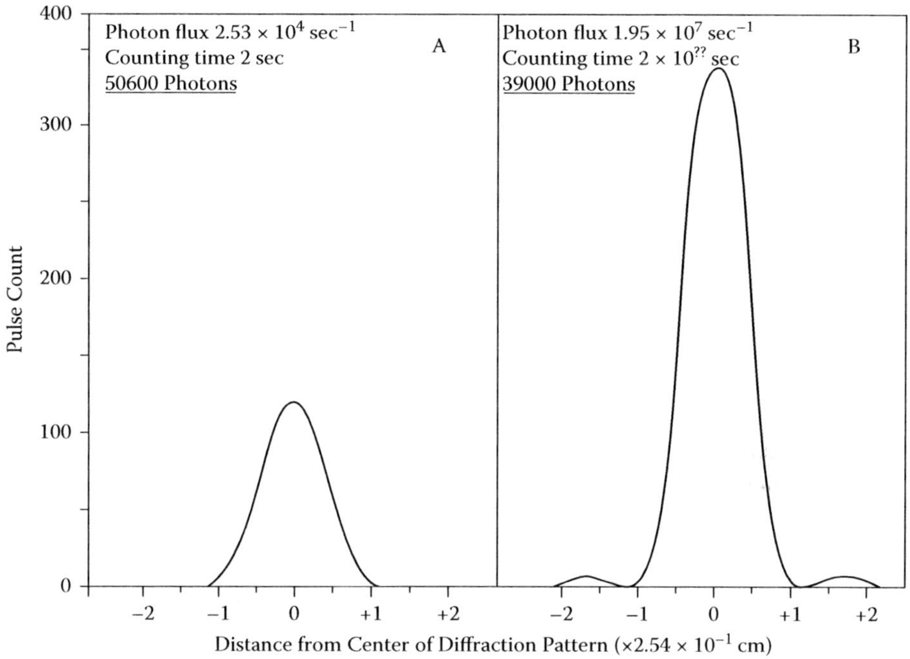

FIGURE 7.8
At regular photon flux (= 1.95×10^7 photons/sec), such that the total number of photons reaching the detector is 39000, we obtain diffraction pattern B which peaks at = 350 counts. When we lower the photon-flux to 2.53×10^4 photons/sec and let a larger number of photons (= 50600) reach the detector, we obtain diffraction pattern A of smaller amplitude (= 130 counts). The nonlinearity is clearly present.

Acknowledgment

The author would like to thank Prof. C. Roychoudhuri and his student, C. Kelley, for their valuable help in creating this summarized reproduction of an earlier publication by the author [12].

References

1. A. Einstein, *Annalen Phys.* **17**, 132 (1905).
2. L. de Broglie, *Compt. Ren. Hebd. Séance Acad. Sci. Paris,* **177**, 506,548, 630 (1930); Thèse de Doctorat (Masson, Paris, 1924).
3. G. I. Taylor, *Proc. Camb. Phyl. Soc. Math. Phys.* **L5**, 114, il9o9).
4. R. Gans and A. P. Miguez, *Ann. Phys.* **52**, 291 (1917).

5. P. Zeeman, *Physica Eindhoven* 325 (November 1925).
6. A. J. Dempster and H. F. Batho, *Phys. Rev.* **30**, 644 (1927).
7. P. A. M. Dirac, *The Principles of Quantum Mechanics* (Clarendon Press, Oxford 1958), p. 9.
8. R. Hanbury-Brown and R. Q. Twiss, *Proc. Roy. Soc.* A242, 300 (1957): A2443, 291 (1958).
9. R. L. Pfleegor and-L.-Mandel, *Phys. Rev.* **159**, 1084 (1967).
10. J. F. Clauser, *Phys. Rev.* **119**, 853 (1974).
11. P. Grangier, G. Roger, and A. Aspect, *Europhys. Lett* **1**, 173 (1986).
12. E. Panarella, Nonlinear behavior of light at very low intensities: the 'photon clump' model, p. 105 in *Quantum Uncertainties: Recent and Future Experiments and Interpretations*, Eds. W. M. Honig, D. W. Kraft and E. Panarella, NATO ASI Series, Series B: Physics Vol. **182**, Plenum Press (1987).
13. R. H. Dicke, *Phys. Rev.* **23**, 99 (1954).
14. E. Panarella, *Ann. Fond. Louis de Broglie* **10**, 1 (1985).
15. L. Levi, *Applied Optics: A Guide to Optical System Design* 4, Vol. I (John Wiley & Sons, New York), 1968, p. 87.
16. W. M. Rosenblum, *Jour. Opt. Soc. Amer.* **58**, 60 (1968).
17. G. L. Polovtseva, A. A. Dybine, and V. V. Lipatov, *Opt. and Spectr.* **33**, 183 (1972).
18. M. Born and E. Wolf, *Principle of Optics*, 3rd ed. (Pergamon Press, Oxford) 1965, p. 397.
19. R. W. Engstrom, *Jour. Opt. Soc. Am.* **37**, 420 (1947).
20. G. T. Reynolds, K. Spartalian, and D. B. Scarl, *Nuovo Cim.* **61B**, 335 (1969).

Section 3

Exploring Photons beyond Mainstream Views

8

What Is a Photon?

C. Rangacharyulu
Department of Physics and Engineering Physics, University of Saskatchewan, Saskatoon, Sakkatchewan S7N 5E2, Canada

CONTENTS

Abstract

The nature of physical objects cannot be clarified independent of our concepts of space and time. We present arguments to show that neither the classical 3D space—1D time nor 4D space-time of special relativity provide a satisfactory theoretical framework to this end, as we encounter non-classical objects. The general relativity is perhaps able to accomplish this task. But, it does so only at the expense of rendering the empty physical space neither isotropic nor homogeneous. Waves are not candidates to represent fundamental objects. We use the celebrated example of Compton scattering to argue that the full description of the experiment makes use of both wave-like and particle-like behavior in the early quantum-mechanical formulations. The later quantum field theoretical descriptions of the same phenomenon abandon causality. We present model arguments from modern particle physics experiments that the photon may be a hadron, at least part of the time.

Key words: Electromagnetic radiation, photons, space-time concepts, ether, particle-wave duality, hadronization.

8.1 Introduction

The quest to describe the physical nature of the universe pervades through all ages and cultures. Among the easily accessible ancient works, a lucid and logical discourse was presented by Lucretius[1], a Roman in the 1st century BC. The ancient natural philosophers had to satisfy themselves with logic and imagination, which they could put forth as proclamations. One sees that there were seeds of future axioms in these assertions. For example, we find the laws of conservation and principle of action and reaction in the statements of Lucretius. It is well known that Democritus advanced the principle of reductionism when he enunciated the atomicity of matter. The main difference between the pre- and post-Renaissance science is that we now insist on experimental verification of the assertions and predictions and we are not simply swayed away by pronouncements. Science in general and physics in particular have become quantitative. Gödel's incompleteness theorem[2], originally intended for mathematical theories, has been found to be of significance for several other disciplines, such as artificial intelligence and information theory etc. Indeed, it would be accurate to say that Gödel's theorem impacts epistemological aspect of each and every discipline of study. As to be expected, the scientific enterprise is also subject to Gödel's theorem.

Implicit in physics theories is the fundamental assumption that the dynamics of physical universe can be discerned as due to interactions among interactants. That is to say, we begin with the idea that there are entities, which have an existence independent of their surroundings. They are the fundamental things and the Universe is made up of some conglomerations of those basic entities. Then, we would attempt to describe all processes and structures as due to interactions among them. The interactions exhibit some universal characteristics independent of the participants. Needless to say, such descriptions rely heavily upon our concepts about space and time, since interactants exists in space and interactions occur in space-time. This fundamental axiom is beyond verification in any current physical theory. At least for the time being, we continue to accept this basic premise. Thus, though the emphasis of this chapter is on the limitations of our knowledge or understanding of what a photon is, we should also address our notions of space and time. The second section is devoted to the concepts of space and time in the frame works of Newtonian and Einsteinian relativity. The third section concerns with wave-particle duality. In this section, we present reasons to show that observer-dependent reality cannot describe the wave-particle duality of photons. For this purpose, we make use of the celebrated example of Compton scattering experiment. In the fourth section, we will address the question of behavior of real and virtual photons and their energy-dependent characteristics as described by high energy physics experiments. The summary and conclusions are presented in the last section.

8.2 Space and Time

An ongoing debate between physicists and philosophers concerns concepts of space and time. The current discourses surround the question whether there is a four-dimensional space-time or if it is a three-dimensional space and one-dimensional time. We should note that this question, referred to as ontology of space and time[3], is of fundamental importance as we deal with the ontology of fundamental entities. The mathematical treatment of relativistic transformations is extensively documented in literature. With regard to space and time, the Lorentz-Einsteinian transformations seemingly reveal features which are beyond everyday experience or common sense. We will recapitulate some well-known simple mathematical results and address the physical meaning.

Say, two observers are in relative motion with respect to each other such that $\beta c = v$ is their relative velocity and $\gamma = 1/\sqrt{1-\beta^2}$. Let us also say that the two observers are each given a meter stick and they measure lengths. Each observer finds that the meter stick of the other is shorter than his own. If we now Lorentz-boost one of the observers to the other's frame, then both observers will find the two rods to be of equal length and that they are of proper length[a].

It is our point that we can reconcile these observations in a simple way by arguing that the length contractions are apparent shortenings of rods and that the rods do not contract. There are two reasons to argue this way. The first reason is the distinction between kinematics and dynamics. One tacitly assumes that special relativity is a kinematical theory just as Galilean relativity. A kinematical theory cannot induce physical changes in the objects or phenomena as it does not involve any forces nor potentials. Clearly, a change in length would amount to a physical transformation and thus a kinematical theory cannot be held responsible for these changes. In the above example, an observer along with his meter stick jumps the frames, while the other one is not affected. Thus, no argument can be presented as to why an unaffected stick will have its length changed.

Another reason is that both observers would see the other's meter shortened rather than one observer seeing a shortened rod and the other seeing a longer one. These observations are interpreted in terms of the idea that simultaneity is relative. Consider the following scenario. In the frame of an observer A, the observer A and his rod are at rest. Another observer B flying along attempts to measure the length of A-rod by determining the coordinates of its two ends. He finds a flying rod. He is confident that he can accomplish this task if he can grab the two ends simultaneously. The two observers disagree on what constitutes simultaneity. However, with the help of Lorentz transformations, we can reinterpret the measurements. This reasoning will

[a] In special relativity, proper length of an object is the length measured in its rest frame. It is the longest.

accommodate the suggested observational results: Measured lengths are always equal to or shorter than proper length of an object. The apparent loss of simultaneity, as measured by two observers, can be accounted for by allowing for the Lorentz transformations and we can again recover the proper length.

If we were to stop here, we may satisfy ourselves that we have clear understanding of these manifestations as simply kinematical, apparent phenomena and that nothing unusual happens to raise this question. However, there are counterexamples in physics. The modern-day experimental observations force us to reconsider this standpoint. Without resorting to non-inertial frames and thus to general relativity, we can look at common experiments at particle physics facilities where secondary particles approaching the speed of light are routinely produced and detected.

It is known that each unstable elementary particle has a characteristic lifetime (τ). It has been verified, on several occasions, that lifetime is an intrinsic property of particle species and it does not depend on external surroundings, such as chemical environment, electromagnetic fields etc. While it is impossible to predict when a particle will decay, we can easily know how many particles decay in a specific time interval. If we have a collection of a species of particles with lifetime τ, we can write the number of particles I(t), surviving after a time "t", as

$$I(t) = I(0)e^{-t/\tau} \tag{8.1a}$$

If we have a beam of particles traveling at a speed close to that of light, the decrease in intensity is found to be governed by the equation

$$I(t) = I(0)e^{-t/\gamma\tau} \tag{8.1b}$$

i.e., time in the laboratory frame is prolonged by a factor of γ. We can write these equations in terms of intensities I(x), at a distance $\times$ from the starting point, as

$$I(x) = I(0)e^{-x/\beta c\gamma\tau} = I(0)e^{-xm/p\tau} \tag{8.1c}$$

where m and p are the rest mass and momentum of the particle in the laboratory, respectively.

What do these equations mean? We may interpret them to say that in the laboratory frame

(i) Characteristic lifetime of a particle has increased from τ to $\gamma\tau$.
(ii) Characteristic length has increased from $c\beta\tau$ to $c\beta\gamma\tau$.

In the laboratory frame, the flying particle has a longer mean life and travels longer distances, increased by a factor of γ. We tell our students that the particle lifetime in the laboratory frame is dilated and that the particle finds the lengths in the laboratory to be contracted. This interpretation is not an idle talk. Experiments in the laboratory are designed to take advantage of this fact. For high energy particles, the experimental layouts can be spread out in the laboratory, while the beamlines of low energy particles will have to be very short.

As an example, consider beams of pions. Pions have a mean life of $\tau = 26$ nsec or $c\tau = 7.8$ meters. If relativistic space-time ideas of increase in the lifetime and characteristic lengths are not correct, the intensity of particles will decrease, independent of the speed of pions, by a factor of 0.368 in each time interval of 26 nsec or a path length of 7.8β meters. In stead, in the laboratory, fast pions travel tens of meters and the flux is quantitatively given by the above equations 8.1b and 8.1c, indicating the increase of lifetime and characteristic distances. Thus, faster pions live longer and get farther. The design, construction and operation of electric and magnetic fields for the transport of charged particles must take these aspects into account.

What would the particle see in the laboratory frame? We can only guess. To get a perspective, let us consider a gedanken experiment. An observer at rest in the laboratory performs two experiments. He measures the flux of the particles flying in the laboratory which are subject to equations 8.1b and 8.1c. He also measures the lifetimes of the same species of particles at rest with him in the laboratory, which obeys the equation 8.1a. Clearly the particle at rest will have a shorter lifetime. One might argue that the flying particle will see the one at rest in the laboratory to have elongated lifetime. Certainly not. If the ones at rest are located at some destination point of the flying particles, they (the ones at rest) would decay before the flying ones reach them. Thus, the flying particles notices the decay of particles at rest in the laboratory. It then has to conclude that the particles at rest would have their life shortened. But more than likely, it would do the Lorentz transformation and account for the seeming time disparity of the rest particles' decay times. Also, the surface of the earth is continually bombarded by cosmic rays. The main component of cosmic rays are muons of proper lifetime of 2.2 μsec, traverse distances of about 15 km before they reach the sea level, corresponding to $\gamma \sim 25$ and lifetime of about 50 μsec. These examples suggest that the elongation of length and time are not simple perceptions but they are real. They are as real as they could get. When one describes this feature as "running clock go slower", we imply it is a real physical phenomenon.

Do space-time concepts of relativity prescribe these changes in characteristic lengths and times? If we insist that the increase of lifetimes is a manifestation of relativity, we have two problems. First, we will find two inertial frames are not equal. In one frame the lifetime is elongated and in the other it is seemingly decreased. For all practical purposes, increase in the characteristic lifetime of flying particle is real. If relativity were a kinematical theory, it could not have caused these real changes in systems. Thus, Einsteinian relativity must be considered as a dynamical theory unlike the Galilean relativity, which is simply a kinematical expression of Newtonian mechanics. If we adopt this view point, at what stage did special relativity become a dynamical theory? Obviously, it has become dynamics at the instant or space-time instant we postulated that speed of light is constant and that it is independent of the motion of observer. We let the features which are purely time-dependent (frequency) or space-dependent (wave length) to be functions of relative motions of the source and observer to render the product of frequency $\times$ wavelength = speed = a constant. The fact that electromagnetic field designs must incorporate the length contractions can be explained away in this manner. Still, the increase in the lifetime of a particle cannot be understood. Is it just a coincidence that the increase in characteristic length and time are given by the Lorentz factor, γ? We don't know the answer. After all, the exponential law is an empirical law and in equations 8.1b and 8.1c, the Lorentz factor "γ" is introduced in an ad hoc manner.

It is often stated, especially in classrooms, that Einsteinian relativity tends to Newtonian relativity at low velocities. While this statement may be mathematically correct, it is not correct in the physics sense. One cannot claim that a four dimensional space-time continuum a la Einstein reorganizes itself to a three dimensional space and one-dimensional time of Newton, at low velocities. There is no evidence of four dimensional space-time at low velocities and/or macroscopic scales. Also, the relativistic speeds are relevant only for microscopic objects. While the special relativity makes rigorous mathematical formulations for the apparent phenomena, it offers an adhoc prescription for the observations of changes in life times. If we adhere to four dimensional space-time, Galilean relativity, though a good mathematical approximation at low velocities, misses out on the important physics. It, then, seems that we are attempting to describe the entities, which inherently rely on four dimensional space-time mathematical logic, with language and concepts based on three dimensional space and one dimensional time. This complicates and leads to confusion for conceptual foundations of microscopic physics, if we insist upon using the language of macroscopic physics to microscopic world. The problem thus seems to stem from the insistence of Copenhagen interpretation of quantum theory[4] that the concepts of classical physics form the language by which we describe the arrangement of our experiments and state the results. However, a few attempts to find alternate terminology have not been successful.

We may not shrug away from this quandary with the above reasoning, since our experimental apparatus function in 3D space and one dimensional time. Clearly, this problem of entities is very closely connected with our concepts of space-time. Is there an empty space in which objects move, which can be found as displacements measured as a function of time? Or, is the physical space a very complicated entity?

Ancient philosophers and also Newton considered an empty space for corpuscular motion. They assigned material media with mechanical properties to sustain wave motions. It is well known that Luminiferous ether, not participating in mechanical motion, was suggested for electromagnetic radiation. Though Einstein set aside the problem of ether in the formulations of special relativity, he addressed this question in General relativity.

However, he was not very helpful in resolving this puzzle. In his readings in 1920[5] and 1925[6], he addresses the question of ether and relativity. At one place, he argued[7] "—if, in fact, nothing else whatever were observable than the shape of the space occupied by the water as it varies in time, we should have no ground for the assumption that water consists of movable particles". Yet, he would not accept the Newton's concept of "action at a distance" and he concludes that empty space is neither homogeneous nor isotropic. While it is understandable that we have no way of knowing "space" without reference to the objects or axes, the transition to general relativity with space defined by ten gravitational potentials obscures the distinction between the entities and space. The mathematical power of extended parametrization allows gravitational field effects to be absorbed into local curvature of space and time. His viewpoint was that the ether of general relativity is determined by connections with the matter and state of ether in the neighboring place. According to him, the ether of general relativity is devoid of all mechanical and kinematical qualities. In a sense, it is not perceptible. Clearly, he drifted from his philosophy that imperceptible entities should not form a part of physical theory.

8.3 Wave-Particle Duality

Lucretius[1] was clear about waves that they cannot be fundamental entities. He reasoned that there must be invisible fundamental particles as basic entities in wind. Some where along the history, we seem to have forgotten it. The basic description of all waves is that they are due to coherent vibrations of atoms, molecules or some such entities which produce perceptible effects such as a wind, sound, etc. We also bring in the examples of water waves in classrooms. One presents an argument that a coherent disturbance in the form of a wave propagates as a pebble is dropped in a lake. This coherent disturbance is a cumulative translation of water molecules, subject to the energy and momentum conservation principles. Thus, at the heart of wave propagations are corpuscular bodies. Elsewhere I argued[8] that the waves cannot be fundamental entities, as they are complex objects of coherent excitations. As we seek structureless objects as candidates to qualify for being classified as elementary particles or quanta, the waves fail this basic criterion.

The phenomena of diffraction and interference are often cited as irrefutable evidence of wave nature of physical entities. Are they? First and foremost, at least mathematically, they are one and the same phenomenon. They are manifestations of superpositions of amplitudes and phases of secondary disturbances, embodied in one mathematical formula, known as Fresnel-Kirchoff integral[9]. The only distinction between these phenomena is our experimental arrangement. The usual question is: how does a photon or an electron or some such object know which slit to pass through, if there is more than one path available? The Fresnel-Kirchoff description tells us "no, it does not know". It simply specifies the field intensity distributions, which are the probabilities that, in a given experimental condition, individual disturbances pass through specific point in space. Does this violate any conservation principle? No. While one is quite happy with this argument for waves in media, one is not satisfied with this reasoning for electromagnetic radiation, since seemingly no medium was found. Einstein side-stepped the medium question in special relativity but his arguments of general relativity will allow him to accommodate this behavior, bending of light and so forth. The ether appears in quantum field theory in the guise of vacuum fluctuations.

8.3.1 Is EM Radiation Wave Phenomenon?

Einstein, in his 1905 seminal paper on photo-electric effect[10], recognized the need to distinguish the propagation of energy which are, in case of light, phenomena such as refraction, diffraction etc. from absorption and emission phenomena which constitute energy and momentum transfers. He clearly states that while the former is described the Hertzian waves a la Maxwellian formulation, one should resort to corpuscular description for the latter. However, scattering phenomena which are subject to energy and momentum conservations principles also affect propagation. Einstein was concerned with absorption phenomenon and thus this subtle aspect was missed by him. Or more likely, it was due to lack of experimental data, that he was not concerned with scattering phenomena. Subsequently, particle-like behavior of electromagnetic radiation has become deeply engrained in physics after the Compton scattering experiment. Also, the particle-wave duality took firm hold after de Broglie's matter wave formalism is developed.

It is attributed to Niels Bohr and his collaborators that wave-particle duality manifests an observer dependent reality, which means that the electromagnetic radiation allegedly behaves according to what our experiment intends to do. The experimental arrangement dictates whether the radiation is corpuscular or waves. Does it? Let us consider the celebrated Compton scattering experiment. In this experiment, electromagnetic radiation scatters off a stationary target. One measures the energy and intensity of scattered radiation as the angle of observation is varied. A basic experimental setup, commonly employed in undergraduate student laboratories looks as below:

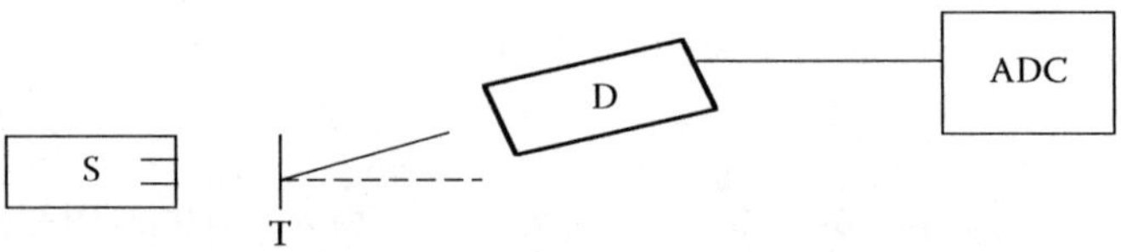

FIGURE 8.1
Compton scattering measurement set-up. S is the radiation source. T is scatterer. The detector D is oriented at an angle θ with respect to the incident beam direction. The signals from the detector are amplified and sent to an ADC, which is a pulse height analyzer.

In this experiment, for a fixed incoming beam energy, the energy of scattered radiation is uniquely determined once the scattering angle is fixed. The kinematical equations are dictated by energy and momentum conservation principles, same as what we employ for the elastic scattering between two billiard balls. We employ relativistic kinematical equations. The pulse height spectrum shows that the energy of scattered photons obeys these principles. Thus, we conclude that the radiation exhibits the corpuscular property. Now, let us ask the question—how many are scattered at the angle θ? Experimentally, we do not need to perform any other measurement. We simply count the number of events registered by the ADC. All this information is stored away in the computer systems somewhere. We simply ask where is the peak of the distribution and what is the area under the peak. The first question is answered by the corpuscular nature of light and the second question by the wave-like property. We can, at our leisure, ask these two questions alternately and find answers for both of them from our stored data of one single experiment.

The first successful theoretical description of scattered intensities was provided by Klein-Nishina formula, which assumes the incoming radiation to be monochromatic waves. Thus, we see both particle-like and wave-like behavior of radiation in one single experiment, contrary to the common assertions that an entity would reveal its particle-like or wave-like behavior but not both of them simultaneously in one single experiment.

It is noteworthy that Klein-Nishina formula was derived in 1929, shortly after the wave mechanics and matrix mechanics have been formulated. Later on, as quantum electrodynamics was being developed to be regarded as "the theory" of electricity and magnetism, Klein-Nishina formula was rederived from the field theoretical arguments. While it restores the idea that the radiation comes as quanta, we pay price in that the causality is lost. We can appreciate this fact, by looking at the Feynman diagrams shown below:

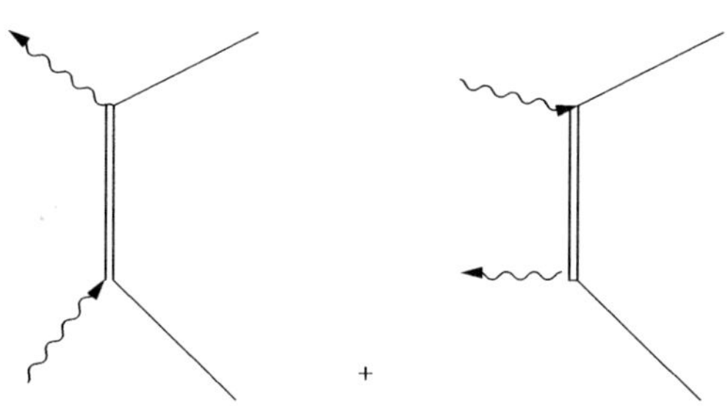

FIGURE 8.2
Feynmann diagrams of the photon Compton scattering off electrons. In each diagram, the spatial coordinates are on left-right axis and the time runs from bottom to the top side of the page. The solid lines are electrons and wiggly ones are photons. The double line is an intermediate state, inaccessible in the laboratory.

The first diagram is to be understood as follows: A photon is incident on an electron, which is raised to a virtual state. The virtual state does not satisfy energy-momentum conservation principles. It is revived to physical state as outgoing electron and photon are emitted, which are subject to energy-momentum conservation principles. This diagram alone is not able to describe the phenomenon and it leads to infinities as solution. We have to make of use a second diagram. Here the electron initially emits an outgoing photon and the electron becomes an unphysical body (it is virtual and does not satisfy the energy-momentum conservations). The virtual particle later absorbs the incident photon and is rendered physical. The overall scattering phenomenon is a superposition of these two processes. Needless to say, the causality is lost. It is important to note that in the second diagram, the initial state when an electron emits a photon should have the precise information of the four momentum vector of the incoming photon, which comes later, for it to be resurrected to physical state. We should leave this issue with individuals to make their decision whether this is an acceptable or palatable physics solution or not.

8.4 Photons as Hadrons[b]

Since the advent of particle accelerators of high energies, it has become possible to produce several unstable particle via interaction of electromagnetic radiation with hadronic matter and/or electromagnetic radiation itself. Examples are production of mesons at electron-positron colliders, routinely performed at Stanford Linear Accelerator Laboratory, Cornell synchrotron etc., or we can consider the experiments at photon beam facilities electron linear accelerators such as the ones at J-Lab and Bates at MIT in the United States and CERN, DESY laboratories in Europe and KEK, SPring-8 in Japan etc.[11]

Within a few years of this enterprise, it was proposed that photon interactions with other entities may be a manifestation that photons appear as vector mesons for some short time intervals. Thus, a high energy photon is neither an electromagnetic quantum, nor a wave, but it can be a meson, which bounces off other hadronic matter and attains a physical status. The early models were inspired by the similarities of interactions of photons and hadrons with nucleons and nuclei. They suggested that the photon acts like a hadron for a small fraction ($\sim\alpha = 1/137$) of the time. Vector meson dominance models[12], based on this physical picture, were quite successful in describing the experimental data. These vector mesons are of about one to a few GeV energies. As we go to higher energies, it is conceivable that photons appear as quark-antiquark pairs. There have been a few models based on these ideas[13]. Figure 8.3 describes the concept.

[b] In particle physics terminology, hadrons are elementary or composite objects which take part in strong interactions. Strong interactions are unique to subatomic physics world. They are of short range and responsible for binding forces of nuclei etc.

FIGURE 8.3
A photon propagating appears as a vector meson for some fraction of time and as quark-antiquark pair for some other fraction of time.

A free photon cannot be a physical vector meson due to energy-momentum imbalance. However, in the field of other material media, the photons may "hadronize" to virtual states and undergo elastic collisions to result in physical vector meson in a final state. Extending this idea to higher energies, one may conceive that a photon lives as quark-antiquark pair for short intervals of time and yields meson in final states.

Mathematically, one represents the photon wavefunction as[12],

$$|\gamma\rangle = \sqrt{Z_3}\,|\gamma_{bare}\rangle + \sum_{v=p^0,\omega,\phi} \frac{e}{f_v}|V\rangle$$

where $|\gamma_{bare}>$ is the electromagnetic quantum that we are accustomed to and $|V>$ are the vector mesons as indicated. They all have the same quantum numbers as photons. As we go to higher energies, we may modify the above equation as

$$|\gamma\rangle = \sqrt{Z_3}\,|\gamma_{bare}\rangle + \sum_{v=p^0,\omega,\phi} \frac{e}{f_v}|V\rangle + \frac{e}{f_{q\bar{q}}}|q\bar{q}\rangle$$

to accommodate the quark degrees of freedom. The last component is the quark-antiquark system.

Summary and Conclusions

So far, we presented a few common notions of electromagnetic radiation in terms of space-time concepts and what we might consider in terms of fundamental constituents of matter and interactions. Einstein was an empiricist, who sought a concise and consistent description of various phenomena. In special relativity, he provided a simple recipe to account for the mathematical formulations of Lorentz, Poincare and hypotheses of Fitzgerald *et al.* He avoided the question of ether in special relativity and he reintroduced it in general relativity. As for the photon, he was concerned with it in the context of absorption and emission processes, relating to energy and momentum conservation principles. We argued that description of Compton scattering makes use of both a particle-like and wave-like picture of the electromagnetic radiation to account for the energies and intensities of radiation, respectively. The quantum field theories seem to avoid wave picture, but it is achieved at the expense of causality. Models, inspired by the particle physics phenomenology, attribute hadronic existence for the photon.

In conclusion, a concrete picture of an elementary "photon" eludes us even today. This problem is not unique to photons. The particle concept of modern physics is not the same as that of Newtonians. Quantum physics, which allows for "Zitterbewegung" or wavy motion of particle-like objects, is not able to describe free-particles in the classical picture. After all, in the classical picture, the space is empty and objects move in straight lines unless forced otherwise. The physical space is not empty in general relativity nor in quantum field theories. In the latter, we find the vacuum is a sea full of objects which, under appropriate conditions, pop out and render the physical phenomena we see. The success of modern physics is the flexibility of formulations to account for the observational data. Alas, it comes at the expense of some conceptual clarity. The physical reality is elusive in these mathematical formulations. I would like to end this article by quoting Henry Stapp[14] "As every physicist knows, or is supposed to have been taught, physics does not deal with physical reality. Physics deals with mathematically describable patterns in our observations".

It was an honor to be able to discuss these conceptual issues of physics openly in the Centennial year of Einstein's landmark publications. I owe sincere appreciation to organizers of the conference for their kind invitation.

References

1. Lucretius, *On the Nature of the Universe*, Translated and introduced by R. E. Latham, Penguin Classics, 1951.
2. Kurt Gödel, *Über formal unentscheidbare Sätze der Principia mathematica und verwandter Systeme I*, Monatshefte für Mathematik und Physik Vol: 38; 173–198 (Leipzig: 1931), English translation can be found at http://home.ddc.net/ygg/etext/godel/godel3.htm.
3. See for example contributions to the 1st International Conference on the Ontology of Spacetime, Concordia University, Montreal, May 11–14, 2004.
4. Werner Heisenberg, The Copenhagen Interpretation of Quantum Theory, page 44, in *Physics and Philosophy*, Prometheus Books, 1999.
5. A. Einstein, *Ether and Relativity*, Address at University of Leiden, English translation in *The Tests of Time*, p. 340, Ed: Lisa M. Dolling, Arthur F. Gianelli and Glenn N. Statile, Princeton University Press, 2003.
6. A. Einstein, Later Comments on General Relativity, English translation "Later Comments on General Relativty" in *The Tests of Time*, p. 347, Ed: Lisa M. Dolling, Arthur F. Gianelli and Glenn N. Statile, Princeton University Press, 2003.
7. See reference 6, page 343.
8. C. Rangacharyulu, No point particles, definitely No Waves, page 329–333, in *New Developments on Fundamental Problems in Quantum Physics*, Eds: Miguel Ferrero and Alwyn van der Merwe, Kluwer Academic Publishers, 1997.

9. This equation is found in any standard optics text book. See, for example, *OPTICS* by Miles V. Klein and Thomas E. Furtak, John Wiley, second edition, 1987.
10. An English translation of Einstein's 1905 paper can be found as Einstein's proposal of the Photon Concept—a translation of the Annalen der Physik Paper of 1905 by A. B. Arons and M. B. Peppard, *American Journal of Physics*. vol. 33, page 367 (1965).
11. A comprehensive list of laboratories may be found at the website http://pdg.lbl.gov
12. Donald R. Yennie, The Hadronic structure of the photon with emphasis on its two-pion constituent, *Reviews of Modern Physics,* Vol. 47, page 311 (1975).
13. Gerhard A. Schuler and Torbjörn Sjöstrand, Towards a complete description of high-energy photoproduction, *Nuclear Physics* vol. B407, page: 539 (1993).
14. H. P. Stapp, EPR and Bell's Theorem: A critical review. *Foundations of Physics,* vol. 21, page 1 (1991).

9

Oh Photon, Photon; Whither Art Thou Gone?

A. F. Kracklauer
P.F. 2040, 99401 Weimar, Germany

CONTENTS

Abstract

A survey of the historically most widely considered paradigms for the electromagnetic interaction is presented along with the conflicts or defects that each exhibited. In particular, problems derived from the concept of the photon and quantum electrodynamics are emphasized. It is argued that a form of direct interaction on the light cone may be the optimum paradigm for this interaction.

Keywords: photon, Quantum Electrodynamics, charged particle interaction.

9.1 The Dilemma

Physics theories comprise at least two elements: a mathematical model and a paradigm. The former encodes the regularities of the phenomena of interest, while the latter provides visual and verbal support for thinking and talking about the phenomena covered by the theory, and, of course, motivation for setting up calculations. Theories about the interaction of charged particles (under circumstances, known as: "light") follow this pattern.

In the case of light, in all its variation of scale considered thus far, from radio to gamma waves, the mathematics seems to be largely in order, at least as far as the needs of radio, optical and electronic engineering are concerned. On the other hand, all paradigms proffered so far throughout history for light, have been found wanting. These include:

- Weber's instantaneous direct interaction (essentially Newton's gravitational force scaled to the strength of electrostatic force). This paradigm was unable to accommodate all dynamic interaction; e.g., magnetism.
- Huygens', Faraday's, Lorentz's and Maxwell's waves. As it is now clear that there is no medium (aether), electromagnetic waves, it seems, can be no more than mental representations for terms in a Fourier series decomposition of the full mathematical expression for the interaction. There are several recognized defects endemic to this paradigm including the infamous divergency of the self energy of the electron (in many guises), "run away" solutions, pre-acceleration, etc.
- Schwarzshild's delayed direct interaction, or Weber's direct interaction evaluated so as to take the time-of-flight of electromagnetic signals into account. Einstein criticized this paradigm because it did not attribute reality to advanced interaction, which, he held, is a valid solution to Maxwell's equations.
- Fokker's mean of advanced and retarded interaction. This paradigm has been faulted for not leading to integrable equations of motion, not to mention the philosophically repugnant concept of "advanced interaction."
- Einstein's "photon." Although very popular at the moment, this paradigm leads to a number of contradictory concepts with respect to interference and severe conflict with General Relativity (more below).
- Second quantized fields. A formalistic elaboration of the photon paradigm; more of a calculational algorithm than a real paradigm.

In addition to the historically well known, but not always emphasized, difficulties with each of the above paradigms, there is a similar set of objections concerned with Special Relativity (SR). It is clear that SR is essentially an application of the fundamentals of electrodynamics; this follows directly from the fact that the core of SR, the Lorentz transforms, contain as an ineluctable parameter, the velocity of *light*. It is not, therefore, unreasonable to speculate that an optimal paradigm for light might render the counter-intuitive aspects of SR less opaque.

9.2 Photons and Quantized Radiation Field

The photon paradigm won advocates by virtue of the simplicity of the motivational imagery it provides for the conservation of energy and momentum involved in calculating e.g., Compton scattering, and its role in the derivation of the Planck blackbody spectrum. In the meantime, is has also acquired critics, because it fails to give a coherent image for interference, as evidenced by the long dispute, to and fro, over whether a single photon can interfere only with itself, or also with other photons.

By far the most pervasive selling point for the photon notion, nevertheless, is the empirical fact, that at very low intensities, radiation in the visible portion of the spectrum is seen to be absorbed always at a single point. This is an artifact, however, of the detection process in this region of the spectrum, which exploits "photo detectors" that convert whatever visible radiation is, to an electron current. Obviously, as an electron current consists of countable electrons which are individually lifted into the conduction band of the detector mass, no matter how radiation arrives, an observer restricted to "seeing" only the photocurrent and limited to inferring the character of whatever stimulated it, is in no position to pass final judgment on the character of the stimulus, in this case, the incoming radiation. Therefore, the majority of the evidence for photons is intrinsically indeterminate.

9.3 Vacuum Fluctuations: Signature of QED

Intimately connected with these issues is a parallel dispute on just how necessary the whole concept of quantized radiation is in fact. Supporters of the so-called neoclassical theory in which matter is quantized but radiation not, have managed to accurately explain too many quantum electrodynamic (QED) effects to allow writing off this line of analysis out of hand. Moreover, their reasoning offers some support for the paradigm proposed below.

The customary approach to quantization of the electromagnetic radiation field leads to the conclusion that there exists a finite ground state with minimal energy, regardless of the existence of charges. As there are no charges to attribute this state to, it is logically attributed to the "vacuum," and discussed as if it had nothing to do with (charge carrying) matter, but is virtually a property of (quantized) "space" itself. Naturally, this evokes the question: is the energy in the ground state "real" or just a formalistic device? If the later, it should appear in the mathematical formalism in just such a way that it requires no physical interpretation, or even precludes it altogether. There are many, however, who argue that the ground state actually is the *physical* cause of phenomena, namely: spontaneous decay, the anomalous magnetic moment of the electron and the Lamb-shift, among others. Indeed, calculations of these effects predicated on this presumption have been so successful, that this view is quite credible, and text books commonly cite this fact as evidence of the fundamental rectitude of QED.

On the other hand, proponents of what has become known as the "neoclassical theory" (NCT) found that a judicious reordering of the terms involved in calculations allows an interpretation of these phenomena in terms of a "source theory" which considers *all* radiation fields as derived from source charges. That is, in plane text: vacuum fluctuations are *not* necessary to explain physical consequences; their use in QED is actually just a device which explains these effects *as if* there were such fluctuations.[1]

As an historical matter, NCT was depreciated in the early 1980's largely on the basis of what were taken then as two capital deficiencies. One of these pertains to the phenomenon of "quantum beats," which was thought at that time to be correctly describable only by QM.[2] It was taken that NCT predicted beats for the case in which two excited levels decay to a single lower state, which is both not observed and not predicted by QED. This writer disputes this argument, however, on the basis of the existence of a model of the experiments fully in accord with NCT, thereby overturning this as an argument against the validity of NCT.[3]

The second large issue speaking against NCT at that time was the then new experimental results from Clauser's group on EPR/Bell type experiments. Although this issue appears not to have been taken up further in the mainline literature, Jaynes himself eventually identified the *lacuna*[4] (See below). Thus, in sum, at the present time there are no unrefuted arguments standing against the NCT.

In view of the current fashionable topics of quantum computing, teleportation and the like, the above statement probably will elicit sharp protest. Proponents of these phenomena, however, have yet to respond to a very elementary and fundamental observation: these phenomena are all described in terms of the algebras of polarization or q-bit spaces, which, unlike phase and quadrature spaces, do not have "quantum" structure. This observation is supported by the fact that whatever noncommutivity, if any, is evident in polarization or q-bit space, is there because of geometric considerations (essentially the $SO(3)$ structure of rotation on a sphere) and not because of Heisenberg Uncertainty, as is the case for quantized phase and quadrature spaces. (Note that for the latter spaces, noncommutivity is an *option*, whereas for the former, it is a geometric ineluctability.) The point here is, that if at all valid, the essence of these phenomena require *no* QM at all for there description, not to mention QED. These effects, from this viewpoint, simply can play no role in the dispute over the necessity of QED.[5]

9.4 Stochastic Electrodynamics: Retrograde QED?

Stochastic electrodynamics (SED) is an attempt to rationalize quantum mechanics (QM) by turning QED on its head. This is done by reversing the sequence of the steps in reasoning underlying QM, i.e.,: starting from QM, one eventually concludes that there exist vacuum fluctuations or a finite ground state of the electromagnetic radiation. For SED, one starts from the basic assumption that there exists a classical, random, background field with the power spectrum of the QM ground state, and tries to show that phenomena otherwise predicted only by QM, result. The origin of the SED background has been justified in two ways. One, is that it is simply admissible initial and boundary data for determining solutions to Maxwell field equations, physically it just emerges from infinity; the other is that it is the average of all interaction with other charges in the universe, and is the dynamic equilibrium of all these separate contributions. In either case, it is argued, it is "visible" as the force that holds up atoms and otherwise is the source of all specifically QM phenomena. [6]

There are two main streams of SED development; one tries to explain QM effects in terms of the stochastic nature of this background, and, in one way or another, calls on some analogy with diffusion processes. It takes certain inspiration from the formal similarity of the Schrödinger equation, having a single time derivative, with the diffusion equation which also has only a single time derivative. The other line of analysis is based on the observation that the single time derivative in the Schrödinger Equation is accompanied by a factor of i, which is for the mathematics in this application equivalent to another time differentiation, so that any analogy with diffusion processes is essentially illusory. This disappointment is compensated by other arguments to the effect that the background can be used to motivate a physical rationalization for De Broglie's pilot waves, thereby bringing the story into the domain of wave phenomena. [7]

It is especially interesting that the two rationalizations for the existence of a SED background field parallel, to some extent, the two approaches to QED, with source theory having virtually a direct link to the dynamic equilibrium model. In any case, however, both rationalizations for the existence of a background can be faulted in that they lead to the same problem that QED introduces, namely a sharp conflict with the "cosmological constant" problem from General Relativity in that each envisions a horrendous quantity of energy resident at every point of space, even where absolutely free of matter.

9.5 Tactic for Remedy

For the purpose of seeking an optimum paradigm for phenomena derived from the interaction of charged particles, two working principles recommend themselves: 1) all notions employed for the paradigm should be as close as possible to incontestably grounded empirical "facts"; and 2) continuously related (in the sense of a correspondence principle) to the successful aspects of those paradigms suggested through history. Arbitrary or not, they seem prudent and reasonable.

9.6 Empirical Base

The following facts seem to constitute the essential optimum supported by observation:

1. Charges come in two genders; likes repel and unlikes attract.
2. The attraction or repulsion is in accord with Gauss' Law, in other words, for static circumstances, it falls off in proportion the inverse square of the separation.
3. When a charge moves, any change in the force it exerts on other charges is delayed by a time lag linearly proportional to the separation. The proportionality constant is called the speed of light.
4. The speed of light is a constant, valid in all inertial frames (see caveats below).

In addition to, or preceeding the pertinent empirical facts, there are a number of more abstract or less material features that might be taken as deeply philosophical in character. The study of such aspects is often associated with Kant, but he may not have had the last word. The basic issue he addressed is: what are space and time?

What they are not is clear. Certainly they are not material objects like stones, atoms or even elementary particles; more likely they are (at least) relational categories and may be further inexplicable. Whatever else they may be, in mechanics they can be seen to be used as organizing relationships among material objects.

As a relationship, space is obviously unique. That is, there is no sense at all to the question: which space? The spacial relationships among material objects are defined strictly with respect to each other. There is no "origin" or other preferred or privileged point; it may be said that in this sense it is "absolute." All the same considerations pertain to time as an ordering parameter, except that, per simple observation, time is flowing in one direction. (As an aside, I emphasize that it is not the point here to enter into philosophical analysis of space or time. The sole point here is to take those features evident directly to observation as given *a priori*. The immediate goal is only to seek the simplest paradigm for charge particle interaction consistent with such features. The analysis of possible deviations from these first order and evident features would be a separate question, deferred to future study.)

To foreclose some misunderstandings, note that the existence of absolute time does not imply the existence of an absolute clock. Such a clock would propagate its "ticking" instantly to the whole universe, and obviously, no known material gadget could do that. Likewise there is no material "absolute meter stick," which would have to be as long as the universe is wide and have no inertia, etc. It can be argued that the non existence of these material items is what really should have been intended by the claim that absolute "time" and "space" do not exist. As ordering abstractions, the latter "exist" as soon as they can be defined in a logically consistent manner, just as is the case for any abstract mathematical concept. The utility of these "absolute" orderings, having been defined, for any purpose within or for a particular theory, is then an internal matter for that theory.

9.7 The Paradigm

The above suggests the following paradigm. The fundamental element is the time-directed interaction link, i.e., the notion that every charge is the source of "action" on every other charge as a sink via delayed attraction or repulsion. Each and every particle is both active or source and passive or sink, where all particles as sinks are functions of a universal parameter conjugate to total energy (or the Hamiltonian of the universe). These links are eternal and primal; that is, they can not be further reduced to more elementary subparts; the charges at each end of a link have no independent existence.

The main difference with, and advantage over, the historical action-at-a-distance on the light cone formulation is that here there is no supposition that any*thing* is being emitted by the source and adsorbed by the sink. This has the consequence that there is no "free," un-targeted energy that eventually will be dissipated at infinity, and no energy to be thought of as "residing" at or passing through an arbitrary, but otherwise vacant point in space. Thus, the calculation of energy thought to be a source of vacuum gravitational energy is preempted from the start, thereby dispatching the "cosmological constant conflict."

For other paradigms, the human predilection for understanding "action" in terms of contact forces has led to the introduction of hypothetical intervening elements such as "fields," and "photons," whose function is to be, in the first instance, an agent of contact. In this way, an association is made with human experience, i.e., lifting, pushing, etc.; the only physiological means of delivering force or interaction with material objects. One might object that the concept of link leaves the essentials of interaction unexplained. But, while this is correct, it is also true of contact forces. The fact that we humans have experience with contact forces makes them intuitively predictable, but still not understandable in any deep sense, just familiar.

A pivotal issue in the historical dispute on the tenability of Maxwell's formation of electrodynamics, is the matter of radiation reaction, or the loss of energy by a charge by cause of its accelerated motion. The classical calculations for this effect have exposed obviously defective understanding of the electric interaction, leading as they do to "preacceleration," or "run-away" (divergent) solutions. This naturally evokes the question for any new proposals for a paradigm: does the new paradigm admit considerations that reasonably avoid such "un-physical" outcomes? Elsewhere this writer has discussed this matter and shown that it appears that radiation reaction may be considered the delayed interaction of a charge with its own induced Debye sheath. Within this paradigm, then, the negative features of solutions to Dirac's equation for radiation reaction can be seen to arise from approximations which neglect the delay time of the reflected signal.[8]

The dynamical aspects of a link require some elaboration. Let us imagine that a charge as source is jerked back and forth to induce a pulse in its links to other charges as sinks. By hypothesis, the pulse will travel up the link at the velocity of light. Such a pulse determines two times, t_q, the source time when the pulse is sent, and t_s, the sink time when the pulse is received. Clearly there can be absolute ambiguity at a sink between two pulses that arrive at the same (sink)-time from the same direction. Reception of pulses from the totality of sources in the universe is a physical realization of projective geometry with two complications: one, the projections are not instantaneous but delayed, and two, the projections are functions of time. For any given sink, including the eye of an experimenter, the totality of incoming pulses from all directions at a given sink-time, t_s, is the 2-dimensional surface of a sphere centered on the sink charge (or eye of an observer), sometimes called the observer's "sky." On this surface, overlapping pulses from sources in the same direction, but at distances such that the transmission delays compensate, can not be distinguished. This is obviously the recipe for the Minkowski metric and, therefore, the justification for the Minkowski space structure with the Lorentz transformations. The only difference with the usual presentation, is that from this viewpoint it is obvious that the Minkowski structure pertains only to the sink times; the source times and positions are interrelated according to the Galilean transformations.

9.8 Universality of Speed of Light

Of the four points considered empirically derived and delineated above, the last, to the effect that the speed of light is the same in all inertial frames, is the weakest. To begin, it is counterintuitive; it was introduced virtually out of desperation by Einstein to make Special Relativity fit together. Moreover, it requires a redefinition of the term "velocity" which was defined originally in terms of its vector character, such that as a matter of syntax, it is to be added according to Galilean transformations. This redefinition induced by Einstein has never really been rationalized by lexicographers, the term has just been given a jargon meaning, distinct from the conventional meaning, solely for discussing electrodynamics and Special Relativity.

In spite of strict taboos, this difficulty re-emerges repeatedly in detailed analysis of certain phenomena. The most convincing to this writer is with respect to the Sagnac effect (Waves sent in opposite directions around a plane figure by means of mirrors, exhibit a interference pattern dependant on the angular velocity of a platform on which the whole experiment, including sources and detectors, is mounted). If one considers the limit of the size of the plane figure as its linear dimensions increase while the angular velocity decreases such that the tangential velocity is constant, then one has a conceptual passage from a circumstance for which there is empirical evidence of the influence of the velocity of the source, to a circumstance where, according to the fourth assumption, the source velocity should have no influence. This conflict is symptomatic of some kind of subtle misunderstanding.[9]

Moreover, there is similar, albeit vague and vanishingly minute, evidence from time-of-flight data for radar signals to distant space vehicles. Because of the many practical effects and defects of equipment that need to be taken into account, this data is not beyond dispute, however. It is less convincing, for this writer at least, but still is a "straw in the wind."

In sum, these complications again render this paradigm too in need of further examination, even whilst overcoming inadequacies found in other paradigms for light.

9.9 Non-Locality: A Banished Bugaboo

One of the most alarming conclusions drawn after analysis of the interpretation of Quantum Mechanics (QM), is that there should be an essential element of non-locality to the natural world. This feature is attributed not only to "light," but implicitly to material particles also. Its ostensible realization, however, has been confined to optical experiments, i.e., to light.

As is widely known, John Bell took up this issue and deduced some inequalities for observable correlation frequencies that, he argued, had to obtain for any theory that might complete QM without reintroducing troublesome features, mostly the non-locality of instantaneous interaction. Experiments showed that these inequalities are violated, so the virtually universally accepted conclusion is: non-locality is an ineluctable fundamental characteristic of nature, specifically to include light.

However, in spite of the acceptance that this conclusion enjoys nowadays, it can be disputed. Evidently the first to identify the source of a misconception (or at least to publish a critique) was Jaynes.[4] He observed that Bell, perhaps mislead by bad notation, misapplied the chain rule for conditional probabilities and, instead of encoding locality into his formula, inadvertently encoded simply statistical independence.

The derivation of a Bell Inequality starts from Bell's fundamental *Ansatz*:

$$P(a,b) = \int d\lambda \rho(\lambda) A(a,\lambda) B(b,\lambda), \tag{9.1}$$

where, per *explicit assumption*: A, a measurement result from one side of a correlated photon pair as envisioned by Einstein, Podolsky, and Rosen, is not a function of b; nor B of a; and each represents the appearance of a photoelectron in its wing, and a and b are the corresponding polarizer filter settings. This is motivated on the grounds that a measurement at station A, if it respects "locality", so argues Bell, can not depend on remote conditions, such as the settings of a remote polarizer.

Jaynes' criticism is that Eq. 9.1 results from a misconstrual of Bayes' formula, or the "chain rule" for conditional probabilities, namely:

$$\rho(a, b, \lambda) = \rho(a|b, \lambda)\rho(b|\lambda)\rho(\lambda), \tag{9.2}$$

where $\rho(a, b, \lambda)$ is a joint probability distribution and $\rho(b|\lambda)$ is a *conditional* probability distribution. Jaynes points out that Bell takes it that the presence of the variable b in the factor $\rho(a|b, \lambda)$ implies instantaneous action-at-a-distance. This is true, however, only for the quantum case for which it is understood according to Von Neumann's measurement theory that wave functions are superpositions of the possible outcomes (even when mutually exclusive) whose ambiguity is resolved by collapse precipitated by the act of measurement. Eq. 9.2, however, for application in non-quantum circumstances implies no more than that there was a *common cause* for a *coincidence* in the past light cones of both measuring stations, a precondition which in QM is preempted by superposition.

The upshot is, that the inequalities that Bell and disciples derived, are valid only for statistically independent events, contrary to the fundamental assumption of the EPR argument, that the systems be correlated. Naturally, then, experimental results have little connection to the widely believed conclusions. (Arguments coming to the same conclusion as Bell's, but not involving inequalities, are also invalidated by error; in this case, not in mathematics, but in, as Barut observed first, the simultaneous application of formula to events that physically cannot be coeval.[10])

Jaynes' point has been rediscovered by various researchers in various styles and considerably extended. This writer, for example, has presented calculations using the *classical* formula for higher order correlation to accurately calculate the intensity curves seen in both EPR (2-fold) and Greenberger, Horne and Zeilinger (GHZ) or (4-fold) correlations. In addition, he has presented a data-point-by-data-point simulation of EPR experiments showing in detail just how the intensity variation as a function of angle arises without non-local interaction being involved in any way. The conclusion from this work is: there is no need whatsoever for quantum concepts to fully explain EPR and GHZ correlations; and, in particular, there is no evidence from any of these experiments for the existence of non-locality (or teleportation) in nature.[11]

Conclusions

Photons, it can be said beyond doubt, present a challenge for contemporary physics. They exhibit two features that call for reconciliation with empirical facts: one, they are an essential element of QED, the paradigm-package which is in drastic conflict with General Relativity; and, two, the issue of their physical extention and interplay with other photons wherever interference comes into play, is not just unclear but contradictory. Nevertheless, as a paradigm, the notion of photon has been fruitful to an astounding degree, so that it can be expected that an improved paradigm will somehow encompass their historical contribution to understanding interaction between charged particles.

One other thing that is beyond doubt is that the classical wave paradigm presents equal challenges. Most of the inadequacies of Maxwell field theory have been known virtually from the start. Many were never attacked thoroughly, as historically the development of QM stole the show leaving research in "classical" E&M as a disparaged step child. But again, the wave paradigm has been, and continues to be, so fruitful that we can be certain that the truth it contains will be retained in an improved story, most probably as the intuitive imagery associated with Fourier analysis of expressions for the full but unwieldy total interaction.

In any case, this writer holds, the optimum tactic to improve the paradigm for "light" is to hew as close as possible to directly experienced, empirical data, without introducing hypothetical constructions. Historically, it has been these hypothetical constructions that eventually led to both contradictions and constraints on imagination impeding progress. Such hypothetical notions in the course of time take on in the folklore a sense of "reality" altogether undeserved but vivid, so that eventually it becomes the implicit goal of science to explain these constructions, in place of nature herself. "Fields" and "photons" are prime examples; both have lead to the idea, now very widely spread, that radiation can detach from its source and exist independently, as if it were a kind of ethereal matter. This is nowhere supported by evidence, however, and is responsible for what can be called "the biggest problem" in physics—that is, the disparity between the minimal ground state energy in the presumed free electromagnetic radiation fields as called for by QED, and the maximum energy level allowed by General Relativity. Taking all acceptable cut-offs into account, puts this at a minimum of 120 orders of magnitude!

The basic facts of the electric interaction seem to point to a permanent, time-directed link between charges, with one serving as, so to speak, a source and the other as a sink; with the complication that each also is linked in the complementary sense, and then with every other charge in the universe. Impugning more to electric interactions of these bare essentials risks reintroducing misleading constructions and irresolvable misconstuctions.

References

1. E. T. Jaynes, "Electrodynamics today," in *Coherence and Quantum Optics IV*, L. Mandel and E. Wolf, eds., pp. 495–509, Plenum Press, New York, 1987.
2. M. O. Scully, "On quantum beat phenomena and the internal consistency of semiclassical radiation theories," in *Foundations of Radiation Theory and Quantum Electrodynamics*, A. O. Barut, ed., pp. 45–48, Plenum Press, New York, 1980.

3. A. F. Kracklauer, "A neoclassical model for quantum beats," *in preparation.*
4. E. T. Jaynes, "Clearing up mysteries: the original goal," in *Maximum Entropy and Bayesian Methods*, J. Skilling, ed., pp. 53–71, Kluwer Academic, Dordrecht, 1989.
5. A. F. Kracklauer, "EPR-B correlations: quantum mechanics or just geometry?," *J. Opt. B: Quantum Semiclass. Opt.* **6**, pp. S544–S548, 2004.
6. L. de la Peña and A. M. Catto, *The Quantum Dice*, Kluwer Academic, Dordrecht, 1996.
7. A. F. Kracklauer, "Pilot wave steerage: a mechanism and test," *Found. Phys. Lett.* **13**, pp. 441–453, 1999.
8. A. F. Kracklauer and P. T. Kracklauer, "Electrodynamics: action-at-*no*-distance?," in *Has the Last Word Been Said on Classical Electrodynamics? –New Horizons*, A. Chubykalo, A. Espinoza, V. Onoochin, and R. Smirnov-Rueda, eds., pp. 118–132, Rinton Press, Princeton, 2004.
9. F. Selleri, "Sagnac effect: end of the mystery," in *Relativity in Rotating Frames*, G. Rizzi and M. I. Ruggiero, eds., pp. 57–74, Kluwer Academic, Dordrecht, 2004.
10. A. F. Kracklauer and N. A. Kracklauer, "The improbability of nonlocality," *Phys. Essays* **15**, pp. 162–171, 2002.
11. A. F. Kracklauer, "Bell's inequalities and EPR experiments: are they disjoint?," in *Foundations of Probability and Physics—3*, A. Khrennikov, ed., *AIP Conf. Proc.* **750**, pp. 219–227, 2005.

10

The Photon Wave Function

A. Muthukrishnan,[1] M. O. Scully,[1,2] and M. S. Zubairy[1]

CONTENTS

Abstract

We review and sharpen the concept of a photon wave function based on the quantum theory of light. We argue that a point-like atom serves as the archetype for both the creation and detection of photons. Spontaneous emission from atoms provides a spatially localized source of photon states that serves as a natural wave packet basis for quantum states of light. Photodetection theory allows us to give operational meaning to the photon wave function which, for single photons, is analogous to the electric field in classical wave optics. Entanglement between photons, and the uniquely quantum phenomena that result from it, are exemplified by two-photon wave functions.

Recently, we wrote an article on the photon concept [1], where we reviewed the wave-particle debate on light and argued for the existence of a wave function description of the photon based on the quantum theory of light [2]. It is interesting to note analogs with both the classical wave theory of light and the quantum mechanics of elementary particles like the electron and neutrino. In the article, we noted that:

> Dual conceptions of light, as wave and particle, have co-existed since antiquity. Quantum mechanics officially sanctions this duality, and puts both concepts on an equal footing (to wit, the quantum eraser).

[1] Institute for Quantum Studies and Department of Physics, Texas A&M University, College Station, TX 77843, USA

[2] Departments of Chemistry and Aerospace and Mechanical Engineering, Princeton University, Princeton, NJ 08544, USA

In this chapter, we revisit the idea of a photon wave function, which exemplifies both the wave and particle aspects of light. The wave aspect is inherent in the phase and amplitude of the wave function, analogous to the electric field in Maxwell theory. The particle aspect is exemplified by the localized nature of the source and detector of photons, in our case, the atoms that act as point-like quantum dipoles that radiate and absorb light. Furthermore, entanglement between photons, as in the quantum eraser [3], can be described using the language of multi-photon wave functions that elucidate fundamental paradigms such as complementarity and two-particle interference.

Historically, the existence of a wave function for photons has been questioned. Bohm raises the issue in his quantum theory book [4], where he argues that there is no quantity for light equivalent to the electron probability density $P(x) = |\psi(x)|^2$ when the region of interest becomes comparable in size to the wavelength of light, and concludes that a precise statement of the conservation of probability cannot be made for the photon. Kramers considers the question in more detail in his quantum mechanics book [5], where he asks whether "it is possible to consider the Maxwell equations to be a kind of Schrödinger equation for light particles." He answers in the negative, for essentially the same reason as mentioned by Power [6], based on the disparity in mathematical form of the two types of equations (specifically, the number of time derivatives in each). The former admits real solutions ($\sin \nu t$ and $\cos \nu t$) for the electric and magnetic waves, while the latter is restricted to complex wave functions ($e^{i\nu t}$ or $e^{-i\nu t}$, but not both).

Nevertheless, recent attempts have continued to motivate the idea of a photon wave function on theoretical grounds [7,8,9,10]. At issue is how well a photon wave packet can be localized in space-time, the connection being that if a wave function $\gamma(\mathbf{r},t)$ exists for the photon, then we can interpret $|\gamma|^2\, d^3r$ as the probability of finding the photon in a volume element d^3r, and realize complete photon localization subject to the uncertainty (Fourier) principle. It is often thought that this is not possible. Mathematical arguments have been advanced against the construction of a position operator for the photon [11], and constraints on the locality of the number/energy density of the photon field have been noted [12,13]. Despite these limitations, we maintain that a physically meaningful photon wave function $\gamma(\mathbf{r},t)$ can indeed be constructed, that is measurably localized in space, everywhere meaningfully defined in both phase and amplitude, and provides a valuable tool for understanding photon interference and correlation experiments.

Our approach is guided by the quantum theory of photodetection pioneered by Glauber [14]. The absorption of a photon by an atom is the basic paradigm that underlies photodetection theory. Consequently, the interaction Hamiltonian for the field and the detector is proportional to the annihilation operator $\hat{E}^{+}(\mathbf{r},t)$, which forms the positive frequency part of the quantized electric field:

$$\begin{aligned}\hat{E}(\mathbf{r},t) &= \hat{E}^{+}(\mathbf{r},t) + \hat{E}^{-}(\mathbf{r},t)\\ &= \sum_{\kappa} \sqrt{\frac{\hbar\nu_k}{\epsilon_0 V}} \left[\hat{a}_k u_k(\mathbf{r}) e^{-i\nu_k t} + \hat{a}_k^{\dagger} u_k^{*}(\mathbf{r}) e^{i\nu_k t}\right],\end{aligned} \tag{10.1}$$

where the $u_k(\mathbf{r})$ are normalized spatial mode functions that satisfy Maxwell's equations and the boundary conditions of the mode volume V. According to Fermi's Golden Rule, the matrix element of the interaction operator between the initial and final states of the field determines the transition probability. If there is only one photon initially in the state $|\gamma\rangle$, then the relevant final state is the vacuum state $|0\rangle$. The probability density of detecting this photon at position $\mathbf{r}$ and time t is then given by

$$G_\gamma^{(1)} = \kappa |\langle 0 | \hat{E}^{+}(\mathbf{r},t) | \gamma \rangle|^2 \equiv |\gamma(\mathbf{r},t)|^2. \tag{10.2}$$

Here, κ is a dimensional constant chosen such that $|\gamma|^2$ has units of inverse volume. The quantity $\gamma(\mathbf{r},t)$ may thus be regarded as a kind of "electric-field wave function" for the photon [2], with $\{\langle 0|\hat{E}^{+}(\mathbf{r},t)\}^{\dagger} = \hat{E}^{-}(\mathbf{r},t)|0\rangle$ analogous to the position state $|\mathbf{r}\rangle$ in the first quantized theory of the electron. The utility of this point of view will be made clear below.

Let us calculate $\gamma(\mathbf{r}, t)$ for the photon spontaneously emitted by an atom when it decays. In the dipole approximation, an atom is like an ideal (point) dipole which radiates light energy, hence it can be regarded as one of Nature's fundamental sources of spatially localized photons. Consider a two-level atom located at $\mathbf{r}_j$, prepared initially in level a, and subsequently decaying at a rate Γ to level b, as shown in Figure 10.1(a). The emitted field state $|\gamma_j\rangle$ is a superposition of Fock states $|1_k\rangle$, summed over all modes κ, given in the Weisskopf-Wigner (or equivalently the Markov) approximation by [2,15]

$$|\gamma_j\rangle = \sum_{\kappa} \frac{g_{ab,k} e^{-i\mathbf{k}\cdot\mathbf{r}_j}}{(\nu_k - \omega) + i\Gamma/2} |1_k\rangle, \tag{10.3}$$

where ω is the atomic frequency, and $g_{ab,k} = \sqrt{\nu_k / 2\epsilon_0 \hbar V}(\mathbf{p}_{ab} \cdot \hat{\varepsilon}_k)$ is an interaction matrix element proportional to the atomic dipole moment $\mathbf{p}_{ab}$ and the polarization of the field mode $\hat{\varepsilon}_k$. The spectrum of the emitted field

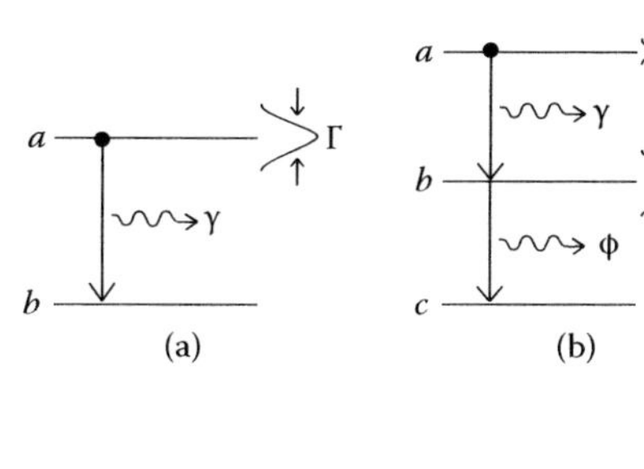

FIGURE 10.1
(a) Two-level atom spontaneously decays from the excited state, emitting a single photon γ. (b) Three-level atom undergoes cascade decay to emit two photons γ and ϕ. Both come from dipole allowed transitions.

is approximately Lorentzian, which corresponds in the time domain to an exponential decay of the excited atom. Calculating $\gamma_j(\mathbf{r},t)$ for this state, we obtain [2]

$$\gamma_j(\mathbf{r},t) = K\frac{\sin\theta_j}{r_j}\theta(t - r_j/c)e^{-i(\omega - i\Gamma/2)(t - r_j/c)}. \tag{10.4}$$

Here, K is for normalization, $r_j = |\mathbf{r} - \mathbf{r}_j|$ is the radial distance from the atom, and θ_j is the azimuthal angle with respect to the atomic dipole moment. The step function $\theta(t - r_j/c)$ is an indication that nothing will be detected until the light from the atom reaches the detector, travelling at speed c. Once the detector starts seeing the pulse, the probability of detection $|\gamma|^2$ decays exponentially in time at the rate Γ, as expected. The angular profile of the pulse mimics the radiation pattern of a classical dipole.

We make several remarks about Eq. 10.4. First, we note that this result can be generalized in two ways: to *vector* fields (**E** and **H**, that depend on the orientation of the dipole), and to higher *multipole* transitions in the same atom or molecule. Second, the correspondence between the photodetector wave function and the classical electric field is not just for the free field (i.e. propagating photons) but can also be made in principle for the source field (e.g. evanescent waves). Third, the photon states $|\gamma_j\rangle$, and their space-time counterparts $\gamma_j(\mathbf{r},t)$, constitute a *causally localized* wave packet basis for the photon, that is localized at the source $(\mathbf{r}_j, t = 0)$, and upon detection $(\mathbf{r},t)$, and capable of spanning a general one-photon state emitted by a collection of dipoles. It is thus instructive to consider the overlap of two states γ_1 and γ_2 that differ in their source location by a variable distance $d = |\mathbf{r}_2 - \mathbf{r}_1|$.

The appropriate quantity to calculate is the inner product $\langle\gamma_1|\gamma_2\rangle$, which translates in real space to an overlap integral of the wave functions (by analogy to the quantum mechanics of an electron):

$$\langle\gamma_1|\gamma_2\rangle \simeq \int d\mathbf{r}^3\gamma_1^*(\mathbf{r},t)\gamma_2(\mathbf{r},t). \tag{10.5}$$

Both the phase and amplitude of the wave function $\gamma(\mathbf{r},t) = |\langle 0|\hat{E}^+(\mathbf{r},t)|\gamma\rangle$ are relevant for determining the inner product. To see the validity of Eq. 10.5, note that for the quasi-monochromatic $(\Gamma \ll \omega)$, one-photon state emitted by an atom, a resolution of identity is given to good approximation by

$$\hat{1} \simeq k\int d^3r\hat{E}^-(\mathbf{r},t)|0\rangle\langle 0|\hat{E}^+(\mathbf{r},t) \tag{10.6}$$

$$= k\sum_k \left(\frac{\hbar\nu_k}{\epsilon_0 V}\right)|1_k\rangle\langle 1_k|. \tag{10.7}$$

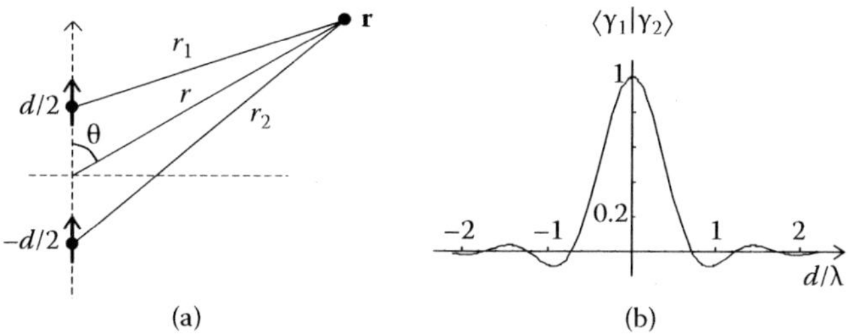

FIGURE 10.2
(a) Geometry for calculating the overlap between the photon states corresponding to atomic dipoles spaced apart by d. The displacement is considered parallel to the dipole vector. (b) Plot of calculated overlap $\langle\gamma_1|\gamma_2\rangle$ versus d/λ.

where $\kappa = e_0/(\hbar\omega)$ is chosen for the mean frequency ω of the atomic transition. Inserting Eq. 10.6 into $\langle\gamma_1|\gamma_2\rangle$ yields Eq. 10.5. This allows us to calculate $\langle\gamma_1|\gamma_2\rangle$ using the wave functions $\gamma_j(\mathbf{r},t)$ in Eq. 10.4. We find that when $\Gamma \ll \omega$, and when the dipoles are displaced parallel to their orientation [see Figure 10.2(a)], we obtain the expression (see Appendix, Ref. [16])

$$\langle\gamma_1|\gamma_2\rangle = \frac{\sin(2\pi d/\lambda) - (2\pi d/\lambda)\cos(2\pi d/\lambda)}{3(2\pi d/\lambda)^2}. \tag{10.8}$$

This is plotted in Figure 10.2(b). Thus, the two emissions are orthogonal only if the atoms are far apart compared to a wavelength: $d = |\mathbf{r}_1 - \mathbf{r}_2| \gg \lambda$.

The non-orthogonality of the wave packet basis $\{|\gamma_j\rangle\}$ is analogous to that of the coherent states of a single-mode field [2]. Furthermore, the wavelength scale of the orthogonality is a reminder of the classical Rayleigh criterion for spatially distinguishing point sources in wave optics [17]. As a bridge between the classical Maxwell theory and the quantum theory of light, the photon wave function has interesting properties that relate to its use in both Hilbert space and real space. In the latter case, just as the electric (or magnetic) field is a localized space-time description of classical light, we contend that our wave function description of the photon is local and real at arbitrarily small length scales. This is nowhere more evident than in the spatially varying phase and amplitude of $\gamma_j(\mathbf{r},t)$ that enables the calculation of the overlap $\langle\gamma_1|\gamma_2\rangle$ in Eq. 10.5.

We note that our definition of the photon wave function can be easily generalized to two or more photons [2]. In particular, for a two-photon state, one can define a joint wave function $\Psi(\mathbf{r},t;\mathbf{r}',t)$ as follows:

$$G_\Psi^{(2)} = \kappa'|\langle 0|\hat{E}^+(\mathbf{r}',t')\hat{E}^+(\mathbf{r},t)|\Psi\rangle|^2 \equiv |\Psi(\mathbf{r},t;\mathbf{r}',t')|^2. \tag{10.9}$$

This is an especially useful tool when the two photons are entangled. An atomic source of entangled photons is the three-level cascade decay shown in Figure 10.1(b), where the frequencies and time of emissions of the photons γ and ϕ are correlated. A calculation of the two-photon cascade wave function yields [2]

$$\Psi(\mathbf{r},t;\mathbf{r}',t') = K'\left(\frac{\sin\theta}{r}\right)\left(\frac{\sin\theta'}{r'}\right)\theta(t-r/c)e^{-(i\omega_{ac}+\Gamma_a/2)(t-r/c)}$$
$$\times\,\theta[(t'-r'/c)-(t-r/c)]e^{-(i\omega_{bc}+\Gamma_b/2)[(t'-r'/c)-(t-r/c)]} \quad (10.10)$$
$$+(1\leftrightarrow 2).$$

where $r = |\mathbf{r} - \mathbf{r}_0|$ and $r' = |\mathbf{r}' - \mathbf{r}_0|$ are the distances from the atom ($\mathbf{r}_0$) to the detection points $\mathbf{r}$ and $\mathbf{r}'$, and θ and θ' are the respective azimuthal angles with respect to the atomic dipole moments for the two transitions.

Historically, inter-particle correlations have played a key role in fundamental tests of quantum mechanics, such as the EPR paradox and Bell inequalities. Indeed, the complementarity of the photon in the quantum eraser scheme [3], where entanglement arises from the spatial separation of the atoms (modeling the double slit in Young's interference experiment), becomes apparent only in the language of wave functions. Through association with photodetection amplitudes, multi-photon wave functions incorporate the phenomenology of quantum-correlated measurement, which makes them qualitatively distinct from classical light waves. Furthermore, many technological advances in quantum metrology are facilitated by the use of two-photon wave functions, in fields such as quantum imaging, quantum microscopy, quantum lithography, and sub-natural spectroscopy.

In conclusion, we argue that a photon wave function can indeed be meaningfully defined based on the quantum theory of photodetection. In our perspective, point-like atomic dipoles serve as a paradigm for the localized creation and detection of photon states, and this provides a natural basis for discussing wave function representations of the quantized electromagnetic field. More than an appealing throwback to classical electromagnetism, or an analog for the quantum mechanics of electrons and neutrinos, the photon wave function is an immensely practical tool for understanding interference and correlation phenomena associated with quantum states of light.

Acknowledgments

The authors thank the support from AFRL, AFOSR, DARPA-QuIST, TAMU TITF initiative, ONR and the Robert A. Welch Foundation. AM thanks Y. Rostostsev for a stimulating discussion.

Appendix: Overlap Integral $\langle\gamma_1|\gamma_2\rangle$ for Photons

Consider the electric field vector of a damped, oscillating dipole located at $\mathbf{r}_j$ and oriented along the z axis, in spherical coordinates:

$$\mathbf{E}_j(\mathbf{r},t) = -\hat{\theta}\varepsilon_0\left(\frac{\sin\theta_j}{r_j}\right)\Theta(t - r_j/c)e^{-i(\omega - i\Gamma/2)(t - r_j/c)} + \text{c.c.}, \tag{10.11}$$

where $r_j = |\mathbf{r} - \mathbf{r}_j|$ and $\varepsilon_0 = (4\pi e_0)^{-1}(|\mathbf{p}_{ab}|^2\omega^2/c^2)$. Apart from the direction vector $\hat{\theta}$, the positive frequency part of the above field $\mathbf{E}^+(\mathbf{r},t)$ is identical to the single-photon wave function $\gamma_j(\mathbf{r},t)$ spontaneously emitted by an atom when it decays [cf. Eq. 10.4]. We wish to calculate the overlap integral for two such fields originating from two dipoles, located independently at $\mathbf{r}_1$ and $\mathbf{r}_2$, as shown in Figure 10.2(a). The quantity to calculate is [cf. Eq. 10.5]

$$\int d^3r\mathbf{E}_1^-(\mathbf{r},t)\cdot\mathbf{E}_2^+(\mathbf{r},t) \simeq \varepsilon_0^2\int_0^{2\pi}d\phi\int_0^{\pi}d\theta\int_0^{\infty}drr^2\sin\theta\left(\frac{\sin\theta}{r}\right)^2 \\ \Theta(t - r_{\max}/c)e^{i\omega(r_2-r_1)/c}e^{-\Gamma[t-(r_1+r_2)/2c]}, \tag{10.12}$$

where $r_{\max}$ is the larger of r_1 and r_2, and we have replaced the quantity $\sin\theta_j/r_j$ with its mean value corresponding to the midpoint of the two dipoles. Assuming that $d \ll r$ (i.e., in the far field), we have $r_2 - r_1 \approx d\cos\theta$ and $(r_1 + r_2)/2 \approx r$ to first order in d/r. Furthermore, the Θ function requires that $r_{\max} \approx r + (d/2)|\cos\theta| \leq ct$, hence we are left with

$$\int d^3r\mathbf{E}_1^-(\mathbf{r},t)\cdot\mathbf{E}_2^+(\mathbf{r},t) \simeq 2\pi\varepsilon_0^2\int_0^{\pi}d\theta\sin^3\theta e^{i(\omega/c)d\cos\theta-(\Gamma/2c)d|\cos\theta|}. \tag{10.13}$$

Here we have assumed the long time limit $t \to \infty$. While it is possible to carry out the θ integral exactly, the result is messy, and it is more instructive to consider the situation where we restrict attention to $\Gamma \ll \omega$. Carrying out the θ integral in this limit and normalizing to $d = 0$ gives us the desired overlap:

$$\langle\gamma_1|\gamma_2\rangle = \frac{\sin(2\pi d/\lambda) - (2\pi d/\lambda)\cos(2\pi d/\lambda)}{3(2\pi d/\lambda)^2} \tag{10.14}$$

where we have used $\omega/c = 2\pi d/\lambda$. This result is plotted in Figure 10.2(b).

References

[1] A. Muthukrishnan, M. O. Scully and M. S. Zubairy, "The concept of the photon revisited," *Optics and Photonics News Trends* **3**, No. 1, S-18, Oct. 2003.
[2] M. O. Scully and M. S. Zubairy, *Quantum Optics*, Cambridge Univ. Press, 1997, ch. 1.
[3] M. O. Scully and K. Drühl, "Quantum eraser: A proposed photon correlation experiment concerning observation and 'delayed choice' in quantum mechanics," *Phys. Rev.* A **25**, 2208 (1982).
[4] D. Bohm, *Quantum Theory*, Constable, London, 1954, p. 98.
[5] H. A. Kramers, *Quantum Mechanics*, North-Holland, Amsterdam, 1958.
[6] E. A. Power, *Introductory Quantum Electrodynamics*, Longman, London, 1964.
[7] E. R. Pike and S. Sarkar, "Spatial dependence of weakly localized single-photon wave packets," *Phys. Rev.* A **35**, 926 (1987); also see by authors, *The Quantum Theory of Radiation*, Oxford UP, London, 1995.
[8] I. Bialynicki-Birula, "On the wave function of the photon," *Acta Phys. Polonica A*, **86**, 97 (1994).
[9] J. E. Sipe, "Photon wave functions," *Phys. Rev. A* **52**, 1875 (1995).
[10] K. W. Chan, C. K. Law and J. H. Eberly, "Localized single-photon wave functions in free space," *Phys. Rev. Lett.* **88**, 100402 (2002).
[11] T. D. Newton and E. P. Wigner, "Localized states for elementary systems," *Rev. Mod. Phys.* **21**, 400 (1949).
[12] L. Mandel, "Configuration-space photon number operators in quantum optics," *Phys. Rev.* **144**, 1071 (1966).
[13] W. O. Amrein, "Localizability for particles of mass zero," *Helv. Phys. Acta* **42**, 149 (1969), and refs. therein.
[14] R. J. Glauber, "The quantum theory of optical coherence," *Phys. Rev.* **130**, 2529 (1963).
[15] V. Weisskopf and E. Wigner, "Calculation of the natural line width on the basis of Dirac's theory of light," *Z. Phys.* **63**, 54 (1930).
[16] A state vector calculation that approximated the angular integral was given by M. Hillery and M. O. Scully, "On state reduction and observation in quantum optics: Wigner's friends and their amnesia," in *Quantum Optics, Experimental Gravity and Measurement Theory*, ed. P. Meystre and M. O. Scully, Plenum, New York, 1983.
[17] J. W. Strutt, "On the manufacture and theory of diffraction gratings," *Philos. Mag.* **47**, 81, 193 (1874); also see Philos. Mag, **8**, 261, 403, 477 (1879).

11

Photons Are Fluctuations of a Random (Zeropoint) Radiation Filling the Whole Space

Emilio Santos
Departamento de Física, Universidad de Cantabria, Santander, Spain

CONTENTS

Abstract

I assume that everywhere in space there is a real random electromagnetic radiation, or zeropoint field (ZPF), which looks similar for all inertial observers, so that the stochastic properties of the field should be Lorentz invariant. This fixes the spectrum except for a single adjustable parameter measuring the scale, which is identified with Planck's constant, so making the ZPF identical to the quantum electromagnetic vacuum. Photons are just fluctuations of the random field or, equivalently, wavepackets in the form of needles of radiation superimposed to the ZPF. Two photons are "classically correlated" if the correlation involves just the intensity above the average energy of the ZPF, but they are "entangled" if the ZP fields in the neighbourhood of the photons are also correlated. These assumptions may explain all quantum optical phenomena involving radiation and macroscopic bodies, provided the latter may be treated as classical. That is, we have an interpretation of quantization for light but not for matter. Detection of photons involves subtracting the ZPF, which cannot be made without a fundamental uncertainty. This explains why photon counters cannot be manufactured with 100% efficiency and no noise (dark rate), which prevents the violation of a genuine Bell inequality (this is the so-called detection loophole). The theory thus obtained agrees very closely with standard quantum optics if this is formulated in the Wigner representation.

Key words: photons, zeropoint field, vacuum fluctuations, Bell's inequality.

11.1 Vacuum Electromagnetic Field

My answer to the question "What is a photon?" derives from a picture of the microworld which rests upon heuristic arguments and will be summarized in the following. That picture provides a *qualitative explanation* for many quantum phenomena. On the other hand the quantum formalism (or, rather, the several physically equivalent formalisms like Hilbert spaces, Feynman path integrals, Wigner function, etc.) provides a set of *calculational rules* which agree *quantitatively* with the experiments, but does not offer a clear picture of the microworld. Matching my picture of the microworld with the quantum calculational rules is not yet achieved, as I shall comment at the end of this chapter.

The starting point of my picture is the problem of the stability of the atom, in particular the hydrogen atom. That is, a negatively charged particle (electron) moving in the static field created by positive point charge at rest (proton). Classical electrodynamics predicts that the electron will move in a spiral orbit due to the energy loss by radiation, so that the atom would be unstable. But the argument is flawed. Because, if there are many atoms in the universe and each atom radiates, then it is more natural to assume that every atom is immersed in some background radiation. Obviously that radiation should be treated as random, which gives rise to randomness of the electron position in the atom. Besides radiating, the atom may absorb energy from the background radiation, so that a dynamical equilibrium could exist where absorption and emission cancel on the average. That state, with a stationary probability distribution of electron positions, will correspond to the quantum ground state of the atom.

The postulated background radiation, or zeropoint field (ZPF), should have Lorentz invariant statistical properties, in order that all inertial observers are equivalent. This constraint fixes the spectrum of the radiation to be

$$\rho = \frac{\hbar}{\pi c^3}\omega^3 = \frac{2\omega^2}{\pi c^3}\frac{1}{2}\hbar\omega, \tag{11.1}$$

where c is the velocity of light, ω the angular frequency and ρ the energy per unit volume and unit frequency interval. In the right hand side of Eq. 11.1

I have written separately the normal modes density (first factor) and the energy per mode (second factor) for later convenience. (The total energy per unit volume of Eq. 11.1 diverges, so that we must assume the existence of a cut-off at high frequencies, but we shall not study this point here). Thus our heuristic arguments lead to a theory containing only one adjustable parameter, $\hbar$, setting the scale of the ZPF. We choose this parameter to be Planck's constant. The sketched theory is known as "stochastic electrodynamics" (SED), a review of which is the book by L. de la Peña and A. Cetto[1]. Thus SED is just classical electrodynamics but replacing the standard boundary condition of Maxwell's equations (no radiation at infinity in the past) by a new boundary condition (ZPF in the past). This boundary condition restores time reversal symmetry in electrodynamics.

The relevance of the ZPF in quantum electrodynamics is widely recognized in phenomena like the Casimir effect or the Lamb shift (see, e.g., the book of Milonni[2]). Specific of SED is the belief that the ZPF is a *real field*. The existence of a real random radiation on the whole space has a lot of consequences which explain—or are related to—most of the characteristic traits of quantum physics. A few examples are the following.

If a charged particle is moving in an one-dimensional potential well, it will interact most strongly with those modes of the ZPF having a frequency close to the typical frequency, ω, of the particle's motion. The stationary state (where absorption cancels emission on the average) will correspond to a mean kinetic energy

$$\frac{1}{2}m\langle v^2\rangle \approx \frac{1}{4}\hbar\omega, \tag{11.2}$$

where $\approx$ means order of magnitude equality. Eq. 11.2 may be rigorously derived from Eq. 11.1 for a harmonic oscillator potential, but we may assume that it is valid in general, at least as a rough estimate. Also a relation exists between the mean square velocity, $\langle v^2\rangle$, and the position variance, $\langle x^2\rangle$, that is

$$\langle v^2\rangle \approx \omega^2\langle x^2\rangle, \tag{11.3}$$

which is again exact for a harmonic oscillator. (We are assuming $\langle v\rangle = \langle x\rangle = 0$.) We may now eliminate the frequency amongst Eqs. 11.2 and 11.3 and get the (Heisenberg) uncertainty relation

$$m^2\langle v^2\rangle\langle x^2\rangle \approx \frac{1}{4}\hbar^2 \tag{11.4}$$

as appropriate for the stationary state in SED. Thus SED predicts correctly the order of magnitude of the ground state energy and size of simple systems. The meaning of the uncertainty relations in SED is transparent: If we confine a particle in a narrow potential, the typical frequency for its motion will be large, the particle will interact mainly with ZPF modes of high energy and it will reach a high average kinetic energy.

One may go a step further and assume that, for relatively long periods, the particle may be coupled mainly to the ZPF via some harmonic, $\omega = n\omega_0$,of the fundamental frequency, ω_0 of the particle's motion. Thus we might write, instead of Eq. 11.2,

$$\frac{1}{2}m\langle v^2\rangle = \frac{1}{4}\hbar\omega = \frac{1}{4}\hbar(n\omega_0), \tag{11.5}$$

n being an integer number. This equality may be also written in the form

$$\oint p dx = \frac{1}{2}nh, \tag{11.6}$$

which is the Sommerfeld-Wilson quantization rule except for a factor 2. This suggests an interpretation of the discrete energy eigenstates as those sustained by a temporary resonance with the ZPF.

The derivation of Eqs. 11.4 and 11.6 presented above provides also a *qualitative* explanation for the connection between energy and frequency in stationary states of atoms. Of course the intuitive picture which emerges does not allow for a precise quantitative agreement with the observations (or the predictions of quantum mechanics). In particular, Eq. 11.6 is valid only in the semiclassical regime, where the action of the ZPF over the particle is relatively weak. Also there are many unanswered questions like: What about the harmonic oscillator where the Fourier expansion of x(t) contains only one frequency and no harmonics?, Why do neutral particles, not interacting with the ZPF, possess quantum behaviour?, What is the explanation of Pauli's principle?, etc. I shall not attempt to rebut all objections here, I say simply that they point towards a limited value of SED, which must be seen as just a starting point to be completed and/or modified. Amongst the needed changes or additions I foresee the following:

1. Including fluctuations in the space-time metric, a kind of gravitational ZPF. This is certainly required, because if we assume the existence of a ZPF electromagnetic field, the idea should be extended to all fields including the gravitational one.
2. Explaining the electron spin and Pauli principle.
3. Explaining the creation and annihilation of particles.

In any case I propose the following general picture: The material world consists of fundamental particles (fermions) plus fields (bosons) as in classical physics. The word "fundamental" is here included in order to take into account that there are "composite systems" consisting of several fundamental particles and fields, for instance atoms, which may behave either as fermions or bosons. I propose that particle properties of Bose fields, like the electromagnetic one, derive from the combination of a random ZPF and the interaction of the field with atoms, as we shall see in more detail below. In contrast wave properties of fundamental fermions, like electrons, or composite systems, like atoms or molecules, would derive from the action on the particles of pervading ZPF's, which are modified by the presence of obstacles, as in the two-slit experiments. This provides a qualitative explanation for the interference of particles, in line with the early de Broglie interpretation of corpuscles guided by waves. The difference is that I do not assume *one* wave associated to every particle, but a background of waves influencing the motion of every particle. An interesting question is whether two particles placed at a small distance will be guided by (i.e., interact most strongly with) the same component of the ZPF's or by different components. My intuition says that this may depend on the external conditions. If the temperature is low enough and the interaction between the particles relatively small it may happen that several, or many, particles are associated to (interact most strongly with) the same wave component of the ZPF. This would give rise to correlated motions of many particles, as is the case in "macroscopic quantum phenomena" like Bose-Einstein condensation or superconductivity. In any other circumstances every particle will be associated to a different wave, which will correspond to the de Broglie "matter wave".

11.2 Understanding Photons

My interpretation of "photons" (i.e., particle properties of light) has been developed in a collaboration with Trevor Marshall, from Manchester University, lasting from 1983 to the present[3]. It derives from SED and we have used the name *stochastic optics* (*SO*) for the approach. It is a pure wave theory where there are no "photons" or, maybe, photons are just wavepackets superimposed to the ZPF. The explanation (or intuitive picture) provided by SO for several quantum phenomena is as follows:

11.2.1 Emission and Absorption of Light

As is well known, Einstein introduced the "quanta" of light in 1905 just as a "heuristic point of view". Only with the work of 1916 Einstein gave particle properties to these quanta, that is momentum in addition to energy. Within SO these properties appear in a natural way as follows. Let us consider an one-electron atom in a quasistationary state of energy E_1. It will make a transition to another state of energy $E_2 < E_1$ when a fluctuation of the ZPF having appropriate frequency arrives at the atom (this frequency is $(E_1 - E_2)/\hbar$, something which I shall not try to explain here, although some hints are provided by the above commented connection between energy and frequency). This means that spontaneous emission may be seen as stimulated by the ZPF, something which is more or less accepted today[2]. Let us fix the origin of the coordinate system at the atom position and assume that the fluctuation of the ZPF may be represented by a plane wave moving in the direction of the OZ axis. The radiation emitted by the atom will have the form of an spherical wave, which will interfere constructively with the incoming plane wave in those points where

$$r - z = r\,(1 - \cos\theta) \lesssim \lambda/2.$$

This means that most of the emitted energy is concentrated within a cone with half angle $\theta \simeq \sqrt{\lambda/r}$. Thus the emitted radiation ("the photon") has the form of a needle rather similar to that foreseen by Einstein in his 1916 paper. Within our approach the probabilistic nature of the time and direction of emission, which so much worried Einstein, has also a simple explanation: it is due to the random character of the ZPF. If the needle of radiation happens to impinge on another atom, an intuitive picture of the absorption of radiation also follows. Furthermore, in case of absorption followed immediately by emission it is easy to understand the conservation of energy and momentum of the "atom plus (incoming and outgoing) photons". This may happen also when the atom is replaced by a free electron, as in the Compton effect.

11.2.2 Anticorrelation after Beam Splitter

As another example of the picture provided by SO, I shall consider the "corpuscular behaviour" of light in an experiment with two detectors after a beam splitter[4]. The experiment seems to prove that a photon is either reflected or transmitted at the beam splitter (BS), never divided, but recombination of the transmitted and the reflected beams at a second BS gives rise to interference, which seems to imply that something goes to every outgoing channel. There is here one of the most "mind boggling" examples of quantum behaviour, which is reinforced if we take into account that our decision to study anticorrelation or recombination may be made after the photon has crossed the BS (*delayed choice experiment*). According to SO (a purely wave theory) when a signal arrives at a BS the radiation intensity is divided, one part being transmitted and the other part reflected, which easily explains interference after recombination. More tricky is explaining anticorrelation, where the ZPF plays an essencial role. Let us label 1 the incoming channel of the BS where the signal arrives and 2 (3) the outgoing channel of the transmitted (reflected) beam. Assuming a balanced BS, the incoming electric field, $E_1(t)$, will be divided so that

$$E_2(t) = \frac{1}{\sqrt{2}} E_1(t), \quad E_3(t) = \frac{i}{\sqrt{2}} E_1(t), \tag{11.7}$$

where we use a representation involving complex quantities and forget about the space dependence and vector character of the electric field. Now, at a difference with conventional classical optics, in SO we shall assume that there is ZPF coming in all channels of the beam splitter. In particular, in the fourth channel there will be some incoming ZPF, represented by $E_0(t)$, so that Eq. 11.7 should be modified taking the form

$$E_2(t) = \frac{1}{\sqrt{2}}[E_1(t) + iE_0(t)], \quad E_3(t) = \frac{1}{\sqrt{2}}[E_0(t) + iE_1(t)]. \tag{11.8}$$

The corresponding intensities (in appropriate units) will be

$$I_2 = |E_2|^2 = \frac{1}{2}(I_1 + I_0) + \mathrm{Im}(E_1E_0^*), \qquad I_3 = \frac{1}{2}(I_1 + I_0) - \mathrm{Im}(E_1E_0^*).$$

The intensity I_0 corresponds to pure ZPF whilst I_1 contains both ZPF and signal, whence $\frac{1}{2}(I_1 + I_0)$ corresponds to the intensity of ZPF plus "half a signal". The term Im $(E_1E_0^*)$ may be positive or negative so that usually either I_2 is below the ZPF level and I_3 is above or vice-versa. If we assume that photon counters have a detection threshold just at the ZPF level, detection will happen only in one of the outgoing channels, so explaining anticorrelation. Of course, if the intensity of the signal is high (it contains many photons, in quantum language) then in both outgoing channels the intensity will quite probably surpass the level of the ZPF, and this is why the corpuscular behaviour of light is exhibited only by weak (single-photon) signals. The unpredictability of the result is a straightforward consequence of the randomness of the ZPF, that is in SO uncertainty derives from noise, as is typical in classical physics, rather than from an "essential randomness" of physical laws, as postulated in quantum theory.

11.2.3 Photon Entanglement

The two beams of light represented in Eq. 11.8 correspond to the quantum state represented, in the Hilbert-space formalism, by

$$|\Psi\rangle = \frac{1}{\sqrt{2}}(|1\rangle|0\rangle + |0\rangle|1\rangle),$$

which is a state where a single photon signal is *entangled* with the vacuum. Thus, in our picture, entangled states of light are situations in which two light beams are correlated, the correlation involving both the signal and a part of the accompanying ZPF. In contrast, classical correlated beams are those where the correlation involves only the part of the radiation field which is superimposed to the ZPF.

Let us develop the argument further. The analysis of the experiment[4] commented in the previous subsection is rather involved because it actually consisted in the measurement of a single, two double and a triple coincidence rates. Thus we shall study a more simple situation of entanglement. We will consider two light beams, arriving at two photon counters, given by the electric fields

$$E_1(t) = E_3(t) + E_4(t),\ E_2(t) = E_5(t) + E_6(t), \tag{11.9}$$

where E_3, E_5, E_4, E_6 are four time-dependent fields which may be treated as stationary stochastic processes. We want to compare the two single detection rates with the coincidence detection rate. In order to make the calculation we shall assume that the single, P_j, and coincidence, P_{12}, detection probabilities during a small time interval are given by

$$P_j = \eta\langle |E_j|^2 - I_{ZPF}\rangle, \qquad P_{12} = \eta^2\langle(|E_1|^2 - I_{ZPF})(|E_2|^2 - I_{ZPF})\rangle, \tag{11.10}$$

η being a constant related to the efficiency of the detectors (assumed identical for simplicity), I_{ZPF} the average intensity of the ZPF entering every detector, and the symbol $\langle\ \rangle$ means either time average or ensemble average. We shall assume that E_3, E_5 correspond to the ZPF, so that $\langle |E_3|^2\rangle = \langle |E_5|^2\rangle = I_{ZPF}$, and E_4, E_6 to the signal. The hypotheses Eq. 11.10 provide a formally simple (but physically absurd, see next subsection) substitute for the assumption that there is a threshold in photon detectors at the level of the ZPF, mentioned in the previous subsection. We may consider two different cases:

1. If E_3, E_5 are uncorrelated to E_4, E_6 and uncorrelated amongst themselves. In this case Eq. 11.10 reduces to

$$P_1 = \eta\langle |E_4|^2\rangle, \qquad P_2 = \eta\langle |E_6|^2\rangle, \qquad P_{12} = \eta^2\langle |E_4|^2 |E_6|^2\rangle, \tag{11.11}$$

   and we may forget about the ZPF. In this case we will say that the correlation of the two beams is "classical", involving only the "radiation above the ZPF level".

2. All four fields are correlated. In this case the detection probabilities may be quite different from the classical prediction Eq. 11.11, in particular much greater. For instance, let us consider that E_3 is correlated to E_5 but both are uncorrelated to E_4 and to E_6. Then we will have, instead of Eq. 11.11,

$$P_1 = \eta\langle |E_4|^2\rangle,\ P_2 = \eta\langle |E_6|^2\rangle,$$
$$P_{12} = \eta^2(\langle |E_4|^2 |E_6|^2\rangle + \langle |E_3|^2 |E_5|^2\rangle - \langle |E_3|^2\rangle\langle |E_5|^2\rangle).$$

In this case we say that the two beams are *entangled*.

We see that SO provides a quite *clear physical picture* of entanglement, which contrasts with the *purely formal definition* of quantum theory ("two particles are entangled if the Hilbert-space vector representing the state cannot be written as a tensor product of single-particle vectors", or an appropriate generalization for mixed states).

11.2.4 Detection Loophole in Tests of Bell's Inequalities

Most books or articles published at present claim that Bell's inequalities have been empirically violated (e.g., in Aspect's experiment). This statement is not true, at least if we consider "genuine" Bell inequalities, derivable from realism and locality without additional assumptions[5]. Incidentally, one of the additional hypotheses used, introduced by Clauser and Home[6] with the name of "no-enhancement", is naturally violated in SO because a light beam crossing a polarizer may increase its intensity, due to the insertion of ZPF in the fourth channel (see arguments leading to Eq. 11.8), which is the possibility excluded by the no-enhancement assumption.

Quantum theory of photon detection rests upon the use of normal ordering of the creation and annihilation of photons, which may be seen to be equivalent to the subtraction of the ZPF average intensity of Eq. 11.10[7]. However, averages like those in Eq. 11.10 cannot correspond to physical processes, where *positive* probabilities (of several possibilities) should be added. In fact the averaged quantity is negative whenever $|E_j|^2 < I_{ZPF}$. For relatively high intensities Eq. 11.10 may be a good approximations to more physical hypotheses like, for instance,

$$P_j \propto \langle (|E_j|^2 - I_{ZPF})_+ \rangle, \qquad P_{12} \propto \langle (|E_1|^2 - I_{ZPF})_+ (|E_2|^2 - I_{ZPF})_+ \rangle \tag{11.12}$$

where $(\,)_+$ means putting 0 if the quantity inside the bracket is negative. Thus Eq. 11.12 amount to assuming a detection threshold at the ZPF level. For low detection efficiencies it is possible to devise physical models of detection which closely approach the quantum detection theory[8], but the possibility does not exist for high enough efficiencies. The reason is that there is always the possibility that high fluctuations of the ZPF are confused, by the detector, with signals, so that at high efficiencies errors are unavoidable (in our wave approach, photon counters are like alarm systems where the combination of few false negative results and few false positive ones is not possible). Thus SO predicts that photon counters (of optical photons) cannot exist with both, high efficiency and good reliability, thus explaining the so-called "low efficiency loophole" in the optical tests of Bell's inequalities.

11.3 Quantum Optics in Wigner Representation

Matching the *qualitative physical picture* of optics presented in this chapter with the *quantitative calculational rules* of quantum optics is not easy. I believe that the matching should be made via the Wigner function formalism. Indeed the

Wigner function of the electromagnetic quantum vacuum may be naturally interpreted as a random radiation field with precisely all the properties of the ZPF assumed in stochastic electrodynamics. Also there is a remarkable property of the Wigner function in quantum optics which I comment in the following.

As is well known, the states of the radiation field which are considered as *classical* are those having a (Glauber-Sudarshan) P-representation which is positive definite. The interesting property is that the Wigner function of such states may be obtained by means of a convolution with the vacuum Wigner function. That is, considering for simplicity a single mode of the radiation, the Wigner function $W(\alpha, \alpha^*)$ of the classical state is

$$W(\alpha,\alpha^*) = \int d\beta\, d\beta^* W_0(\beta-\alpha, \beta^*-\alpha^*) P(\beta,\beta^*), \tag{11.13}$$

where $P(\alpha, \alpha^*)$ is the P-function of the classical state and $W_0(\alpha, \alpha^*)$ is the Wigner function of the vacuum state. This result allows for a natural interpretation of the classical states of the radiation field. They are those states where some radiation exists on top of the ZPF, but are uncorrelated with it. Indeed, Eq. 11.13 is just the standard formula for the probability of a random variable which is the sum of two uncorrelated ones. In sharp contrast, the socalled non-classical states of the radiation field (e.g., squeezed states or entangled states) are those states where the ZPF has been modified.

However there are some problems for an interpretation in terms of the Wigner function, which I just summarize in the following:

1. The Wigner function is not positive definite. This well kown fact prevents us form interpreting the Wigner function as a probability distribution. However, at a difference with what happens in elementary quantum mechanics, the Wigner function is very frequently positive definite in quantum optics. Indeed this is the case for all experiments involving parametric down conversion[7]. We have made the conjecture that, when one takes into account all sources of uncertainty in the representation of field states in quantum optics, the Wigner function will be indeed positive[3].
2. The evolution of the Wigner function guarantees that the positivity is maintained only if the evolution equations are linear (the Hamiltonian quadratic) in the creation and annihilation operators. This is the case in all PDC experiments[7], but not in general.
3. The most frequent criticism to the reality of the ZPF is the fact that it does not give rise to activation of photodetectors. This problem is eliminated by the assumption, mentioned in section 1, that the ground state of physical systems (say detectors) correspond to a dynamical equilibrium with the ZPF. Thus photon counters should be activated only when they receive radiation above the level of ZPF.

References

[1] L. de la Peña and A. M. Cetto, *The quantum dice. An introduction to stochastic electrodynamics*. Kluwer, Dordrecht, 1996.

[2] P. W. Milonni, *The quantum vacuum,* Academic Press, Boston, 1994.

[3] T. W. Marshall and E. Santos, Found. *Phys.* **18**, 185 (1988); *Recent Res. Devel. Optics,* **2**, 683 (2002).

[4] P. Grangier, G. Roger and A. Aspect, *Europhys. Lett.* **1** (1986), 173.

[5] E. Santos, Found. *Phys.* **34**, 1643 (2004); quant-ph/0410193 (2004).

[6] J. F. Clauser and M. Home, *Phys. Rev.* **D 10**, 526 (1974).

[7] A. Casado et al., *Phys. Rev.* **A 55**, 3879 (1997); **A 56**, 2477 (1997); *J. Opt. Soc. Amer.* **B 14**, 494 (1997); **B 15**, 1572 (1998); *Eur. Phys. J.* **D 11**, 465 (2000); **D 13**, 109 (2001).

[8] E. Santos, *Quant-ph*/0207073 (2002).

12

Violation of the Principle of Complementarity and Its Implications

Shahriar S. Afshar

Department of Physics, Harvard University, Cambridge, Massachusetts 02138, USA, and Department of Physics and Astronomy, Rowan University, Glassboro, New Jersey 08028, USA

CONTENTS

Abstract

Bohr's principle of complementarity predicts that in a *welcher weg* ("which-way") experiment, obtaining fully visible interference pattern should lead to the destruction of the path knowledge. Here I report a failure for this prediction in an optical interferometry experiment. Coherent laser light is passed through a dual pinhole and allowed to go through a converging lens, which forms well-resolved images of the respective pinholes, providing complete path knowledge. A series of thin wires are then placed at previously measured positions corresponding to the dark fringes of the interference pattern upstream of the lens. No reduction in the resolution and total radiant flux of either image is found in direct disagreement with the predictions of the principle of complementarity. In this chapter, a critique of the current measurement theory is offered, and a novel nonperturbative technique for ensemble properties is introduced. Also, another version of this experiment without an imaging lens is suggested, and some of the implications of the violation of complementarity for another suggested experiment to investigate the nature of the photon and its "empty wave" is briefly discussed.

Key words: Photon, complementarity, wave-particle duality, *welcher weg*, which-way experiments, parametric down-conversion, Afshar experiment, measurement theory, empty wave, wavefunction collapse.

12.1 Introduction

The wave-particle duality has been at the heart of quantum mechanics since its inception. The celebrated Bohr-Einstein debate revolved around this issue and was the starting point for many illuminating experiments conducted during the past few decades. Einstein believed that one could confirm both wave-like and particle-like behaviors in the same interferometry experiment. Using a movable double-slit arrangement, he argued that it should be possible to obtain *welcher-Weg* or which-way information (WWI) for an electron landing on a bright fringe of an interference pattern (IP) "to decide through which of the two slits the electron had passed".[1] Although Einstein ultimately failed to achieve this goal, his logical consistency argument (LCA) was the initial motivation behind Bohr's Principle of Complementarity (PC).[1] The general formulation of LCA, in the context of the double-slit experiment, could read as follows:

(1) Perfectly visible IP implies that the quantum passed through *both* slits (sharp wave-like behavior).

(2) Complete WWI implies that the quantum passed through only *one* of the slits (sharp particle-like behavior).

(3) Satisfaction of both (1) and (2) in a *single* experimental setup is a logical impossibility, since (1) and (2) are mutually exclusive logical inferences.

Bohr famously avoided the logical impasse mentioned in (3) by applying Heisenberg's uncertainty principle to the experimental setup,[2] showing that under any *particular* experimental configuration one can only achieve (1) or (2), and *never* both.

In Bohr's own words: "... we are presented with a choice of *either* tracing the path of the particle, *or* observing interference effects...we have to do with a typical example of how the complementary phenomena appear under *mutually exclusive* experimental arrangements".[1] Several recent experiments,[3–9] however, suggest independence of the interferometric complementarity from the uncertainty principle; hence, we shall only discuss the limitations of PC in this chapter. A quantitative formulation for which-way detection has been developed on the basis of theoretical[10–15] and experimental[9, 16–19] investigations of PC during the past two decades, leading to a wave-particle duality relation covering both sharp and intermediate stages expressed as:

$$V^2 + K^2 \leq 1, \tag{12.1}$$

where the two complementary measurements are $0 \leq V \leq 1$, the visibility or contrast of the IP, and $0 \leq K \leq 1$ the which-way knowledge corresponding to WWI. The visibility is given by

$$V = (I_{max} - I_{min})/(I_{max} + I_{min}), \tag{12.2}$$

where I_{max} is the maximum intensity of a bright fringe and I_{min} is the minimum intensity of the adjacent dark fringe, so that $V = 1$ when the fringes are perfectly visible (sharp wave-like behavior), and $V = 0$ when there is no discernible IP. By analogy, for the which-way knowledge $K_1 = (I_1 - I_2)/(I_1 + I_2)$, so $K = 1$ when the WWI is fully obtained (sharp particle-like behavior), and $K = 0$ when the origin of the quantum cannot be distinguished.

It is noteworthy to mention that quantum mechanics does not forbid the presence of *non-complementary* wave and particle behaviors in the same experimental setup. What is forbidden is the presence of *sharp complementary* wave and particle behaviors in the same experiment. Such complementary observables are those whose projection operators *do not commute*.[20]

In this paper we shall only investigate sharp complementary wave and particle behaviors explicitly forbidden by PC in the same experiment. Therefore, intermediate conditions, where $0 < V < 1$, and $0 < K < 1$ shall not be covered. We assume full validity for quantum mechanical formalism, and make use of it to test the predictions of PC as a particular interpretation of quantum mechanics. Finally, although in our experiments we have not used a coherent *single-photon* source, it is expected that exactly the same results would be obtained if such a source is used.

12.2 Conventional Measurements of Complementary Observables

12.2.1 Modern Version of Principle of Complementarity

We can take advantage of the recent developments in the debate over the PC to update the definition of interferometric complementarity. Based on Eq. 12.1 a *modern* version of the orthodox PC—the contemporary principle of complementarity (CPC)—can be formulated as follows in any *particular* experimental arrangement:

(i) If V = 1, then K = 0.

(ii) If K = 1, then V = 0.

It is clear from CPC (i) that in any *welcher weg* experiment, obtaining full visibility for the IP should lead to a complete loss of the WWI for the quanta. Let us pay homage to orthodoxy by applying its tenets to two experiments.

12.2.2 Destructive Measurement of IP Visibility

In the first experiment, we test the validity of CPC(i) in a *conventional* manner. As shown in Figure 12.1(a), coherent and highly stable laser light of wavelength $\lambda = 650$ nm impinges upon a *dual pinhole* with a center-to-center distance of $a = 2000$ μm and pinhole diameters of $b = 250$ μm. Two diffracted beams represented by wave functions ψ_1 and ψ_2 emerge. The overlapping diffraction patterns of the beams caused by the corresponding pinholes are apodized (see Appendix A), by passing the light through an aperture stop (AS) permitting only the maximal Airy disks of radius $s = 10.4$ mm to pass, thus eliminating higher order diffraction rings. A photosensitive surface is placed at plane σ_1 at a distance $l = 400$ cm from the dual pinhole, and a fully visible IP ($V = 1$), with peak-to-peak distance of $u = 1.4$ mm for the consecutive fringes, is observed as shown in Figure 12.1(b).

Assuming that Ψ_1 and Ψ_2 are the *apodized* wave functions, the probability density, or its classical equivalent, the irradiance, for the *coherent* superposition state $\psi_{12} = \psi_1 + \psi_2$, is given by

$$I_{12} = |\psi_{12}|^2 = |\psi_1|^2 + |\psi_2|^2 + \Gamma, \tag{12.3}$$

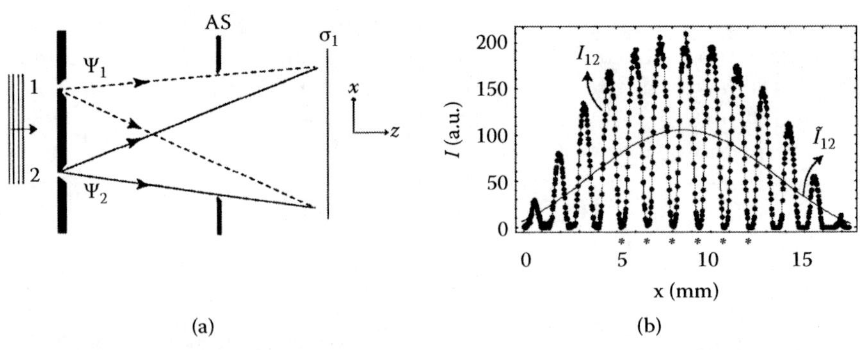

FIGURE 12.1
(a) Laser light impinges upon a dual pinhole and two diffracted beams ψ_1 and ψ_2 emerge. The beams are apodized by an aperture stop AS. (b) The interference pattern I_{12} is observed at plane σ_1. Here $V = 1$, and $K = 0$. The red curve shows the theoretical decoherent irradiance profile $\tilde{I}_{12}$. The irradiance is measured in arbitrary units a.u. of grey-level intensity.

where $\Gamma = \psi_1^*\psi_2 + \psi_1\psi_2^*$ is the usual interference term. It is clear that observing the IP in this configuration leads to a complete loss of WWI, because the photosensitive surface at σ_1 *destructively* absorbs all of the incoming light and no further analysis can take place, hence $K = 0$. Here, in *conformity* with Eq. 12.1 the complementary measurements are $V = 1$, and $K = 0$. For comparison, the red curve shown in Figure 12.1(b) depicts the theoretical irradiance profile for the case $V = 0$, where

$$\tilde{I}_{12} = |\psi_1|^2 + |\psi_2|^2 \tag{12.4}$$

is the irradiance for the *decoherent* state, which clearly lacks any interference fringes.

12.2.3 Destructive Measurement of Which-Way Information

The application of a converging lens for which-way detection has a long history and is already implicit in the classic "Heisenberg's microscope" proof of the uncertainty principle, where the spatial resolution of the lens Δx, enters directly into the uncertainty relation $\Delta p_x \cdot \Delta x \sim h$.[2,21,22] Wheeler has used the lens explicitly for which-way detection in a proposed *welcher weg* experiment,[23] such that photons registered at each image of the two slits are assumed to have passed through the corresponding slit, thus providing WWI.

In the second experiment, as shown in Figure 12.2(a), we remove the photosensitive surface at σ_1, and allow the light to pass through a suitable converging lens (L), here, with a focal length $f = 100$ cm and effective diameter of $d = 30$ mm, placed at a distance $p = 420$ cm from the pinholes, which then forms two well-resolved images (1′ and 2′) of the corresponding pinholes (1 and 2) at the image plane σ_2 at a distance of $q = 138$ cm from the lens. The

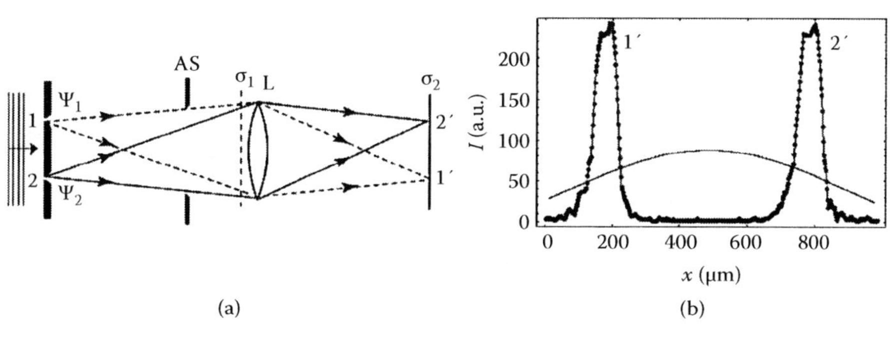

FIGURE 12.2
(a) A converging lens L placed in front of σ_1 produced two well-resolved images of the pinholes. (b) The irradiance profile of the images 1′ and 2′. The photons landing in 1′ originate in pinhole 1, and those landing in 2′ originate in pinhole 2. $V = 0$, and $K = 1$. The curve shows a theoretical irradiance profile for a $K = 0$ case.

image data collected at σ_2 is shown in Figure 12.2(b) in black. The theoretical spatial resolution of the lens in this experiment is $R \approx 30$ μm, which matches well with the observation. Less than 10^{-6} of the peak value irradiance from either image is found to enter the other channel, essentially providing $K = 1$. For comparison, the red curve in Figure 12.2(b) shows the theoretical irradiance profile for a $K = 0$ case (no WWI), where a single unresolved peak instead of the two well separated peaks would be observed.

Again in this experiment, the photons are destructively detected at σ_2, and no further analysis can take place *afterwards*. However, Eq. 12.1 in conjunction with LCA(3) predicts a visibility of $V = 0$ for the IP in this experiment, which entails a decoherent state for the two wave functions Ψ_1 and Ψ_2 at σ_1 with a corresponding *decoherent* irradiance distribution $\tilde{I}_{12} = |\psi_1|^2 + |\psi_2|^2$ as shown in Figure 12.1(b). In contrast to I_{12}, in this case the resulting irradiance $\tilde{I}_{12}$ lacks the interference term Γ.

In this experiment, the decoherence of the wave functions *prior* to entering the lens is a counter-intuitive conclusion dictated by PC, as it implies that the potential future act of obtaining WWI (the detection of the pinhole images at σ_2) leads to the loss of the IP at an *earlier* stage (at σ_1) in a non-local manner. As Feynman puts it, this situation "has in it the heart of quantum mechanics" and "contains the only mystery" of the theory.[24]

12.3 Theoretical Digression: Measurement Theory Revisited

12.3.1 Critique of Orthodox Concept of "Measurement"

Before we discuss the main experiment, let us momentarily take an uncustomary digression to theory to elucidate the motivation behind the experiment. Measurement in general, can be defined as *a physical process by which quantitative knowledge is obtained about a particular property of the entity under the study.* Most orthodox measurements of quantum systems involve the interaction of a microscopic quantum particle with a macroscopic classical measuring apparatus, which inevitably leads to an *irreversible* and *destructive* change in the property we want to measure. For instance, the energy of a particle can be measured by bringing it to a halt in a scintillator. This process irreversibly "destroys" the particle's energy, i.e., the particle no longer carries the initial energy after the measurement process. Although in the so-called quantum nondemolition measurements we can preserve a particular property after successive measurements, this is achieved at the expense of introducing *irreversible* perturbation to the particle's other physical properties. What these types of destructive measurements have in common is that they are performed at the level of a *single particle* and *lead to an irreversible change in the final quantum state of the detector.* It is indeed *impossible* to obtain quantitative knowledge about a particular physical property of a single particle in a non-destructive and non-perturbative manner. Unfortunately, in his reasoning for the necessity of the principle of complementarity, Bohr erroneously applies destructive measurement schemes for establishing the wave-like behavior of photons in a *welcher weg* experiment, as discussed in section 2.2.[1] However, as we shall demonstrate in the next section, the measurement of a *multi-particle* or *ensemble* property *need not be destructive.*

12.3.2 Coherence and Wave-Like Behavior

Formation of an IP is aptly considered as evidence for coherent wave-like behavior of quantum particles. However, whereas in classical electromagnetism a *continuous* IP would be formed no matter how weak the source, in contrast quantum mechanics disallows such a state due to the fact that upon arriving at the observation plane, each quantum produces only a single dot. Figures 12.3(a–c) show the theoretical buildup of an IP from a coherent single-photon source over progressively extended periods of time, with 30, 300, and 3000 photons registered respectively. For comparison, Figures 12.3(d–f) show the decoherent photon distribution of the same number of photons respectively. It is *impossible* from the data in Figures 12.3(a) and 12.3(d), with only 30 photons registered, to discern which of the two show a coherent distribution (i.e., an IP) or a decoherent one. It is only as larger and larger numbers of photons arrive that one can recognize the lack or presence of an IP. In other words, ***evidence for coherent wave-like behavior is not a single-particle property, but an ensemble or multi-particle property.***

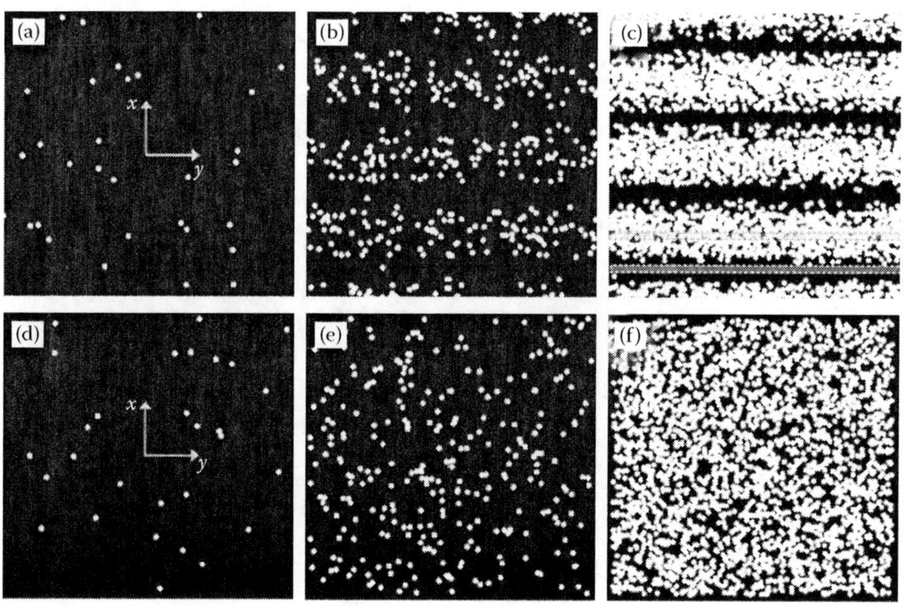

FIGURE 12.3
The interference pattern produced by a single-photon source with (a) 30, (b) 300, and (c) 3000 photons registered. In contrast, the decoherent distribution of (d) 30, (e) 300, and (f) 3000 photons lacks the dark fringes.

In contrast to single-particle properties such as the arrival of a single photon at a particular pinhole image, which immediately provides WWI as discussed in section 2.3, evidence for coherence *essentially* involves multiple measurements. The other important feature of coherent behavior is that there exist "forbidden" regions in space corresponding to the dark fringes, where no photons can be found. This avoidance of the dark fringe region is essential for the definition of an IP and its visibility.

12.3.3 Nondestructive Measurement of IP Visibility

The conventional method of obtaining the visibility of an IP involves two separate measurements: (1) destructive measurement of the maximum radiant flux at a bright fringe in order to obtain I_{max} and (2) destructive measurement of the minimum radiant flux at a dark fringe in order to obtain I_{min}. By substituting the values for I_{max} and I_{min} in Eq. 12.2, $V = (I_{max} - I_{min})/(I_{max} + I_{min})$, the visibility is calculated. The above process is necessary *if* $V < 1$, however, if the IP is perfectly visible ($V = 1$), then step 1 would be *entirely* superfluous. This is because in a perfectly visible IP, $I_{min} = 0$, and under such a condition, Eq. 12.2 is reduced to $V = \frac{I_{max}}{I_{max}} \equiv 1$, regardless of the actual value of I_{max}. *Therefore, as long as the total radiant flux of the dual pinhole output is nonzero (thus ensuring* $I_{max} \neq 0$*), all we need to establish perfect visibility is to determine* $I_{min} = 0$.

We can obtain $I_{min} = 0$ in two different ways: (i) by directly measuring the flux by placing a very thin detector array at the dark fringe, making sure it does not obstruct the bright fringes, or (ii) by placing an opaque obstacle such as a thin wire at the middle of a dark fringe and comparing the total radiant flux before and after the obstacle. Due to the technical impracticality of method (i), in our experiment, we opt for method (ii).

Figure 12.4(a) shows the schematics of method (ii) where the wire is shown as a small dark disk in the cross-section view, and σ_0 and σ_1 are parallel planes immediately before and after the wire. Assuming a coherent behavior, if we denote the distance between the centers of the pinholes as a, the diameter of the pinholes as b, the distance between the dual pinholes and σ_0 as l, and the wavelength of the laser as λ, then the IP is bounded within an Airy disk of radius

$$s = 3.833\, l\, \lambda/b, \tag{12.5}$$

and the distance between the peaks of each neighbouring bright fringe within the disk is

$$u = l\, \lambda/a \tag{12.6}$$

The *coherent* irradiance is given by

$$I_{12} = |\psi_{12}|^2 = [2\cos\alpha\, J_1(\beta)/\beta]^2, \tag{12.7}$$

$$\alpha = \pi\, x/u, \tag{12.8}$$

$$\beta = 1.22\pi\, x/s, \tag{12.9}$$

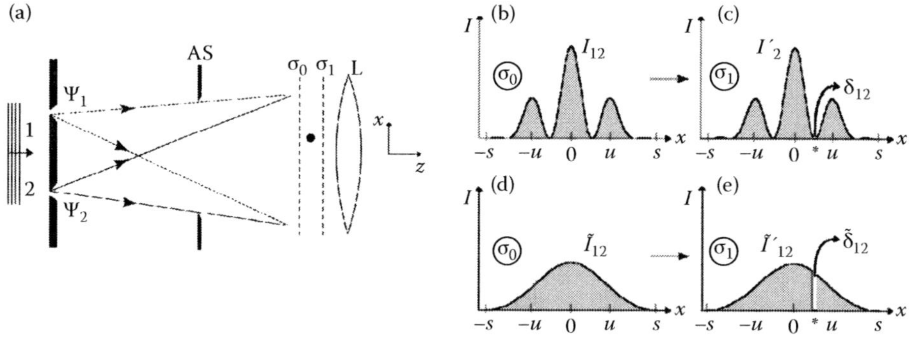

FIGURE 12.4
The effect of an opaque obstacle placed at the dark fringe of an interference pattern. (a) The planes σ_0 and σ_1 are located immediately before and after the obstacle, which is a wire shown as the small black disk. The irradiance profile I_{12} of the coherent superposition state $|\psi_{12}\rangle$, at (b) plane σ_0, and (c) plane σ_1. The irradiance profile $\tilde{I}_{12}$ of a decoherent state, at (d) plane σ_0, and (e) plane σ_1.

and $J_1(\beta)$ is the Bessel function of first order and first kind.[25] For clarity, we have selected an IP with three bright fringes as shown in Figure 12.4(b). Here we assume that the thickness of the wire is $e = u/10$ and is placed at the position $x = u/2$, in the middle of the right centermost dark fringe shown as an asterisk in Figure 12.4(c) depicting the irradiance I'_{12} at σ_1 immediately after the wire. It is clear that for the coherent case, the wire does not reduce the transmitted light appreciably, since it receives virtually no incident light such that

$$\int_{-s}^{s} I_{12}\, dx = \int_{-s}^{s} I'_{12}\, dx + \delta_{12}, \tag{12.10}$$

$$\delta_{12} = \int_{x_1}^{x_2} |\psi_{12}|^2\, dx \approx 0, \tag{12.11}$$

where $x_1 = (u - e)/2$ and $x_2 = (u + e)/2$.

Therefore, denoting $\Phi = \|\psi\|^2 = \int_{-s}^{s} |\psi|^2\, dx = \int_{-s}^{s} I\, dx$ for the total radiant flux (see A.4) we can rewrite Eq. (12.9) as

$$\Phi_{12} = \Phi'_{12} + \delta_{12}. \tag{12.12}$$

In contrast, the situation for a *decoherent* distribution, where $V = 0$ is quite different. As shown in Figure 12.4(d), the decoherent irradiance

$$\tilde{I}_{12} = 2[J_1(\beta)/\beta]^2, \tag{12.13}$$

also bound within the same Airy disk as the coherent state,[25] suffers a reduction in total radiant flux of

$$\tilde{\delta}_{12} = \int_{x_1}^{x_2} \tilde{I}_{12}\, dx \neq 0. \tag{12.14}$$

Therefore,

$$\tilde{\Phi}_{12} = \tilde{\Phi}'_{12} + \tilde{\delta}_{12} \tag{12.15}$$

Clearly $\tilde{\delta}_{12}$ is a significant fraction of the initial decoherent total radiant flux as shown in Figure 12.4(e). We know that

$$\int_{-s}^{s} [2\cos\alpha\, J_1(\beta)/\beta]^2\, dx = \int_{-s}^{s} 2[J_1(\beta)/\beta]^2\, dx, \tag{12.16}$$

and using Eqs. (12.5–12.16), the relationship between the coherent and decoherent states, can be expressed as

$$\Phi_{12} = \Phi'_{12} = \tilde{\Phi}'_{12} + \tilde{\delta}_{12}. \tag{12.17}$$

Eq. 12.17 simply restates the fact that for the *coherent* state, the presence of the wire makes no significant difference in the total radiant flux entering the lens ($\Phi_{12} = \Phi'_{12}$), and that it is the *same* as in the case when there is no wire present. This leads to the conclusion that the total radiant flux of the pinhole images 1′ and 2′ are not affected by the presence of the wire, if the light is in a *coherent* state at σ_1. In contrast, the same cannot be said about the decoherent state, since in this case the presence of the wire leads to a loss of $\tilde{\delta}_1 = \tilde{\delta}_2 = \tilde{\delta}_{12}/2$ in the total radiant flux of each image.

12.3.4 Impossibility of Interaction/Attenuation-Free Diffraction by Opaque Obstacle According to QM

In the discussion of diffraction, textbooks often fail to mention that the initial wavefunction is *always* attenuated after interaction with the opaque obstacle which produces the diffraction pattern in the transmitted wave function perhaps because the relative intensities within a distribution is of interest and thus normalization is justified. An optically opaque obstacle is an impenetrable barrier which has a cross section $e >> \lambda$. The interaction of a wave function with such an obstacle is a completely *local* process governed by Schrödinger equation, for which a *non-zero* amplitude must be present at the surface of the obstacle. Figures 12.5(a–c) depict the quantum mechanical simulation of a Gaussian wave packet directly hitting an obstacle

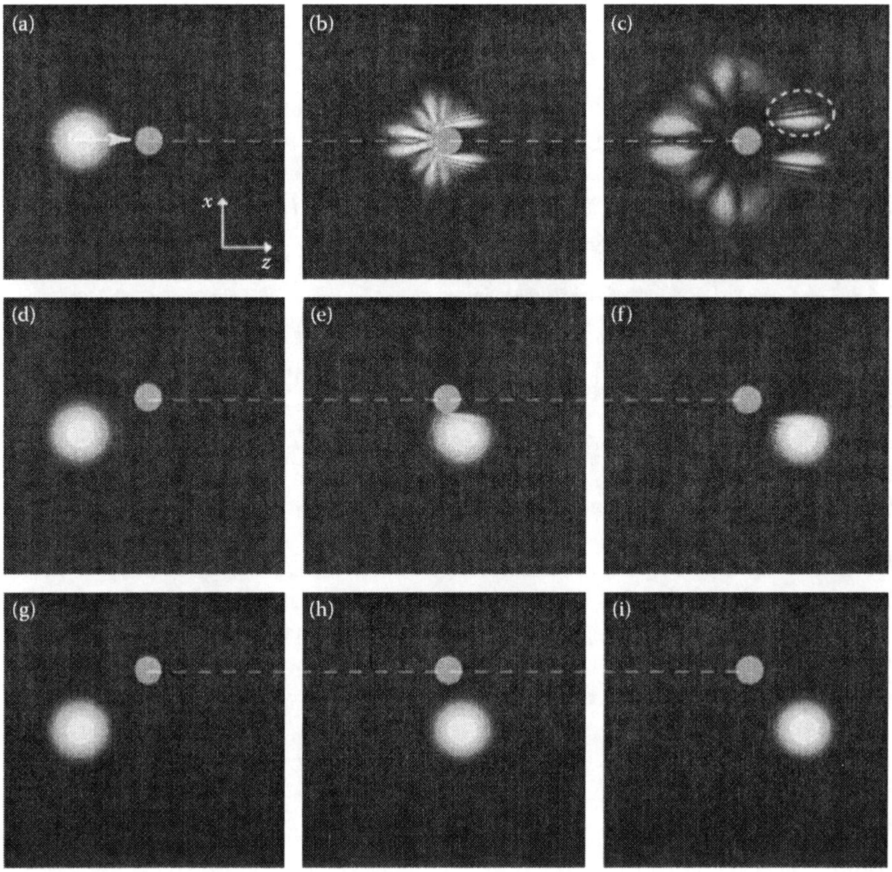

FIGURE 12.5
Theoretical simulation of the quantum-mechanical effect of an opaque obstacle on the evolution of a Gaussian wave packet for three different positions of the obstacle. (a–c) The wave packet directly hits the obstacle, producing significant attenuation and diffraction in the transmitted light. (d–f) The wave packet interacts with only the lower surface of the obstacle. (g–i) The wave packet nearly misses the obstacle.

(here $e = 30\lambda$) and consequently being partly reflected backwards, and partly diffracted in the direction of initial motion. In our simulation, the obstacle satisfies the Dirichlet boundary condition and is assumed to be a perfect mirror, reflecting the incident wave function without any damping.[26] It is clear that the transmitted part of the wave function is greatly attenuated and contains the telltale diffraction "lobes", enclosed within the dashed ellipse in Figure 12.5(c).

In contrast, Figures 12.5(d–f) show the same initial wave packet nearly missing the obstacle. In this scenario, the wave function interacts with only the lower surface of the obstacle, and therefore the reflected and diffracted portions of the wave function are dramatically reduced.

Finally, Figures 12.5(g–h) depict the same initial wave packet, this time completely missing the obstacle. It is clear that the wave function continues to move *undisturbed*, and no diffraction takes place. This is essentially a unitary time development during which the norm of the wave function remains unchanged. Therefore, we can make the following statement: ***If a wave function is not attenuated after passing a region within which a fully opaque obstacle is placed, it is not diffracted by the obstacle, and vice-versa: attenuation ⇔ diffraction.***

12.3.5 Formal Proof of Interference

Now we shall proceed to formally discuss the condition in which the incident wave function has a large enough lateral extent along the x-axis to completely cover the obstacle, yet after passing the obstacle, it is not attenuated (see Figure 12.6.) We show that: ***the lack of attenuation of the transmitted wave function is a necessary and sufficient condition for the existence of destructive interference at the position of the obstacle.***

Theorem 1. Suppose an *apodized* wave function $\psi(x, z, t_1)$ localized along the x-axis within $-s \leq x \leq s$ (see Appendix A) is immediately incident on an *opaque* obstacle of thickness $e >> \lambda$ placed at position $x = u$, $-s \leq u \leq s$. Immediately after the obstacle, the transmitted wave function $\psi'(x, z, t_2)$ continues to move along the z-axis. The following relation holds:

$$\|\psi\|^2 = |\psi'\|^2 \neq 0 \Leftrightarrow \delta = \int_{x_1}^{x_2} |\psi|^2 \, dx = 0 \tag{12.18}$$

where $x_1 = (u - e)/2$ and $x_2 = (u + e)/2$.

PROOF The interaction of ψ with the obstacle can be written as

$$|\psi\rangle \otimes |\varphi\rangle \xrightarrow{T} |\psi'\rangle, \tag{12.19}$$

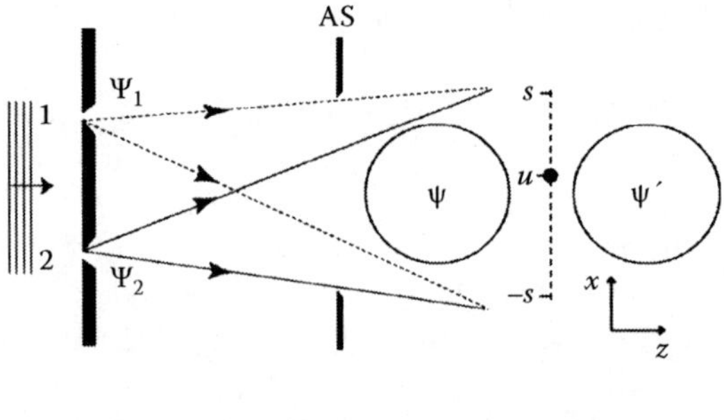

FIGURE 12.6
Apodized wave function ψ moving along the z-axis impinges upon an opaque obstacle placed at $x = u$. The transmitted wave function ψ' would have the same norm as ψ, if and only if there is a destructive interference at $x = u$, establishing the presence of an interference pattern.

where $|\varphi\rangle$ represents the obstacle, and T is the unitary time development operator.

We know that $\|\psi\|^2 = \|\psi'\|^2 \neq 0$, therefore

$$\Phi_{12} = \Phi'_{12} > 0. \tag{12.20}$$

But according to Eq. 12.11 we have $\Phi_{12} = \Phi'_{12} + \delta_{12} > 0$. Therefore, we have

$$\delta = \int_{x_1}^{x_2} |\psi|^2 \, dx = 0. \tag{12.21}$$

Theorem 2. For any wavefunction $\psi(x)$, and a given value $x = u$ the following holds:

$$|\psi(u)|^2 = 0 \Leftrightarrow \psi(u) = 0. \tag{12.22}$$

PROOF Since ψ is a complex wave function, we have for any given point within the wavefunction a complex vector $\psi(u) = Ae^{i\theta}$, where A is the modulus of the complex number $\psi(u)$. Since $|\psi(u)|^2 = A^2 = 0$ therefore $A = 0$, which necessarily leads to $\psi(u) = 0$. Therefore $|\psi(u)|^2 = 0 \Leftrightarrow \psi(u) = 0$.

Theorem 3. For any wave function $\psi(x, y)$, and a given value $x = u$, and $y = v$, the following holds:

$$\|\psi(x,y)\|^2 > 0 \wedge \psi(u,v) = 0 \Rightarrow \psi(u,v) = \psi_1(u,v) + \psi_2(u,v) = 0. \tag{12.23}$$

PROOF The wave function has a nonzero norm, and the particular complex vector for a point within the wave function is given as $\psi(u, v) = A\, e^{i\theta} = 0$.

A can be written as $A = \sqrt{B^2 + C^2}$, $B = B_1 + B_2 = 0$ and $C = C_1 + C_2 = 0, B_n \wedge C_n \neq 0$. We can thus construct at least two complex numbers $\psi_1(u,v) = \sqrt{B_1^2 + C_1^2}\, e^{i\theta} \neq 0$, $\psi_2(u,v) = \sqrt{B_2^2 + C_2^2}\, e^{i\theta} = -\sqrt{B_1^2 + C_1^2}\, e^{i\theta} \neq 0$. It is clear that the sum of these two nonzero complex vectors can be written as $\psi(u, v) = \psi_1(u, v) + \psi_2(u, v) = 0$, which is the superposition of two complex vectors with a phase difference of π.

12.4 Experimental Test of Complementarity

12.4.1 Experimental Verification of Nondestructive Measurement: Methodology

Now that we are theoretically motivated, let us get back to that most important tool of a physicist's trade, the experiment. Figure 12.7(a) depicts the essential parts of a configuration that can test the validity of PC. In this experiment, we use the absence of photons at the dark fringes (due to total

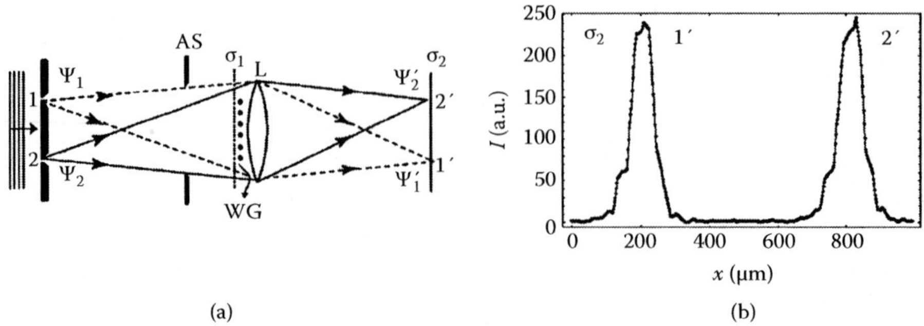

FIGURE 12.7
(a) The configuration testing the effect of the wires in the wire grid (WG). (b) Data representing the images of pinholes 1 and 2. No reduction in the resolution of the images is found at the image plane σ_2. This implies that no diffraction is produced by the WG and thus WWI is still complete (see text for theoretical justification) so that $K = 1$.

destructive interference), *as opposed to* their arrival at bright fringes (due to total constructive interference), as an equally valid evidence for the coherent wave-like behavior. In order to increase the "shadowing" effect of the wire, we place a series of six equidistant, and parallel thin wires (shown as black dots in the cross-section view of the setup) of thickness $e = 127\ \mu m \approx 0.1u \approx 200\ \lambda$ in front of the lens, at previously measured positions depicted by the asterisks in Figure 12.1(b), corresponding to the minima of the six most central dark fringes. Each wire is independently placed at the middle of the selected dark fringe with an alignment and positional accuracy of $\pm 1.6\ \mu m$. These wires can be considered as a wire grid (WG) with the same periodicity as the IP. Figure 12.7(b) shows the irradiance profile of the images at σ_2, while the WG is present. A comparison with the data in Figure 12.2(b) immediately demonstrates that the presence of the WG has not affected either the resolution, or the total radiant flux of the images.

The placement of the CCD directly at σ_2 leads to relatively large errors in the total radiant flux measurement. This is because the diameter of each pinhole image is quite small and few CCD elements receive the incident light, leading to saturation and blooming into the nearby pixels. In order to increase the accuracy, we used the configuration shown in Figures 12.8(a–c), where mirrors placed at the image plane σ_2 further separate the incident beams from each pinhole and direct them into different high resolution CCDs 1 m away from the image. Naturally, this reflected beam is distributed over a larger number of CCD elements, reducing the local irradiance and thus avoiding the blooming-related errors.

Figure 12.8(a) depicts the control run, where no WG is present and both pinholes are open. The total radiant flux Φ_C of this run for image 2′ is used to normalize the measurements in the next two experiments. Figure 12.8(b) shows the configuration and data for the simulation of a *decoherent* distribution of light at σ_1. One of the pinholes is closed and therefore there would be

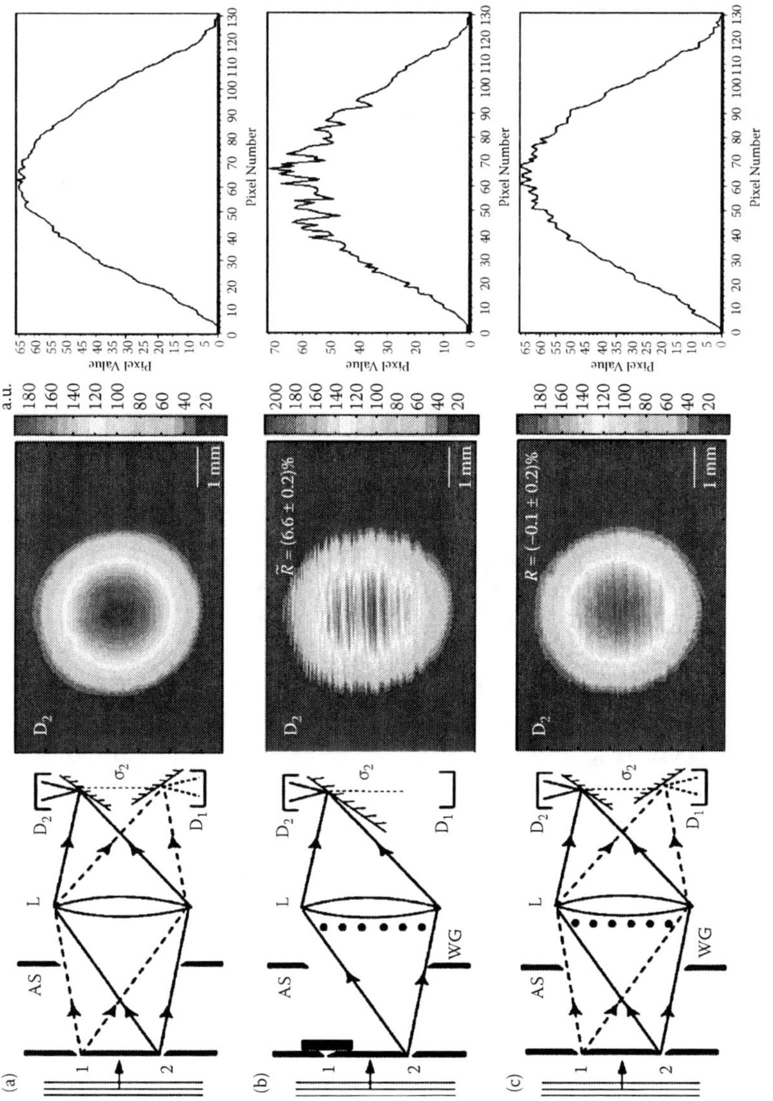

FIGURE 12.8

Test of complementarity. (a) Control configuration, with both pinholes open and no WG in place. The light from image 2′ is directed to detector D_2. (b) Simulation of decoherent state at σ_1 is achieved by closing pinhole 1, and placing the WG in the path of ψ_2. The total radiant flux is reduced by $\tilde{R} = (6.6 \pm 0.2)\%$. Compared to control data the loss of resolution of the image due to diffraction caused by the WG is clear. (c) Both pinholes are open and WG is placed at the dark fringes of the IP. The attenuation of the radiant flux of 2′ is found to be $R = (-0.1 \pm 0.2)\%$, which is negligible. Also the resolution of the image is only slightly reduced compared to control, since no diffraction takes place by WG. Here in violation of PC, $V = 1$, and $K = 1$, in the same experimental configuration.

incident photons on the WG, which attenuates and diffracts the transmitted light gathered by detector D_2. Using Eqs. 12.14 and 12.15, the normalized reduction in the total radiant flux of image of pinhole 2 for the *decoherent* case is given by

$$\tilde{R} = 100\tilde{\delta}_2/\Phi_C. \tag{12.24}$$

The loss of the radiant flux due to the WG in this case is theoretically calculated to be $\tilde{\delta}_2 = \Sigma_6\tilde{\delta}_2 = \Sigma_6\tilde{\delta}_{12}/2 = 6.5\%$ of Φ_C. The normalized radiant flux blocked by the wires is found to be $\tilde{R} = (6.6 \pm 0.2)\%$ by the analysis of the data, which matches the above theoretical value very well. Also, as expected, it is evident from the density plot of the D_2 output that the *resolution* of image 2′ has been significantly *reduced* in comparison to that of the control case.

12.4.2 Test of PC

In similar fashion to Eq. 12.24, using Eqs. 12.11 and 12.12, the normalized reduction in the total radiant flux of image of pinhole 2 for the *coherent* case is given by

$$R_{\text{Coherent}} = 100\,\delta_2/\Phi_C. \tag{12.25}$$

Figure 12.8(c) shows the configuration in which both pinholes are open, and the WG is present. The data show that the attenuation of the transmitted light in this case is negligible, $R = (-0.1 \pm 0.2)\%$ indicating that the WG has not absorbed or reflected a measurable amount of light within the margin of error, thus establishing the presence of dark fringes at σ_1, so that $V = 1$. It is also evident that the loss of the resolution of the image compared to the decoherent case is negligible. There is a very good agreement between the theoretical value of $R_{\text{Coherent}} = 0$, and the observed value R. This is compelling evidence for the presence of a perfectly visible IP ($V = 1$) just upstream of WG.

12.5 Discussion and Conclusion

Using Eq. 12.24 and the observed value for R, we can define a new parameter:

$$\eta = \frac{\tilde{R} - R}{\tilde{R} + R}, \quad 0 \le \eta \le 1. \tag{12.26}$$

If PC is correct, then in any experiment, we *must* find $\eta = 0$ since the observed value for R must be that of the decoherent case $\tilde{R}$, due to the fact that we find no reduction in the resolution of the images as shown in Figure 12.7(b), so that $K = 1$. The presence of a perfect IP would result in a $R = 0$, and therefore would lead to an ideal result of $\eta = 1$. Bearing in mind the margins of error in our measurements, in this experiment we find that $0.97 \leq \eta \leq 1.1$, again confirming a clear violation of PC. It is expected that this result can be improved upon by reducing the thickness e of the wires in the WG, yet maintaining the condition for opacity ($e >> \lambda$), and increasing the resolution and sensitivity of the CCDs.

I have endeavoured here to introduce a novel, non-destructive measurement process for the visibility of the IP which can be generalized to any ensemble property, be it spatial, temporal, or otherwise. In the last experiment shown in Figure 12.8(c), no attenuation of the transmitted light, and no significant reduction in the resolution of the image of pinhole 2 (it could as well have been pinhole 1) is found, although the WG is present in the path of the light. ***It is concluded therefore, that the coherent superposition state at the IP plane σ_1 persists ($V = 1$) regardless of the fact that the WWI is obtained ($K = 1$) at the image plane σ_2 in the same experiment.***

One might be tempted to argue that the reliability of the WWI is lost due to the presence of the WG. However, as discussed at length in sections 12.4 and 12.5, since the diffraction by WG could be the only reason for the reduction of K, we have established no such diffraction takes place, since no attenuation in the transmitted light is observed. *This simply means there was no light incident on the wires in the WG to diffract.* Therefore, since no diffraction takes place, no reduction in K is possible. Thus it is established that in the *same* experiment, sharp complementary wave and particle behaviors can coexist so that $V^2 + K^2 \approx 2 > 1$, violating Eq. 12.1 and the PC.

It is worth mentioning that since the so-called "delayed-choice" class of experiments[23] rely primarily on the validity PC, the results of this experiment demonstrate that there is really no "choice" to be made, as the coherent superposition state remains intact although WWI is obtained. Since the arguments presented in this chapter are valid for all quantum particles, it is plausible that equivalent experiments could be performed involving electrons or neutrons with identical results to this experiment.

12.6 Corollary

Since the initial results of the experiment were made available publicly in March 2004,[27] numerous critiques of the interpretation of the experiment were offered by the physics community. It would be impossible to discuss all those criticisms due to the page limitation of this publication, however, I would like to suggest three new experiments that may go a long way in answering most of the critics.

The first suggested experiment is a modified version of Wheeler's original delayed-choice experiment, in which two mutually coherent beams simply cross each other. Figure 12.9 depicts two beams crossing each other at plane Σ_1 and unitarily evolving unto well-separated beams further downstream at

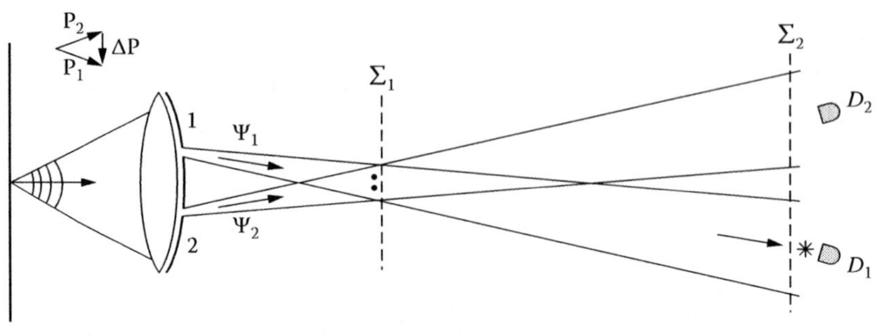

FIGURE 12.9
Configuration for first suggested experiment.

Σ_1. It is clear that at Σ_1 the beams will interfere and by the passive placement of the wires at the minima we can gain information about the visibility of the interference there. A single-photon detector, say D_1 registers a photon in Σ_2. Since the linear momentum of the photon is conserved, we cannot accept the proposition that this photon could have originated in pinhole 2, due to the fact that it must have changed its direction of motion at some point. We know that the wires cannot exchange momentum with the photon since they do not intercept it, and thus complete WWI is obtained, thus violating PC again.

The second experiment is based on the assumption that PC is indeed violated. The take-home message of such a violation is that the so-called collapse of the wavefunction does not take place. If so, the question is whether "empty waves" could help produce interference at the last beam splitter in a Mach-Zehnder type interferometer. This experiment is a modified version of the empty wave experiment of Mandel *et al.*[28] conducted in 1991 to investigate whether empty waves can induce coherence. As shown in Figure 12.10, the pump laser is incident on a beam splitter and equally irradiates two identical down-conversion crystals NL1, and NL2. The idler beam from NL1 is aligned such that its optical path overlaps with the idler beam from NL2. The signal beams from both crystals are brought together before detector D_s and a first order IP with visibility of about 33% is obtained. Now, I modify their experiment in two critical ways: (1) allow all of i_1 to enter NL2 to ensure maximum induced coherence. (2) place two identical 50–50 beam splitters BS_1 and BS_2 just before the final beam splitter. Step (2) gives us the opportunity to investigate the effects of the wavefunction collapse by observing say the upper beam before (A), at (B), and after (C) detection of a photon at D_s. This means we can now compare the resulting first order spatial IP at D_s with and without the beam splitters and with and without the collapse of the wavefunction for s_1. If we observe no reduction in the visibility of the IP (given we allow the same number of photons to

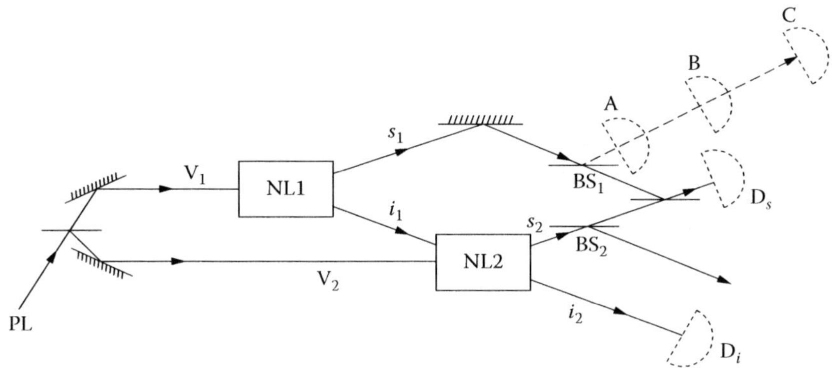

FIGURE 12.10
Configuration for second suggested experiment.

accumulate), then we can at least claim that the empty waves are capable of guiding a real photon to allow it to participate in an IP.

Should this second experiment prove positive, the next step would be to isolate the empty waves and observe their dynamical properties by perhaps accumulating large numbers of such waves within a carefully controlled optical cavity and looking for any changes in its temperature. Figure 12.11 depicts a possible setup. The isolation is achieved by opening a delayed Optical Gate (OG)—e.g., a Pockels Cell, only after detector D has detected the single photon emerging from the beam splitter. From the point of view of quantum mechanics, upon such detection, the wavefuntion should collapse, and the other channel must be considered as completely empty. If we observe any physical properties for this beam, we will have discovered a new form of electromagnetic field and would have to revise all our theories of radiation and detection.

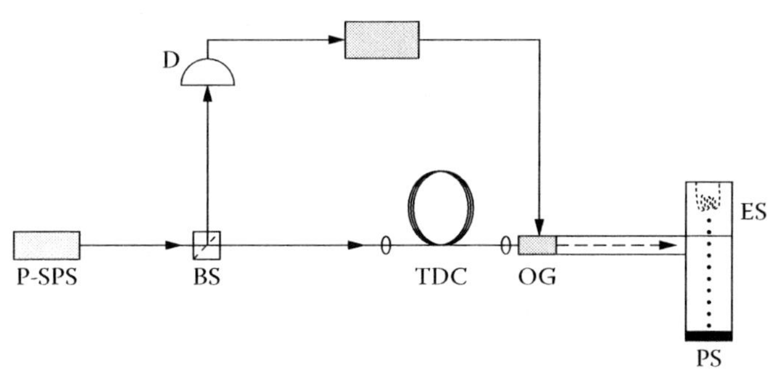

FIGURE 12.11
Configuration for third suggested experiment.

Acknowledgments

I would like to thank G. B. Davis, D. W. Glazer, J. Grantham, and other financial supporters of this research, to all of whom I am deeply grateful. I also would like to thank Christopher Stubbs and the Harvard University Department of Physics for their hospitality and the opportunity to duplicate and verify the experiment initially conducted at IRIMS. I also want to thank Eduardo Flores and Ernst Knoesel for many delightful conversations and Rowan University's Department of Physics and Astronomy for their ready assistance and support.

Appendix A

The total probability of finding a photon with wave function $\Psi(x, y, z, t)$ somewhere in space is given by

$$\|\Psi(x,y,z,t)\|^2 = \int_{-\infty}^{\infty}\int_{-\infty}^{\infty}\int_{-\infty}^{\infty} |\psi(x,y,z,t)|^2 \, dx\,dy\,dz. \tag{A.1}$$

In this appendix, we use the *one-dimensional* notation $\Psi(x)$ for simplicity of argument without any loss of generality and use the equivalence of the classical notion of irradiance and quantum mechanical probability distribution such that we have

$$\Phi = \|\psi(x)\|^2 = \int_{-\infty}^{\infty} |\psi(x)|^2 \, dx = \int_{-\infty}^{\infty} I(x)dx, \tag{A.2}$$

where Φ is the total radiant flux, and $I(x)$ is the classical irradiance at position x. Due to the practical impossibility of scanning the entire space, we employ apodization in our experiment for the wave functions Ψ_1 and Ψ_2 so that only the maximal Airy disks are allowed to go through the aperture stop AS and the resulting apodized wave functions ψ_1 and ψ_2 emerge. These wave functions are bounded within $-s \le x \le s$, where $s = 3.833\, l\, \lambda/b$, l is the distance of plane σ_1 from the dual pinhole, and b is the diameter of each pinhole [25]. Therefore, we have

$$\psi_i(x) = \begin{cases} 0 & \text{for } x > s \\ \psi_i(x) & \text{for } -s \le x \le s \\ 0 & \text{for } x < -s \end{cases} \tag{A.3}$$

where $i = 1, 2$.

Bearing in mind that both ψ and the irradiance I are functions of x, for apodized wave functions, the total radiant flux in (A.2) is reduced to

$$\Phi_i = \|\psi_i\|^2 = \int_{-s}^{s} |\psi_i|^2 \, dx = \int_{-s}^{s} I_i \, dx. \tag{A.4}$$

References

1. N. Bohr, in *Albert Einstein: Philosopher-Scientist,* P. A. Schilpp, Ed. (Library of Living Philosophers, Evanston, IL, 1949).
2. N. Bohr, *Nature* 121, 580 (1928).
3. H. Rauch *et al., Phys. Lett. A* 54 (1975) 425.
4. S. Haroche, in: *High Resolution Laser Spectroscopy,* K. Shimoda, Ed. (Springer, New York, 1976).
5. X. Y. Zou, L. J. Wang and L. Mandel, *Phys. Rev. Lett.* 67 (1991) 318.
6. T. J. Herzog, P. G. Kwiat, H Weinfurter and A. Zeilinger, *Phys. Rev. Lett.* 75 (1995) 3034.
7. U. Eichmann *et al., Phys. Rev. Lett.* 70 (1993) 2359.
8. G. Badurek, H. Rauch, and D. Tuppinger, *Phys. Rev. A* 34 (1986) 2600.
9. S. Dürr, T. Nonn and G. Rempe, *Nature* 395, (1998) 33.
10. W. Wooters and W. Zurek, *Phys. Rev. D.* 19 (1979) 473.
11. R. Glauber, *Ann. (N.Y.) Acad. Sci.* 480 (1986) 336.
12. D. Greenberger and A. Yasin, *Phys. Lett. A* 128 (1988) 391.
13. L. Mandel, *Opt. Lett.* 16 (1991) 1882.
14. G. Jaeger, A. Shimony and L. Vaidman, *Phys. Rev. A* 51 (1995) 54.
15. B.-G. Englert, *Phys. Rev. Lett.* 77 (1996) 2154.
16. H. Rauch and J. Summhammer, *Phys. Lett. A* 104 (1984) 44.
17. J. Summhammer, H. Rauch, and D. Tuppinger, *Phys. Rev. A* 36 (1987) 4447.
18. P. Mittelstaedt, A. Prieur and R. Schieder, *Found. Phys.* 17 (1987).
19. F. De Martini *et al., Phys. Rev. A* 45 (1992) 5144.
20. G. Kar, A. Roy, S. Ghosh, and D. Sarkar, Los Alamos National Laboratory e-print (xxx.lanl.gov), quant-ph/9901026; S. Bandyopadhyay, *Phys. Lett. A* 276 (2000) 233.
21. W. Heisenberg, in *The Physical Principles of the Quantum Theory,* (Dover, New York, 1949).
22. V. B. Braginsky and F. Y. Khalili, in *Quantum Measurement,* K. S. Thorne, Ed. (Cambridge University Press, Cambridge, UK, 1992).
23. J. A. Wheeler, in *Mathematical Foundations of Quantum Theory,* A. R. Marlow, Ed. (Academic Press. 1978); *ibid,* in *Some Strangeness in the Proportion,* H. Woolf, Ed. (Addison-Wesley, Reading, MA, 1980).
24. R. P. Feynman, in: *The Feynman Lectures on Physics,* (Addison-Wesley, Reading, MA, 1963).
25. M. Born, E. Wolf, in: *Principles of Optics,* (Cambridge University Press, 1999).
26. B Thaller in: *Visual Quantum Mechanics,* (Springer, New York, 2000).
27. S. S. Afshar, http://irims.org/quant-ph/030503/, M. Chown in New Scientist, July 24, 2004.
28. X. Y. Zou, L. J. Wang, and L. Mandel, *Phys. Rev. Lett.* 67, (1991) 318.

13

The Bohr Model of the Photon

Geoffrey Hunter,[a] Marian Kowalski,[b] and Camil Alexandrescu[c]

CONTENTS

Abstract

The photon is modeled as a monochromatic solution of Maxwell's equations confined as a soliton wave by the principle of causality of special relativity. The soliton travels rectilinearly at the speed of light. The solution can represent any of the known polarization (spin) states of the photon. For circularly polarized states the soliton's envelope is a circular ellipsoid whose length is the observed wavelength (λ), and whose diameter is λ/π; this envelope contains the electromagnetic energy of the wave ($h\nu = hc/\lambda$). The predicted size and shape is confirmed by experimental measurements: of the sub-picosecond time delay of the photo-electric effect, of the attenuation of undiffracted transmission through slits narrower than the soliton's diameter of λ/π, and by the threshold intensity required for the onset of multiphoton absorption in focussed laser beams. Inside the envelope the

[a] Chemistry Department, York University, Toronto, Ontario M3J 1P3, Canada
[b] Optech, Inc., Toronto, Canada
[c] Physics Department, York University, Toronto, Ontario M3J 1P3 Canada

wave's amplitude increases linearly with the radial distance from the axis of propagation, being zero on the axis. Outside the envelope the wave is evanescent with an amplitude that decreases inversely with the radial distance from the axis. The evanescent wave is responsible for the observed double-slit interference phenomenon.

13.1 Introduction

The Bohr model of the photon was first published in 1989 [1]; here the theory and supporting experimental evidence presented in [1] are is summarized and augmented by recent developments. It is a Bohr model in the sense that it is a solution of the classical equations of motion that is subsequently quantized. In Bohr's well-known model of the hydrogen atom the classical equations are Newton's equations for the motion of an electron within the field of a proton, whereas for the photon (light regarded as electromagnetic radiation) the appropriate classical equations are Maxwell's equations in vacuum.

In Bohr's model of the hydrogen atom the quantization makes the angular momentum of the electron an integer multiple of Planck's constant, $\hbar = h/2\pi$. In the Bohr model of the photon the quantization of the photon's angular momentum arises as an appropriately chosen solution of Maxwell's equations; in addition, the energy of the oscillating electromagnetic field is quantized to be $h\nu$—the known energy of the photon; this energy quantization is actually generalized to be $nh\nu$ with $n > 1$ representing a multiphoton.

The solution of Maxwell's equations was chosen to be a monochromatic traveling wave having the observed angular momentum of the photon; i.e., a spin of $\pm\hbar$; constant parameters multiplying each of these spin states allows for representation of all the known polarization states of light.

The chosen solution of Maxwell's equations is confined within a finite space-time region by the principle of causality of Special Relativity; i.e., that causally related events must be separated by time-like intervals. With the idea that a photon is self-causing as it propagates, causality imposes the condition that events within the wave having the same phase must be separated by time-like intervals. In the limit where the interval becomes null (light-like), causality leads to the inference that the length of the photon along its axis of propagation is the wavelength, λ.[1] In addition, for circularly polarized states the causally connected field is contained within a circular ellipsoid with maximum diameter (transverse to the axis of propagation) of λ/π; the length of the ellipsoid (along the axis of propagation) is the wavelength.[2]

---

[1] Or equivalently in time, the period of oscillation $\tau = \nu{-}1$.

[2] The ellipsoidal soliton can be visualized as an egg, or as a rugby or American football.

This modeling of the photon as an ellipsoidal soliton arises from the imposition of causality upon the solution of Maxwell's equations (which are linear and homogeneous) whereas non-relativistic solitons arise as solutions of non-linear equations; this is considered further in §4.2.1.

Derivation of the size and shape of the soliton allowed for the quantization of the energy; the wave's electromagnetic energy, $\mathbf{E}^2 + \mathbf{H}^2$, was integrated over the volume of the ellipsoid and set equal to $h\nu$.[3] This determined the amplitude of the wave and led to an expression for the average intensity within the soliton [1, eqn.57]:

$$I_p = \frac{4\pi hc^2}{\lambda^4} \tag{13.1}$$

which we regard as the photon's intrinsic intensity.

13.2 Experimental Confirmation of Soliton

Several distinct experimental measurements confirm the predicted size, shape and intrinsic intensity of the photon:

- its length of λ is confirmed by:
  - the generation of laser pulses that are just a few periods long;
  - for the radiation from an atom to be monochromatic (as observed), the emission must take place within one period, τ, [2];
  - the sub-picosecond response time of the photoelectric effect [3];
- the diameter of λ/π is confirmed by:
  - the attenuation of transmission of circularly polarized light through slits narrower than λ/π: our own experiments with microwaves ([1, p.166]) confirmed this within the experimental error of 0.5%;
  - the resolving power of a microscope (with monochromatic light) being "a little less than a third of the wavelength" [4];
- The predicted intrinsic intensity (given by eqn. 13.1) is the threshold (minimum) intensity to which a laser beam must be focussed in order to produce multiphoton absorption: 2 distinct experimental confirmations of this are cited in [1, p.165].

[3] Or in general, to $nh\nu$.

13.3 Solution of Maxwell's Equations

Maxwell's equations [5] relate the first derivatives of the six components of the electromagnetic field; they comprise eight partial differential equations which must be satisfied simultaneously.[4] The key to the rather daunting task of finding appropriate solutions, is obtained by further differentiation to produce second derivatives followed by elimination of common terms between the resulting equations to yield the result that each Cartesian component of the field (E_x, E_y, E_z, H_x, H_y, H_z) separately satisfies d'Alembert's wave equation [5].[5]

For a wave traveling parallel to the z-axis at the speed of light, c, the solution must be any function of $z - ct$ [6], and if this wave is monochromatic the functional form is simply:

$$S(z - ct) = \exp\{i(z - ct)\} \tag{13.2}$$

When this form is assumed to be a factor of the solution, insertion into d'Alembert's equation causes a complete separation of z and t from the transverse coordinates ($x = r\cos\phi$, $y = r\sin\phi$),[6] plane polar coordinates (r, ϕ) being chosen in preference to the Cartesian coordinates (x, y) in view of the axial symmetry of the direction of propagation.

Separation of the radius, r, from the polar angle, ϕ, produces the two ordinary differential equations:

$$\frac{1}{\Phi(\phi)}\frac{d^2\Phi(\phi)}{d\phi^2} = m^2 = -\frac{1}{R(r)}\left\{\frac{d^2R(r)}{dr^2} + \frac{1}{r}\frac{dR(r)}{dr}\right\} \tag{13.3}$$

where m^2 is the real separation constant introduced to separate r from ϕ.

The simplest solution of eqn. 13.3 is the plane wave ($m^2 = 0$); i.e., $R(r)$ and $\Phi(\phi)$ both being constants.[7] However, this solution was rejected as unphysical because light is observed to travel along very narrow beams.[8]

The next simplest solution of eqn. 13.3 is for $m^2 = 1$: i.e., a factor of r or $1/r$, with an angular factor of $\exp\{i(\phi)\}$ or $\exp\{-i(\phi)\}$.

These angular factors are eigenfunctions of the z-component of angular momentum, $\mathbf{L_z} = \frac{\hbar}{i}\frac{\partial}{\partial\phi}$, in Schrödinger quantum mechanics [9, p.217], the eigenvalues of $\pm\hbar$ being those observed for the spin angular momentum of the photon; thus these solutions for $m^2 = 1$ are appropriate to represent

[4] The equations are linear and homogeneous with constant coefficients.

[5] The separate satisfaction of d'Alembert's wave equation only obtains for the *Cartesian* components of the field; it does not prevail for the spherical or cylindrical components.

[6] The separation is complete in the sense that there is no separation constant between the z, t and the r, ϕ differential equations.

[7] Plane waves are widely used in the quantum field theory of light [7, 8].

[8] A plane wave has field components that have the same value throughout any plane perpendicular to the axis of propagation, and thus it is completely non-localized.

the photon:

$$\psi(r, \phi, z - ct) = (\alpha r + \beta/r)\,(A \exp\{i\phi\} + B \exp\{-i\phi\}) \exp\{i(z - ct)\} \qquad (13.4)$$

Having determined this as the appropriate solution of d'Alembert's equation, each of the 6 field components (E_x, E_y, E_z, H_x, H_y, H_z) will have this form, the coefficients (α, β, A, B) being different in each component. The relationships between the coefficients of different components were determined by Maxwell's equations. This produced the results:

$$\begin{aligned} &E_z = H_z = 0 \\ &E_x = (\alpha r + \beta/r)\,(A \exp\{i\phi\} + B \exp\{-i\phi\}) \exp\{i(z - ct)\} = \mu_0 c H_y \qquad (13.5) \\ &E_y = i(\alpha r - \beta/r)\,(A \exp\{i\phi\} - B \exp\{-i\phi\}) \exp\{i(z - ct)\} = -\mu_0 c H_x \end{aligned}$$

Imposition of the causality condition led to the result that if A or B is zero, then the field must be contained within a circular ellipsoid of length λ and cross-sectional diameter λ/π [1, §2.5].

Since Maxwell's equations are linear and homogeneous they do not determine the amplitude of the solutions. Thus it was proposed to determine the amplitude by integration of the energy of the wave, $\mathbf{E}^2 + \mathbf{H}^2$.[9] This proposal led to the realization that the form $1/r$ would cause a divergent contribution to the energy at $r = 0$, while the form r would cause a similar divergence as $r \to \infty$. Thus, in view of the causality condition limiting the domain of the field to an ellipsoid along the axis of propagation, it was decided to discard the $1/r$ form and retain the r form in order to produce a finite integrated energy. This discarding of the $1/r$ term (i.e., $\beta = 0$ in eqn. 13.5) was concordant with the need to make the field an eigenfunction of L_z [1, §2.6].

This normalization of the amplitude of the photon's field yielded:[10]

$$A^2 + B^2 = 1 \quad \text{and} \quad \alpha^2 = 120nhc\pi^4/(\varepsilon_0\lambda^6) \qquad (13.6)$$

13.4 Soliton's Evanescent Wave

An evanescent wave outside the ellipsoid is necessary as an adjunct to the theory presented in [1], because while the relativistic principle of causality confines the wave within the ellipsoid, the radial dependence of the wave

[9] This is analogous with Bohr's quantization of the electron's angular momentum in his model of the hydrogen atom.

[10] In [1] the amplitude squared ($\alpha^2 = S_0^2$ in [1, eqn. 47]) was given as, $\alpha^2 = 64nhc\pi^4/(\varepsilon_0\lambda^6)$, which corresponds to integration over a cylinder (length λ and diameter λ/π) rather than the ellipsoid; the factor of 120 in eqn.(13.6) is correct for integration over the ellipsoid.

within the soliton is simply r, which is a maximum at the surface of the ellipsoid; physically the wave cannot sharply cut-off to zero at this surface; it must smoothly decay towards zero outside the ellipsoid; an evanescent wave will decay in this way.

The radial dependence of the evanescent wave is postulated to be $1/r$; i.e., the apposite solution of Maxwell's equations (eqn. 13.5) with $\alpha = 0$. The intensity of this wave decreases as $1/r^2$ as the radial distance, r, from the soliton increases.

J. J. Thomson derived the same solution (eqn. 13.5) of Maxwell's equations in 1924 [10]; he noted that a radial dependence of r is appropriate near $r = 0$, with $1/r$ being appropriate as $r \to \infty$, but he didn't pursue his analysis as far as deducing an ellipsoidal soliton, with the wave having the r form within the ellipsoid, and the $1/r$ form outside the ellipsoid.

The r dependence within the ellipsoid and the $1/r$ dependence outside the ellipsoid, makes the r-derivative of the wave discontinuous on the surface of the ellipsoid. While this may appear to be unphysical, it is the same discontinuity exhibited by the gravitational force due to the mass of the Earth: on the assumption of a uniform density, inside the Earth, the gravitational force is proportional to the radius, r, whereas outside the Earth it decreases like $1/r^2$ [11]. In reality the Earth's mass-density is greatest at its centre, while the mass-energy density of the photon-soliton is greatest just inside the surface of the ellipsoid at its maximum diameter (mid-way along its length) of λ/π; i.e. at $r = \lambda/(2\pi)$.

13.4.1 Matching Soliton and Evanescent Waves

While the gradient of the wave (w.r.t. r) has a cusp at $r = \lambda/(2\pi)$ (noted above), the amplitude must be continuous at $r = \lambda/(2\pi)$; this equating of the soliton and evanescent wave amplitudes at $r = \lambda/(2\pi)$ is expressed by:

$$\alpha r = \beta/r \qquad \text{for} \qquad r = \lambda/(2\pi) \tag{13.7}$$

and since α^2 is given by eqn. 13.6 it follows that:

$$\beta^2 = [\lambda/(2\pi)]^4 \times 120nhc\pi^4/(e_0\lambda^6) = 7.5nhc/(\varepsilon_0\lambda^2) \tag{13.8}$$

13.4.1.1 Orthogonality of Radial Gradients

The radial gradient of the soliton wave is simply the normalization constant, a, while that of the evanescent wave is $-\beta/r^2$. Thus at the cusp where the two waves join (at $r = \lambda/(2\pi)$) the ratio of these gradients is:

$$\text{ratio of gradients} = -\frac{\beta}{\alpha r^2} = -1 \;\; \text{at} \;\; r = \lambda/(2\pi) \tag{13.9}$$

Thus where the soliton and evanescent waves meet (at $r = \lambda/(2\pi)$) they are orthogonal to each other—independent of the wavelength, λ.

13.4.2 Evanescent Wave Characteristics

The polar components of the evanescent field are given by eqn. 13.8 of [1] for $\alpha = 0$ and β given by eqn. 13.8, which show that none of these components have any dependence upon the polar angle (ϕ), and that E_r and H_ϕ are real, while H_r and E_ϕ are imaginary:

$$E_r = \frac{\beta}{r}[A+B] = \mu_0 cH_\phi \quad E_\phi = -i\frac{\beta}{r}[A-B] = -\mu_0 cH_r \tag{13.10}$$

This independence of the angle, ϕ, means that the evanescent wave carries none of the angular momentum of the photon,[11] and hence none of the photon's energy; it is a truly evanescent wave [12].

13.4.2.1 Caveat

The matching of the soliton and evanescent waves in §4.1 was made at the soliton's maximum diameter of λ/π; this raises the question of their matching at values of z other than $z = 0$; i.e., at other points on the ellipse:

$$(2\pi r)^2 + (2z)^2 = \lambda^2$$

$$\text{i.e. when} \quad r = \frac{1}{2\pi}\sqrt{(\lambda)^2 - (2z)^2} \quad \text{for} \quad -\frac{\lambda}{2} < z < +\frac{\lambda}{2} \tag{13.11}$$

It might appear natural to apply the matching condition of eqn. 13.7 for all values of r specified in eqn. 13.11 to produce:

$$\begin{aligned}\beta^2 &= \left[\frac{1}{2\pi}\sqrt{(\lambda)^2 - (2z)^2}\right]^4 \times 120 nhc\pi^4/(\epsilon_0\lambda^6) \\ &= [(\lambda)^2 - (2z)^2]^2 \times 7.5 nhc/(\epsilon_0\lambda^6)\end{aligned} \tag{13.12}$$

This would have the effect of making the amplitude of the evanescent wave, β, become smaller as z changes from $z = 0$ towards $z = \pm\frac{\lambda}{2}$, with β actually being zero at these limits (the ends of the ellipsoid). Physically this is what would be expected.

However, this conjecture would make β a function of z (as in eqn. 13.12) rather than a constant, and hence the evanescent field (eqn. 13.5 for $\alpha = 0$, $\beta \neq 0$)

---

[11] Because the operator for the z-component of angular momentum is $\mathbf{L}_z = \frac{\hbar}{i}\frac{\partial}{\partial\phi}$.

would no longer be a solution of Maxwell's equations, but rather of some similar, non-linear equations. The resolution of this physical vs. mathematical paradox may be found within the framework of General Relativity, in which the photon's local energy produces a non-Lorentzian metric.

13.4.3 Diffraction and Interference

The evanescent wave is believed to be responsible for the phenomena of diffraction and interference. As a photon-soliton passes close to the edge of, or through a slit in, a material obstacle placed within the beam of light, the interaction between the electrons within the obstacle and the photon's evanescent wave will cause its path to bend as it passes by, the angle of bending (diffraction) being dependent upon the impact parameter of the soliton's axis with the edge or slit.

Double slit interference can be understood by the soliton itself (like the C_{60} molecules in Zeilinger's experiment [13]) going through one slit or the other, while its evanescent wave extends over both slits. The evanescent wave is like a classical continuous wave in extending throughout all space, and hence the interference minima and maxima will appear at the same positions as predicted by Huygens' theory. However, the soliton model predicts that:

- the individual photons will arrive at local positions in the detection plane, whereas the classical continuous wave model predicts a uniformly visible interference pattern: that the former (rather than the latter) is actually observed supports the soliton model [13];
- the visibility of the interference pattern[12] will decrease with slit separation (because the intensity of the evanescent wave decreases like r^{-2}, r being the distance from the soliton's axis of propagation), whereas the classical continuous wave model predicts a visibility independent of slit separation. This seems not to have been investigated experimentally.

A double-slit experiment by Alkon [14] exhibits the expected interference pattern even though the individual photons are constrained to pass through one slit or the other by an opaque barrier extending from the source (a laser) up to the mid-point between the slits.[13] This experiment demonstrates that the particle-like photon (the Bohr model soliton) passes through one slit or the other, and yet its passage through this slit (and the subsequent diffraction) is affected by the presence of the other slit; this effect of the other open

[12] Visibility, V, is defined by: $V = \frac{I_{max} - I_{min}}{I_{max} + I_{min}}$, I_{max} and I_{min} being the measured intensities at the interference maxima and minima respectively; it has the range: $0 \leq V \leq 1$.

[13] Alkon's experiment is the experimental proof that the continuous wave concept that "the photon goes through both slits and interferes with itself" is not correct.

slit is evidence for the existence of the evanescent wave surrounding the soliton.[14]

A causal model of diffraction has been proposed by Gryzinski [16]; it is based upon the photon being a particle-like (localized) electromagnetic wave that interacts with the array of positive atomic nuclei and negative electrons within a solid, as it passes:

- through a crystal (Bragg diffraction of X-rays), or
- adjacent to an edge of a sheet of the solid (an edge of a slit).

Gryzinski's model of diffraction does not specify (does not need to specify) the size or shape of the soliton, but it quantitatively explains both Bragg diffraction and double-slit interference; his concept of the latter is that while the localized photon goes through one slit, its wave extends to the other slit. His theory is concordant with the Bohr model's evanescent wave, specifically because his localized model involves the concept that "the photon's electric field decreases when distance [from its center] increases".

Gryzinski pertinently cites Zeilinger's observation that each photon manifests its particle (localized) nature in each detection event: the distribution of detection events[15] only becomes manifest after a large number ($\geq 10^4$) of detection events have been recorded [13]; each photon detection is a localized event.

The evanescent wave explanation for diffraction and interference is not readily invoked for the Mach-Zender type of interferometer, because the two alternative paths for the photon are typically separated by distances over which the evanescent wave's intensity would have become negligible; a small difference (of the order of the wavelength) between the lengths of the two paths determines the observed interference pattern. However, just as has already been proven for diffractive "interference" (discussed above), the continuous wave concept that the wave goes along both paths of the interferometer and interferes with itself, is unlikely to be the true explanation for Mach-Zender interferometry.

Acknowledgments

A discussion between the author and Chandra Roychoudhuri at the 2005 Quantum Optics conference (Snowbird, Utah) initiated this presentation.

[14] Interaction between the evanescent waves of collaterally moving photon-solitons could be the cause of the very small (but finite) divergence of a laser beam [15, p.6].

[15] Attributed in the continuous wave model to the wave going through both slits and self-interfering.

References

[1] Geoffrey Hunter and Robert L.P. Wadlinger, Photons and Neutrinos as Electromagnetic Solitons, *Phys Essays*, 2, 158–172 (1989).

[2] Marian Kowalski, Photon Emission from Atomic Hydrogen, *Phys Essays*, 12, 312–331 (1999).

[3] M.C. Downer, W.M. Wood, and J.I. Trisnadi Comment on Energy Conservation in the Picosecond and Subpicosecond Photoelectric Effect, *Phys.Rev.Lett.* **65**, 2832 (1990).

[4] S.G. Starling and A.J. Woodall, *Physics*, second edition, Longmans Green and Co., London, UK, 1957, p. 706.

[5] J.H. Fewkes and J. Yarwood, *Electricity and Magnetism*, University Tutorial Press, London 1956, p. 509.

[6] C.A. Coulson, *Waves* , 7th edition. Oliver and Boyd, London, 1955, pp. 103,111.

[7] W. Heitler, *Quantum Theory of Radiation*, Oxford University Press, Oxford, UK, 1954.

[8] Marlan O. Scully and M. Suhail Zubairy, *Quantum Optics*, Cambridge University Press, Cambridge, UK, 1997.

[9] D.A. MacQuarrie, *Quantum Chemistry*, University Science Books, Sausalito, CA (1983).

[10] J.J. Thomson, *Philos.Mag Ser.* 6, **48**, 737 (1924); *Nature.* **137**, 23 (1936).

[11] http://abel.math.harvard.edu/~knill/math21a/hell.pdf

[12] M.V. Berry Evanescent and Real Waves in Quantum Billiards and Gaussian Beams, *J.Phys.A.* **27** L391–L398 (1994).

[13] Olaf Nairz, Markus Arndt, and Anton Zeilinger, Quantum Interference Experiments with Large Molecules, *Am.J.Phys*, **71**, 319–325, (2003).

[14] Daniel L. Alkon, *"Either-Or"* Two-Slit Interference: Stable Coherent Propagation of Individual Photons Through Separate Slits, *Biophys J*, **80**, 2056–2061 (2001).

[15] T.P. Softley, *Atomic Spectra*, Oxford University Press, Oxford, UK, 1994.

[16] M. Gryzinski, Spin-Dynamical Theory of the Wave-Corpuscular Duality, *Int., J. Theoretical Phys*, **26**, 967–979 (1987); http://www.iea.cyf.gov.pl/gryzinski/teor1ang.html

14

The Maxwell Wave Function of the Photon

M. G. Raymer and Brian J. Smith
Oregon Center for Optics and Department of Physics, University of Oregon, Eugene, Oregon 97403, USA

CONTENTS

Abstract

James Clerk Maxwell unknowingly discovered a correct relativistic, quantum theory for the light quantum, forty-three years before Einstein postulated the photon's existence. In this theory, the usual Maxwell field is the quantum wave function for a single photon. When the non-operator Maxwell field of a single photon is second quantized, the standard Dirac theory of quantum optics is obtained. Recently, quantum-state tomography has been applied to experimentally determine photon wave functions.

Key words: photon, wave function, Wigner function.

> "But to determine more absolutely what light is, after what manner refracted, & by what modes or actions it produceth in our minds the Phantasms of colours, is not so easie. And I shall not mingle conjectures with certaintyes."*
>
> — **Isaac Newton**

* A Theory Concerning Light and Colors, *Cambridge University Library Add MS 3970.3 ff. 460-66, http://www.newtonproject.sussex.ac.uk/prism.php?id=1*

14.1 Maxwell Photon Wave Function

In about 1862, James Clerk Maxwell determined mathematically from his then-new equations, that electromagnetic waves travel at a speed very nearly equal to the measured value of the speed of light. In 1864 he wrote [1],

> "This velocity is so nearly that of light that it seems we have strong reason to conclude that light itself (including radiant heat and other radiations) is an electromagnetic disturbance in the form of waves propagated through the electromagnetic field according to electromagnetic laws."

In 1862 he wrote in On Physical Lines of Force [1],

> "We can scarcely avoid the inference that light consists in the traverse undulations of the same medium which is the cause of electric and magnetic phenomena."

Maxwell's equations are, for a source-free region of space (in Gaussian units),

$$\begin{aligned} &\frac{\partial}{\partial t}\vec{E}(\vec{r},t) = c\vec{\nabla}\times\vec{B}(\vec{r},t), \quad \frac{\partial}{\partial t}\vec{B}(\vec{r},t) = -c\vec{\nabla}\times\vec{E}(\vec{r},t) \\ &\vec{\nabla}\cdot\vec{E}(\vec{r},t) = 0, \quad \vec{\nabla}\cdot\vec{B}(\vec{r},t) = 0. \end{aligned} \tag{14.1}$$

Max Planck said [1], on the centenary of Maxwell's birth in 1931, that Maxwell's theory "... remains for all time one of the greatest triumphs of human intellectual endeavor."

Planck was correct—even more so than he realized. For, just a year earlier, in 1930, Paul Dirac had shown the way to formulate dynamical equations for relativistic elementary particles. It is now understood that Dirac's particle approach, when applied to massless spin-one particles, leads directly to Maxwell's equations. This means that Maxwell unknowingly discovered a correct relativistic, quantum theory for the light quantum, forty-three years before Einstein postulated the photon's existence! In this theory, the (non-operator) Maxwell field is the quantum wave function for a single photon. When the non-operator Maxwell field of a single photon is quantized, the standard Dirac theory of quantum optics is obtained.

Here we review the derivation of Maxwell's equations from relativistic, quantum particle dynamics, which in recent times was expounded on in detail by Bialynicki-Birula [2] and by Sipe [3], and later by Kobe [4]. We follow [2] and [3], while trying to present a simpler version of the derivation.

In modern terms, a photon is an elementary excitation of the quantized electromagnetic field. If it is known a priori that only one such excitation

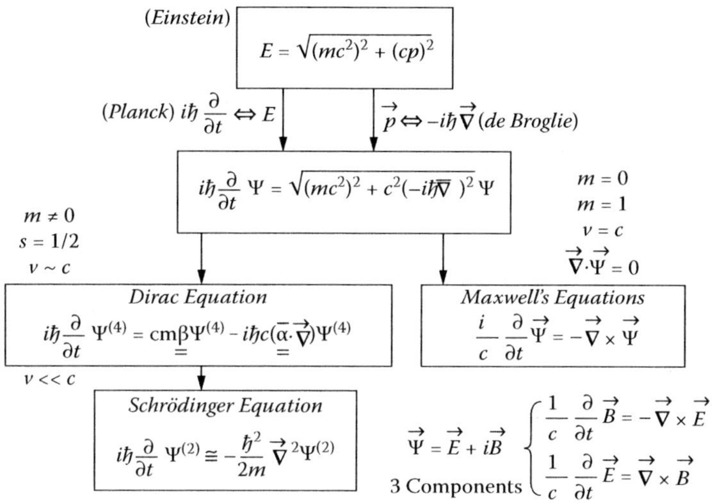

FIGURE 14.1
Flow chart for derivations of electron and photon wave equations, m = rest mass, s = spin, v = velocity.

exists, it can be treated as a (quasi-) particle, roughly analogous to an electron. It has unique properties, arising from its zero rest mass and its spin-one nature. In particular, there is no position operator for a photon, leading some to conclude that there can be no properly defined wave function, in the Schroedinger sense, which allows localizing the particle to a point. On the other hand, it is known that even electrons, when relativistic, don't have properly defined wave functions in the Schroedinger sense [2,3], and this opens our minds to broader definitions of wave functions. In relativistic quantum theory, one distinguishes between charge-density amplitudes, mass-density amplitudes, and particle-number density amplitudes. These can have different localization properties. Photons, of course, are inherently relativistic, so it is not surprising that we need to be careful about defining their wave functions.

Dirac's theory of a particle is based on the kinematic equation for energy E, momentum $\vec{p} = (p_x, p_y, p_z)$, and rest mass, m,

$$E = \sqrt{(mc^2)^2 + c^2 \vec{p} \cdot \vec{p}}. \tag{14.2}$$

Define a multicomponent amplitude function $\tilde{\psi}(\vec{p}, E)$ obeying the normalization condition

$$(2\pi\hbar)^{-3} \int d^3p \, \tilde{\psi}^*(\vec{p}, E) \cdot \tilde{\psi}(\vec{p}, E) = 1, \tag{14.3}$$

where the dot indicates a vector dot product. Multiplying Eq. 14.2 by this function gives

$$E\tilde{\psi}(\vec{p},E) = \sqrt{(mc^2)^2 + c^2\vec{p}\cdot\vec{p}}\,\tilde{\psi}(\vec{p},E). \tag{14.4}$$

For a spin-one-half particle with non-zero rest mass, this equation gives, upon recognizing that the wave function $\tilde{\psi}(\vec{p},E)$ must have two components (for the positive-energy solutions), and transforming to space-time variables $\vec{r},t$, the Dirac equation for the electron. The electron wave function is a two-spinor, $\psi(\vec{r},t) = (\psi_{1/2}, \psi_{-1/2})$, where the components $\psi_{\pm 1/2}$ are amplitudes for the states of plus- and minus-1/2 spin-projection onto the quantization axis. In this case, Eq. (14.3) represents normalization of the probability to find particular values of the electron's momentum.

Because a photon is massless, its wave function should obey

$$E\tilde{\psi}(\vec{p},E) = c\sqrt{\vec{p}\cdot\vec{p}}\,\tilde{\psi}(\vec{p},E), \tag{14.5}$$

Since the photon is a spin-one particle, its wave function should have three components, forming a (non-operator) three-component vector field $\tilde{\psi}(\vec{p},E) = (\tilde{\psi}_x, \tilde{\psi}_y, \tilde{\psi}_z)$. To represent the square-root operator $\sqrt{\vec{p}\cdot\vec{p}}$ we look for a vector operator $\hat{A}$ with the property $(\hat{A})^2\tilde{\psi} = \vec{p}\cdot\vec{p}\tilde{\psi}$. Such an operator can be found by elementary means, by trying $\hat{A} = i\vec{p}\times$, where $\times$ is the cross-product operator. Then a well-known vector identity gives

$$\hat{A}\hat{A}\tilde{\psi} = -\vec{p}\times(\vec{p}\times\tilde{\psi}) = (\vec{p}\cdot\vec{p})\tilde{\psi} - \vec{p}(\vec{p}\cdot\tilde{\psi}). \tag{14.6}$$

Any vector field can be written as the sum of two linearly independent parts, $\tilde{\psi} = \tilde{\psi}_T + \tilde{\psi}_L$, where the transverse part obeys $\vec{p}\cdot\tilde{\psi}_T = 0$, and the longitudinal part obeys $\vec{p}\times\tilde{\psi}_L = 0$. Identifying the transverse part as the relevant field for the photon, we derive the equivalent of Eq. 14.5,

$$E\tilde{\psi}_T(\vec{p},E) = c\, i\vec{p}\times\tilde{\psi}_T(\vec{p},E). \tag{14.7}$$

This deceptively simple-looking equation is actually equivalent to Maxwell's equations. To see this, first note that $\tilde{\psi}_T$ must be a complex-valued vector if Eq. 14.7 is to be satisfied. Next, Fourier transform the amplitude function $\tilde{\psi}(\vec{p},E)$ from momentum space to coordinate space, and from energy to time, accounting for the constraint between energy and momentum $(E = c\,|\vec{p}|)$ by including a delta function. This allows E to be considered as an independent variable, and gives

$$\tilde{\psi}(\vec{r},t) = (2\pi\hbar)^{-3}\iint dE\,d^3p\;\delta(E - c|\vec{p}|)\exp(-iEt/\hbar + i\vec{p}\cdot\vec{r}/\hbar)\,f(E)\tilde{\psi}(\vec{p},E). \tag{14.8}$$

The momentum-space weight function $f(E)$ has been included to allow different forms of normalization of the coordinate-space function $\tilde{\psi}_T(\vec{r},t)$.

(In the case of the electron, discussed above, the standard choice is $f(E) = 1$.) For the photon, we adopt the choice advocated by Sipe [3], $f(E) = \sqrt{E}$, which gives for the coordinate-space normalization,

$$\int d^3r \vec{\psi}(\vec{r}, t)^* \cdot \vec{\psi}(\vec{r}, t) = (2\pi\hbar)^{-3} \int d^3p\, E(p)\tilde{\psi}^*(\vec{p}, E(p)) \cdot \tilde{\psi}(\vec{p}, E(p)) = \langle E \rangle. \quad (14.9)$$

where we defined $E(p) = c\,|\vec{p}|$, and $\langle E \rangle$ denotes the expectation value of the photon's energy. This choice of normalization reflects the fact that a photon has no mass that can be localized at a point; rather it has only helicity and energy, and the energy cannot strictly be localized at a point. The function $(\vec{\psi}(\vec{r}, t)^* \cdot \vec{\psi}(\vec{r}, t))/\langle E \rangle$ is the probability density for energy, not particle location [2,3].

Equations 14.7 and 14.8 together give the "complex Maxwell equations,"

$$i\frac{\partial}{\partial t}\vec{\psi}_T(\vec{r}, t) = c\vec{\nabla} \times \vec{\psi}_T(\vec{r}, t). \quad (14.10)$$

Notice that $\hbar$ acts only as a scaling factor in the Fourier transform functions, and cancels in Eq. 14.10. Also note that we did not have to postulate the de Broglie relation, $\vec{p} = -i\hbar\vec{\nabla}$; rather it emerges naturally from the Fourier transform. Further note that the transverse part of the field defined in Eq. 14.8 has zero divergence, $\vec{\nabla} \cdot \vec{\psi}_T = 0$, and the longitudinal part has zero curl, $\vec{\nabla} \times \vec{\psi}_L = 0$.

Now write the complex wave function as a sum of real and imaginary parts $\vec{E}_T(\vec{r})$ and $\vec{B}_T(\vec{r})$.

$$\vec{\psi}_T(\vec{r}, t) = 2^{-1/2}(\vec{E}_T(\vec{r}, t) + i\vec{B}_T(\vec{r}, t)). \quad (14.11)$$

Using Eq. 14.10, the real and imaginary parts $\vec{E}_T(\vec{r},t)$ and $\vec{B}_T(\vec{r},t)$ are found to obey Maxwell's equations, Eq. 14.1. Therefore, to paraphrase Maxwell's quote above, we can scarcely avoid the inference that the photon's quantum wave function consists in the traverse undulations of the same medium which is the cause of electric and magnetic phenomena. That is, the classical Maxwell equations are the wave equation for the quantum wave function $\vec{\psi}_T$ of a photon. Evidently, the longitudinal part of the $\vec{\psi}$ function corresponds to longitudinal electric and magnetic fields, which are non-propagating.

As a check, calculate the space normalization integral,

$$\int d^3r \vec{\psi}(\vec{r}, t)^* \cdot \vec{\psi}(\vec{r}, t) = \int d^3r \frac{1}{2}(\vec{E}_T \cdot \vec{E}_T + \vec{B}_T \cdot \vec{B}_T) = \langle E \rangle, \quad (14.12)$$

which has the proper meaning that $\frac{1}{2}(\vec{E}_T \cdot \vec{E}_T + \vec{B}_T \cdot \vec{B}_T)$ is the local energy density.

The above derivation is for a particular helicity (handedness) of the photon angular momentum. The opposite helicity is described by changing Eq. 14.11 to $\vec{\psi}_T(\vec{r},t) = 2^{-1/2}(\vec{E}_T(\vec{r},t) - i\vec{B}_T(\vec{r},t))$, and multiplying the right-hand side of Eq. 14.10 by –1.

14.2 Measuring the Maxwell Photon Wave Function

If a single-photon state of the electromagnetic field is created, then to know its quantum state means to know its electric and magnetic field distributions in space and time. Such a state is a single-photon wave-packet state, and its generation is an important goal in quantum-information research.

Recently, a technique has been developed to measure the transverse spatial quantum state of an ensemble of identically prepared photons [5, 6]. The single-photon light beam is sent into an all-reflecting, out-of-plane Sagnac interferometer, which performs a relative rotation of 180° and a mirror inversion on the wave fronts of the counter-propagating beams. The Sagnac performs a two-dimensional parity operation on one of the beams relative to the other. The fields are recombined at the output beam splitter and are interfered on a photon-counting photomultiplier tube (PMT), allowing the emerging beams to be detected at the single-photon level. The mean photocount rate is directly proportional to the transverse spatial Wigner function at a phase-space point that is set by the tilt and translation of a mirror external to the interferometer.

The situation becomes even more interesting when the joint spatial wave function of a pair of photons is considered. In the case that the two photons' spatial and momentum variables are described by an entangled state, such a state measurement will provide the maximal-information characterization of the entanglement. By sending two entangled photons into two parity-inverting interferometers, one can measure the joint two-photon transverse spatial Wigner function, and completely characterize the transverse entanglement of this system [5,6]. The two-photon wave function exists in six spatial dimensions, and its equation of motion can be called the two-photon Maxwell equations.

To conclude, the usual (classical) Maxwell field is the quantum wave function for a single photon. That it transforms like a three-dimensional vector arises from the spin-one nature of the photon. (In contrast, the electron transforms like a two-dimensional spinor.) When two photons are present, the joint wave function "lives" in a higher dimensional space. These observations imply the interpretation of the Maxwell field as akin to the Schrödinger wave function, which evolves probability amplitudes for various possible quantum events in which the electron's position is found to be within a certain volume, rather than being a realistic description of the electron as being here or there. In this sense, the Maxwell equation evolves the probability amplitudes for various possible quantum events in which the photon's energy is found within a certain volume. In addition, quantum-state tomography methods have been devised for determining spatial states of one- and two-photon fields.

Note: We have reviewed in detail the treatment given in [7] and have extended it to the case of two photons, [8], and many photons, [7].

Acknowledgments

We thank Cody Leary for helpful discussions about field theory. This research was supported by the National Science Foundation's ITR Program, Grant 0219460.

References

1. From the web site Light through the ages: Relativity and quantum era, http://www-groups.dcs.st-and.ac.uk/~history/HistTopics/Light_2.html
2. I. Bialynicki-Birula, *Acta Phys. Pol.* **34**, 845 (1995); and "Photon wave function," in *Progress in Optics XXXVI*, E. Wolf, ed. (Elsevier, Amsterdam, 1996); and *Phys. Rev. Lett.* **80**, 5247 (1998).
3. J. E. *Sipe, Phys. Rev. A* **52**, 1875–1883 (1995).
4. D. H. Kobe, *Found. Phys*, **29**, 1203 (1999).
5. E. Mukamel, K. Banaszek, I. A. Walmsley and C. Dorrer, *Opt. Lett.* **28**, 1317–1319 (2003).
6. Brian J. Smith, Bryan Killett, Andrew Nahlik, M. G. Raymer, K. Banaszek and I. A. Walmsley "Measurement of the transverse spatial quantum state of light at the single-photon level," *Opt. Lett.* **30**, 3365–3367 (2005).
7. Brian J. Smith and M. G. Raymer, "Photon wave functions, wave-packet quantization of light, and coherence theory, *J. New Phys.* 9, 414, (2007).
8. Brian J. Smith and M. G. Raymer, "Two-photon wave mechanics," *Phys. Rev. A*, 74, 062104 (2006).

15

Modeling Light Entangled in Polarization and Frequency: Case Study in Quantum Cryptography

John M. Myers
Gordon McKay Laboratory, Harvard University, Cambridge, MA 02138, USA

CONTENTS

Abstract

With the recognition of a logical gap between experiments and equations of quantum mechanics comes: (1) a chance to clarify such purely mathematical entities as probabilities, density operators, and partial traces—separated from the choices and judgments necessary to apply them to describing experiments with devices, and (2) an added freedom to invent equations by which to model devices, stemming from the corresponding freedom in interpreting how these equations connect to experiments.

Here I apply a few of these clarifications and freedoms to model polarization-entangled light pulses called for in quantum key distribution (QKD). Available light pulses are entangled not only in polarization but also in frequency. Although absent from the simplified models that initiated QKD, the degree of frequency entanglement of polarization-entangled light pulses is shown to affect the amount of key that can be distilled from raw light signals, in one case by a factor of 4/3.

Open questions remain, because QKD brings concepts of quantum decision theory, such as measures of distinguishability, mostly worked out in the context of finite-dimensional vector spaces, into contact with infinite-dimensional Hilbert spaces needed to give expression to optical frequency spectra.

Key words: Quantum cryptography, polarization entanglement, frequency spectrum.

15.1 Introduction

In physics, every now and then some big shift in theory happens (think Planck) or some big invention in devices changes the landscape. A striking feature is that a revolution on the blackboard of theory can leave lenses on the optics bench unchanged, and, similarly, a new light source need not change a theory. Blackboard and bench have a certain independence, as everybody knows. What is not so well known is that this independence is no flaw in the current practicalities, but is a feature of quantum mechanics.

Although quantum states nicely express interference effects, outcomes of experimental trials show no states directly; they indicate properties of probability distributions for outcomes. In a previous paper,[1] it is proved categorically that probability distributions leave open a choice of quantum states and operators and particles, resolvable only by a move beyond logic, which, inspired or not, can be characterized as a guess. In contrast to any hope for a seamless, unique blackboard description of devices on a laboratory bench, no matter what experimental trials are made, if a quantum model generates calculated probabilities that match given experimentally determined relative frequencies, there are other quantum models that match as well but that differ in their predictions for experiments not yet performed.

That means that quantum physics stands not only on "serious, careful experimentation and analysis,"[2] but also on a third leg of irreducible improvisation and guesswork, needed to link the experimentation to the analysis. With the recognition of a logical gap between experiments and equations of quantum mechanics, come two areas of opportunity: (1) a chance to clarify such purely mathematical entities as probabilities, density operators, and partial traces—separated out from the choices and judgments necessary to apply them to describing experiments with devices, and (2) a certain freedom to invent equations by which to model devices, stemming from the corresponding freedom in interpreting how these equations connect to experiments.

Section 2 of this report exploits these opportunities to develop equations by which to model faint light and its detection in quantum key distribution (QKD). This story is a drama of negotiating between concepts of quantum decision theory, such as distinguishability, mostly worked out in the context of finite-dimensional vector spaces, and the concept of an infinite-dimensional Hilbert space needed to give expression to optical frequency spectra.

Section 3 steps back to offer a perspective, emerging from the study presented in Section 2, that makes probability distributions the central mathematical objects of interest, pushing into a subsidiary role the various entropies and information measures that can be used to prove theorems about these distributions. Open questions are posed.

15.2 Frequency Spectra of Light in Quantum Key Distribution

A collaboration of BBN Technologies, Boston University, and Harvard University is fielding several varieties of quantum key distribution (QKD) over a fiber-optic network. For each link of the QKD network, the system distributes a crypto-graphic key that the users, say Alice and Bob, share. The objective is to provide a high level of security for Alice and Bob against undetected eavesdropping attacks. Part of the work has been to choose equations by which to model QKD that uses the BB84 protocol[3] with polarization-entangled light pulses.

Assuming, among other things, the imperfect distinguishability of certain light states from others, various simplified models of QKD suggest how some classes of eavesdropping attacks disturb the key in ways that Alice and Bob can detect. These models express the degree of ignorance of an eavesdropper in relation to eavesdropping-induced disturbances in detection probabilities for Alice and Bob.[4] Taken from quantum decision theory, the measure of distinguishability of any two quantum states is their trace distance, and if trace distance increases, security drops.

Available polarization-entangled light pulses (produced by parametric down conversion) are entangled not only in polarization but also in *frequency*, and they act not as what are called "single-photon states" but include also "multi-photon components," and hence involve a "mean photon number μ." John Schlafer of BBN Technologies asked me how increasing μ contributes to errors in the key. The first step toward an answer was to choose a definition. For purposes of answering John's question, I defined "photon number" and "mean photon number" mathematically, in terms of weighted integrals over products of creation operators.[5] As a step toward an answer, I wanted to find out: how do the trace distances between QKD light states (which affect QKD error probabilities), depend on the mean photon number μ, and how is this dependence on μ modulated by the degree of *frequency* entanglement?

To find this, I had to choose a QKD system model, starting by laying out what is assumed.

15.2.1 Simplifying Assumptions

To simplify the discussion, I make the (dubious) assumptions that

1. The eavesdropper refrains from modifying Alice's and Bob's devices, for which kindness I call her Evangeline for "angelic Eve".[6]
2. Evangeline limits her attacks to one bit at a time (individual attacks).
3. Memory effects in detectors (and all other devices) are negligible.

With these assumptions, I model each attempt to generate a single raw key bit as a *trial* (in contrast to joint attacks, for which the transmission of a string of raw bits would be modeled by a single trial).

15.2.2 System Description

Consider a source of polarization-entangled light pulses (Fig. 15.1) from which light propagates along single-mode fiber to Alice and to Bob, assuming that Alice controls the source of entangled light pulses, so that only the fiber from the source to Bob is open to eavesdropping. Alice and Bob each have four detectors. I ignore memory and dead time in detectors to view each detector as responding to each trial with a 0 (no detection) or a 1 (detection). The four detectors for Alice and the four for Bob are polarized at angles 1 through 4, corresponding to 45° increments, as shown in Fig. 15.2. Any combination of Alice's detectors can fire, so there are sixteen possibilities for Alice's component of the measurement event. If one and only one of Alice's detectors fires, we code her event component i by the corresponding polarization label

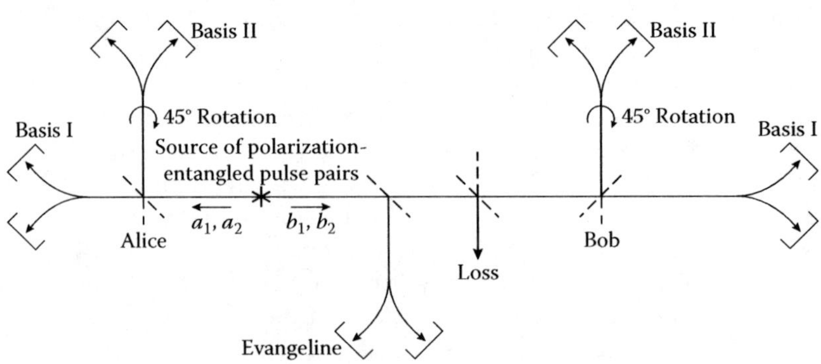

FIGURE 15.1
Polarization-entangled QKD system subjected to eavesdropping attack.

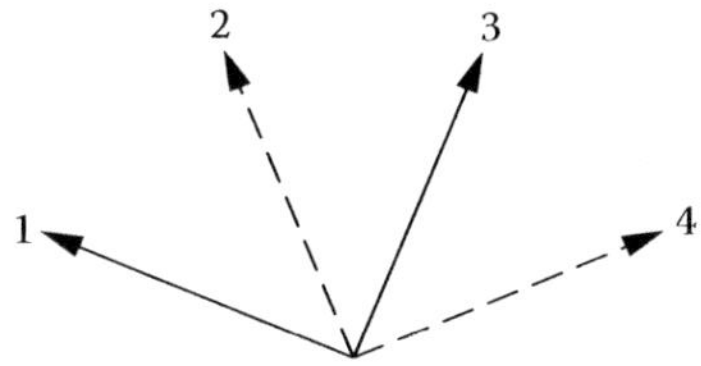

FIGURE 15.2
Four polarizations used in BB84. Basis I: solid. Basis II: dashed.

1, ... , 4; otherwise we code her measurement event component by some integer greater than 4. The same holds for Bob's event component j.

In analogy with the form of BB84 in which Alice transmits light to Bob (no entangled source),[7,8] polarizations 1 and 3 are called "Basis I" and polarizations 2 and 4 are called "Basis II." Following Ref. 7, I limit my analysis to the probabilities pertaining to sifted bits, assuming the following sifting rule: sift out trials except those for which exactly one of Alice's detectors and exactly one of Bob's detectors register a detection and the bases match.

15.2.3 Form of Model

By definition, quantum modeling invokes equations expressing probabilities of measurement events for trials, with the probabilities expressed by the trace rule applied to appropriate operators on a Hilbert space.[9] For a Hilbert space $\mathcal{H}$, let $\mathcal{B}(\mathcal{H})$ denote the set of bounded operators on $\mathcal{H}$. A trial consists of "preparing a state" as expressed by some density operator $\rho \in \mathcal{B}(\mathcal{H})$ and "measuring a state" as expressed by some positive operator-valued measure (POVM) M; these can be interspersed by a temporal-evolution expressed by a unitary operator U. The probability of a measurement event X is then

$$\Pr(X) = \mathrm{Tr}[U \rho U^{\dagger} M(X)]. \tag{15.1}$$

For entangled-state BB84, I take the Hilbert space to be the tensor product space of three factors, each infinite-dimensional: $\mathcal{H}_E \otimes \mathcal{H}_B \otimes \mathcal{H}_A$, where $\mathcal{H}_A$ is the Hilbert space for light detected by Alice, using a POVM M_A. At each trial Evangeline prepares a probe state $\rho_E \in \mathcal{B}(\mathcal{H}_E)$ and the entangled light source prepares a fixed entangled state ρ_{BA} on $\mathcal{H}_B \otimes \mathcal{H}_A$. The total prepared state is $\rho_E \otimes \rho_{BA}$. The eavesdropping interaction by which Evangeline probes light propagating to Bob is modeled by U_{EB} acting on $\mathcal{H}_E \otimes \mathcal{H}_B$. There are three components of a measurement event, corresponding to a tensor product of three POVM's, M_E for Evangeline, M_B for Bob, and M_A for Alice.

For a trial in which Alice and Bob match in bases, Evangeline selects a POVM according to that basis.[7] The joint probability distribution for measurement-event components for Alice, Bob, and Evangeline (which determines all that our modeling can say about individual eavesdropping attacks) becomes, in the notation defined in Appendix A:

$$\begin{aligned}\begin{pmatrix} E & B & A \\ k & j & i \end{pmatrix} &= \mathrm{Tr}_{EBA}[M_{\mathrm{EI,II}}(k)M_B(j)M_A(i)U_{EB}(\rho_E \otimes \rho_{BA})U_{EB}^{\dagger}] \\ &= \mathrm{Tr}_{EBA}\{M_{\mathrm{EI,II}}(k)M_B(j)U_{EB}(\rho_E \otimes \mathrm{Tr}_A[M_A(i)\rho_{BA}])U_{EB}^{\dagger}\} \\ &= \mathrm{Tr}_{EB}\{M_{\mathrm{EI,II}}(k)M_B(j)U_{EB}(\rho_E \otimes \sigma_B(i))U_{EB}^{\dagger}\},\end{aligned} \tag{15.2}$$

where the scaled reduced density operator $\sigma_B(i)$ is defined by

$$\sigma_B(i) \stackrel{\mathrm{def}}{=} \mathrm{Tr}_A[M_A(i)\rho_{BA}]. \tag{15.3}$$

By Bayes' rule, the conditional probability, given that Alice obtains event component i, is

$$\begin{pmatrix} E & B & \big| A \\ k & j & \big| i \end{pmatrix} = \mathrm{Tr}_{EB}\{M_E(k)M_B(j)U_{EB}(|e_1\rangle\langle e_1| \otimes \rho_B(i))U_{EB}^{\dagger}\}, \tag{15.4}$$

where a *reduced density operator* has been defined by

$$\rho_B(i) = \rho_B(i)/\mathrm{Tr}_B[\rho_B(i)]. \tag{15.5}$$

So far as detection by Bob and by Evangeline is concerned, models of this form display QKD system behavior exactly as if Alice simply transmitted a light state $\rho_B(i)$ to Bob with probability

$$\begin{pmatrix} A \\ i \end{pmatrix} = \mathrm{Tr}_B[\sigma_B(i)]. \tag{15.6}$$

As explained in Appendix A, to explore frequency effects on probabilities, the first step is to calculate the frequency dependence of the trace distances for states in distinct bases, for example $\mathrm{Tr}|\rho_B(1) - \rho_B(2)|$. Before that, however, the modeler must specify ρ_{BA} and the POVM M_A in order to determine the reduced density operators $\rho_B(i)$.

15.2.4 Formulating Quantum Optics for Fiber-Optic QKD

To model any QKD system, we must specify the density operators and POVM's. Appropriately for groundbreaking work, the modeling equations that originated QKD were simplified; they omitted the frequency spectrum of the key-carrying light (allowing one to say "photon" while pointing to vector in a two-dimensional vector space). But modeling that accounts for frequency spectra is needed, to expose vulnerabilities that occur if the eavesdropper has finer frequency filters than do Alice and Bob.[10,11] Frequency spectra involve infinite-dimensional Hilbert spaces, and in the context of these models *photon* refers to an infinite-dimensional vector or operator. To specify these operators, we follow an earlier framework[5] to arrive at equations that express the frequency spectra of quantized light in single-mode optical fiber. (For a fuller description, see Ref. 10.)

Although called "single-mode," each fiber propagates both vertical and horizontal polarizations. Let $a_1^\dagger(\omega)$ denote the creation operator for vertically polarized light at angular frequency ω, propagating from the source toward Alice, and let $a_2^\dagger(\omega)$ denote the creation operator for horizontally polarized light. Similarly introduce $b_1^\dagger(\omega)$ and $b_2^\dagger(\omega)$ for light from the source to Bob. For the quantization, adapt from Yuen and Shapiro[12] the commutation rule

$$[b_j(\omega), b_k^\dagger(\omega')] = \delta_{j,k}\delta(\omega-\omega'), \text{ along with } b_j(\omega)|0\rangle = \langle 0| b_j^\dagger(\omega) = 0 \text{ and } \langle 0|0\rangle = 1, \tag{15.7}$$

with the same equations holding when b's are replaced by a's; in addition, all a-operators commute with all b-operators. Polarization-entangled light, such as that demonstrating violations of Bell inequalities, can exhibit an interesting invariance under SU(2) polarization transforms of both a-modes and b-modes. For this chapter, I limit my attention to such light. The most general such state has the form:

$$|\psi_{BA}\rangle = \sum_{n=0}^{\infty} C_n |\psi_{BA,n}\rangle, \quad \sum_{n=0}^{\infty} |C_n|^2 = 1; \tag{15.8}$$

and the normalized state vector $|\psi_{BA,n}\rangle$ signifies n photons transmitted to Bob (and to Alice); it is built from an operator of form

$$f_n\!: (a_1b_2 - a_2b_1)^{\dagger n} \stackrel{\text{def}}{=} \int \cdots \int d\omega_1 \ldots d\omega_n d\tilde{\omega}_1 \ldots d\tilde{\omega}_n f_n(\omega_1, \ldots, \omega_n, \tilde{\omega}_1, \ldots, \tilde{\omega}_n)$$
$$\times \prod_{j=1}^{n} [a_1(\omega_j) b_2(\tilde{\omega}_j) - a_2(\omega_j) b_1(\tilde{\omega}_j)]^\dagger. \tag{15.9}$$

Then we have

$$|\psi_{BA,n}\rangle = \mathcal{N}(n) f_n\!:(a_1b_2 - a_2b_1)^{\dagger n}|0\rangle, \tag{15.10}$$

where the coefficient $\mathcal{N}(n)$ is needed to assure $\langle\psi_n|\psi_n\rangle = 1$.

To express the coefficient $\mathcal{N}(n)$ and for other calculations to come, it is convenient to use the abbreviations:

$$\vec{\boldsymbol{\omega}}_j \text{ for } (\omega_j, \tilde{\omega}_j), \quad \vec{\boldsymbol{\omega}}_n \text{ for } (\omega_1, \ldots, \omega_n, \tilde{\omega}_1, \ldots, \tilde{\omega}_n), \quad d\vec{\boldsymbol{\omega}}_n \text{ for } d\omega_1 \ldots d\omega_n\, d\tilde{\omega}_1 \ldots d\tilde{\omega}_n. \tag{15.11}$$

The coefficient $\mathcal{N}(n)$ needed to assure unit norm is

$$\mathcal{N}(n)=\left[n!^2\sum_{k=0}^{n}\int d\vec{\boldsymbol{\omega}}_n f_n^*(\vec{\boldsymbol{\omega}}_n)\mathcal{S}(\vec{\boldsymbol{\omega}}_n)\mathcal{S}(\omega_1,\ldots,\omega_k)\mathcal{S}(\omega_{k+1},\ldots,\omega_n)f_n(\vec{\boldsymbol{\omega}}_n)\right]^{-1/2} \tag{15.12}$$

where for $\boldsymbol{L}_k$ any list of k arguments, the operator $\mathrm{S}(\boldsymbol{L}_k)$, important to spectra in the quantum context, is a symmetrizing operator over the k arguments in the list $\boldsymbol{L}_k$; in particular $\mathcal{S}(\vec{\boldsymbol{\omega}}_n)$ is to be understood as

$$\begin{aligned}\mathcal{S}(\vec{\boldsymbol{\omega}}_n)f_n(\vec{\boldsymbol{\omega}}_n) &= \mathcal{S}(\vec{\omega}_1,\ldots,\vec{\omega}_n)f_n(\vec{\omega}_1,\ldots,\vec{\omega}_n)\\ &= \frac{1}{n!}\sum_{\pi\in S_n} f_n(\vec{\omega}_{\pi 1},\ldots,\vec{\omega}_{\pi n}) = \frac{1}{n!}\sum_{\pi\in S_n} f_n(\omega_{\pi 1},\ldots,\omega_{\pi n},\tilde{\omega}_{\pi 1},\ldots,\tilde{\omega}_{\pi n}).\end{aligned} \tag{15.13}$$

Remark: (1) Although S is not a quantum operator, it is a projection operator on a function space, so that, whatever arguments it averages over, $S^2 = S$. For this reason, if f is a function with arguments operated on by S then Sf is invariant under the group of permutations over which S averages. In particular, Sf is invariant under the swapping of any two of the arguments listed in S. (2) A transposition of $\vec{\omega}_j$ and $\vec{\omega}_k$ followed by a transposition of ω_j and ω_k is a transposition of $\tilde{\omega}_j$ and $\tilde{\omega}_k$. Because the permutation group is generated by transpositions, $S(\vec{\omega})\, S(\omega_1,\ldots,\omega_k)\, f(\vec{\omega})$ is invariant under the action of $S(\tilde{\omega}_1,\ldots,\tilde{\omega}_k)$.

With the abbreviated notation, I condense the right-hand side of Eq. 15.9 to obtain the expression

$$\begin{aligned} f_n\colon (a_1b_2-a_2b_1)^{\dagger n} &\overset{\text{def}}{=} \int d\vec{\omega}_n f_n(\vec{\omega}_n)(a_1b_2-a_2b_1)^{\dagger}(\vec{\omega}_n)\\ &= \int d\vec{\omega}_n [S(\vec{\omega}_n)f_n(\vec{\omega}_n)](a_1b_2-a_2b_1)^{\dagger}(\vec{\omega}_n)\\ &= \int d\vec{\omega}_n [S(\vec{\omega}_n)f_n(\vec{\omega}_n)]\sum_{k=0}\binom{n}{k}[(a_1b_2)^{\dagger k}(\vec{\omega}_1,\ldots,\vec{\omega}_k)]\\ &\quad\times[(-a_2b_1)^{\dagger n-k}(\vec{\omega}_{k+1},\ldots,\vec{\omega}_n)].\end{aligned} \tag{15.14}$$

To highlight the effect of frequency entanglement in the distinguishability of states, we choose a case in which $\mathrm{Tr}_B[\sigma_B(i)] = \mathrm{Tr}_B[\sigma_B(j)]$; then we can study the trace distance for the corresponding (unit-trace) reduced density operators $\rho_B(i) = \sigma_B(i)/\mathrm{Tr}_B[\sigma_B(i)]$. The essential effect can be seen in a simplified design for entangled-light production and detection, illustrated in Fig. 15.3; instead of using four detectors, Alice uses only two detectors, orthogonally polarized, one oriented at an angle θ to the vertical, the other oriented at $\sigma\perp = \theta - \pi/2$. Let $a_\theta(\omega) = \cos\theta\, a_1(\omega) + \sin\theta\, a_2(\omega)$. Ignoring dark counts, inefficiencies, misalignments, and loss between the source and Alice's receiver, I model

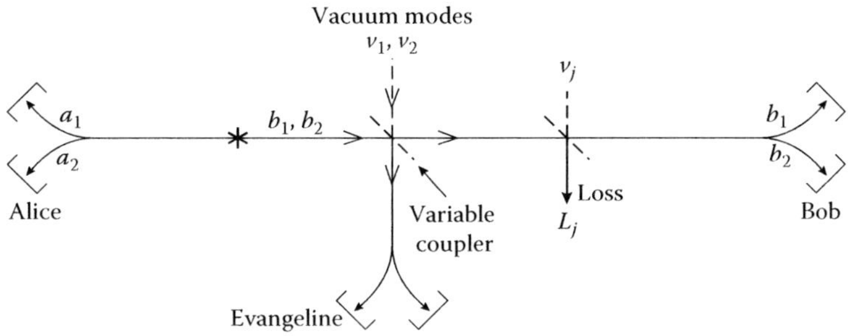

FIGURE 15.3
Simplified network.

detection by frequency-independent projections. Let $P_0(a_\theta)$ be the projection that discards those terms that have no creation operators $a_\theta^\dagger$. Then a detector that responds to all number states of one or more photons polarized along $\theta\perp$ is expressed by $[1 - P_0(a_{\theta\perp})]$. I assume a rule for sifting bits that works for the general light states discussed here essentially as discussed by Slutsky *et al.*[8]: sifted bits are limited to those for which exactly one of Alice's detectors fires. When the detector for $\theta\perp$ fires while the detector for θ does not, the scaled reduced state contains b-creation operators the other way around: it contains operators for θ and none for $\theta\perp$. I name the detection operator for this situation by its effect on b-operators as $M_A(\theta, \bar{\theta}\perp) = [1 - P_0(a_{\theta\perp})]P_0(a_\theta)$.

With this detection operator, we have

$$\sigma_B(\theta) = \mathrm{Tr}_A[M_A(\theta, \bar{\theta}\perp)\rho_{BA}] = \mathrm{Tr}_A([1 - P_0(a_{\theta\perp})]P_0(a_\theta)|\psi_{BA}\rangle\langle\psi_{BA}|). \quad (15.15)$$

From this follows[10]

$$\sigma_{B,n}(\theta) \stackrel{\text{def}}{=} n!\mathcal{N}^2(n)\int d\boldsymbol{\omega}_n d\tilde{\boldsymbol{\omega}}_n d\tilde{\boldsymbol{\omega}}'_n [\mathcal{S}(\boldsymbol{\omega}_n)\mathcal{S}(\tilde{\boldsymbol{\omega}}_n) f_n(\boldsymbol{\omega}_n, \tilde{\boldsymbol{\omega}}_n)][\mathcal{S}(\boldsymbol{\omega}_n)\mathcal{S}(\tilde{\boldsymbol{\omega}}'_n) f_n^*(\boldsymbol{\omega}_n, \tilde{\boldsymbol{\omega}}'_n)]$$
$$\times b_\theta^{\dagger n}(\tilde{\boldsymbol{\omega}}_n)|0_B\rangle\langle 0_B| b_\theta^n(\tilde{\boldsymbol{\omega}}'_n). \quad (15.16)$$

15.3 Trace Distances for Frequency-Dependent Light States

While on the one hand the originating equations for QKD were simplified, on the other hand they invoked concepts of quantum decision theory little used in quantum optics. In quantum decision theory, the main job is to calculate what is learned about an unknown state $\rho(i)$ from measuring that state, assuming the measurement is expressed by a positive operator-valued measure (POVM).[13] While problems in this field are mostly challenging and open, Helstrom and Holevo have shown how to express the least possible probability of error for a binary decision between density operators in terms of the trace distance between these.[13] This gives trace distance a special importance, in contrast to various other measures of distance between operators, such as those constructed from fidelity. In particular, in QKD models, trace distances play a big part in answering John Schlafer's question. This leads to rephrasing the question as: how do the trace distances relevant to QKD depend on the mean photon number μ, and how is that dependence modulated by the degree for *frequency* entanglement?

While trace distances critical to QKD are simple to calculate in finite-dimensional models that omit frequency, here we calculate them by drawing on equations of quantum optics in a form that expresses frequency spectra of light propagating in optical fiber. In formulating quantum-mechanical equations for frequency-dependent light in fiber, there is the complication of dispersion, absent in vacuum, but also a simplification in that only a few spatial modes propagate along a fiber. A preliminary formulation of single-photon and multi-photon, frequency-dependent light states propagating in fiber modes can be found in Ref. 5. When frequency enters, trace distances between density operators on infinite-dimensional Hilbert spaces involve convolution integrals.[10]

$$\mathcal{D}(\theta) \stackrel{\text{def}}{=} \frac{1}{2}\mathrm{Tr}_B|\rho_B(\theta/2) - \rho_B(-\theta/2)| = \frac{\sum_{n=1}|C_n|^2\sqrt{1-\cos^{2n}\theta}\,\kappa_f(n)}{\sum_{n=1}|C_n|^2\kappa_f(n)}, \tag{15.17}$$

where

$$\begin{aligned}\kappa_f(n) &= \frac{1}{2}n!\int d\boldsymbol{\omega}_n h_n(\boldsymbol{\omega}_n, \boldsymbol{\omega}_n) = n!^2\,\mathcal{N}^2(n)\int d\boldsymbol{\omega}_n d\tilde{\boldsymbol{\omega}}_n |\mathcal{S}(\boldsymbol{\omega}_n)\mathcal{S}(\tilde{\boldsymbol{\omega}}_n) f_n(\boldsymbol{\omega}_n, \tilde{\boldsymbol{\omega}}_n)|^2 \\ &= \mathrm{Tr}_B[\sigma_{B,n}(\theta)],\end{aligned} \tag{15.18}$$

which is independent of θ.

Example

We study an example in which

$$f_n(\boldsymbol{\omega}_n, \tilde{\boldsymbol{\omega}}_n) = \prod_{j=1}^{n} g(\boldsymbol{\omega}_j, \tilde{\boldsymbol{\omega}}_j) \Rightarrow \mathcal{S}(\boldsymbol{\omega}_n)\mathcal{S}(\tilde{\boldsymbol{\omega}}_n) f_n(\boldsymbol{\omega}_n, \tilde{\boldsymbol{\omega}}_n) = \mathcal{S}(\boldsymbol{\omega}_n)\prod_{j=1}^{n} g(\omega_j, \tilde{\omega}_j). \tag{15.19}$$

For this case, $\kappa_f(n)$ defined by Eqs. 15.18 and 15.12 becomes

$$\kappa_f(n) = \frac{1}{n!}\Xi_g(n)\left(\sum_{k=0}^{n}\frac{\Xi_g(k)\Xi_g(n-k)}{k!(n-k)!}\right)^{-1}, \tag{15.20}$$

with

$$\Xi_g(n) \overset{\text{def}}{=} \sum_{\pi \in S_n} \int d\boldsymbol{\omega}_n d\tilde{\boldsymbol{\omega}}_n \prod_{j=1}^{n} [g^*(\omega_{\pi j}, \tilde{\omega}_j) g(\omega_j, \tilde{\omega}_j)], \tag{15.21}$$

where, as before, 8_n is the permutation group of order n. In this function a variety of convolution integrals appear, for example for $n = 2$, we have

$$\begin{aligned} \Xi_g(2) = \int d\omega_1 d\omega_2 d\tilde{\omega}_1 d\tilde{\omega}_2 \{ & g^*(\omega_1 \tilde{\omega}_1) g(\omega_1 \tilde{\omega}_1) g^*(\omega_2 \tilde{\omega}_2) g(\omega_2 \tilde{\omega}_2) \\ & + g^*(\omega_2 \tilde{\omega}_1) g(\omega_1 \tilde{\omega}_1) g^*(\omega_1 \tilde{\omega}_2) g(\omega_2 \tilde{\omega}_2) \}. \end{aligned} \tag{15.22}$$

This function Ξ_g is discussed in detail in Ref. 5, in which Sect. 10 and Appendix C describe properties of Ξ_g and Appendix F lists MATLAB programs for it.

15.3.1 Examples of Frequency Functions

We consider a family of functions $g_\zeta(\omega, \tilde{\omega})$ and show two limiting cases. For any real-valued functions $\phi(\omega)$ and $\tilde{\phi}(\tilde{\omega})$ and positive real parameters ϕ and $\tilde{\sigma}$, let

$$g_\zeta(\omega, \tilde{\omega}) = \frac{1}{\sqrt{\sigma \tilde{\sigma}}} e^{i\phi(\omega)} e^{i\tilde{\phi}(\tilde{\omega})} F\left(\zeta; \frac{\omega - \omega_0}{\sigma}, \frac{\tilde{\omega} - \tilde{\omega}_0}{\tilde{\sigma}} \right), \tag{15.23}$$

where we define

$$\begin{aligned} F(\zeta; x, y) &\overset{\text{def}}{=} \sqrt{\frac{2}{\pi}} \exp\left\{ -\frac{1}{2} \left[\left(\sqrt{\zeta^2 + 1} + \zeta \right)(x + y)^2 + \left(\sqrt{\zeta^2 + 1} + \zeta \right)^{-1} (x - y)^2 \right] \right\} \\ &= \sqrt{\frac{2}{\pi}} \exp\left\{ -\left[\sqrt{\zeta^2 + 1}(x^2 + y^2) + 2\zeta xy \right] \right\}. \end{aligned} \tag{15.24}$$

Regardless of the value of ζ, $\int_{-\infty}^{\infty} dx\, dy |F(\zeta; x, y)|^2 = 1$. Thus for any choice of center frequencies ω_0 and $\tilde{\omega}_0$, bandwidth parameters σ and $\tilde{\sigma}$, and phase functions $\phi(\omega)$ and $\tilde{\phi}(\tilde{\omega})$, we get a family of g_ζ's. For any such family, consider two limiting cases as follows.

Case I: No Frequency Entanglement. For this case $\zeta = 0$, and Eq. 15.24 shows a product of a function of ω times a function of $\tilde{\omega}$, so there is no frequency entanglement. For this case, as shown in Sect. 10 of Ref. 5, $\Xi_g(n)/n! = 1$ so that $\kappa_f(n) = 1/(n + 1)$. Then letting $|C_n|^2$ be given by a Poisson distribution with mean photon number μ, one evaluates Eq. 15.17 for small μ to show

$$\mathcal{D}_{\text{I}}(\theta) = |\sin\theta| \left[1 + \frac{\mu}{3} \left(\sqrt{1 + \cos^2\theta} - 1 \right) \right] + O(\mu^2). \tag{15.25}$$

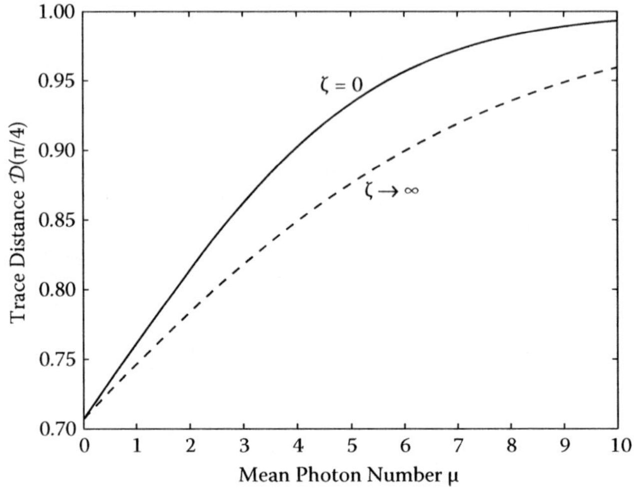

FIGURE 15.4
Dependence of trace distance $D(\pi/4)$ on μ, showing effect of frequency entanglement for Case I, $\zeta = 0$ (no frequency entanglement), and Case II, $\zeta \to \infty$ (extreme frequency entanglement).

Case II: Extreme Frequency Entanglement. In the limit as $\zeta \to \infty$, as shown in Ref. 5, $\Xi_g(n)/n! \to 1/n!$ and $\kappa_f(n) = 2^{-n}$, in which case Eq. (15.17) evaluated for small μ shows

$$\mathcal{D}_{\mathrm{II}}(\theta) = |\sin\theta| \left[1 + \frac{\mu}{4}\left(\sqrt{1+\cos^2\theta} - 1\right) \right] + O(\mu^2). \tag{15.26}$$

For $\theta = \pi/4$, Fig. 15.4 illustrates both cases over a larger range of μ.

15.4 Food for Thought

The beachhead into the modeling of frequency-dependent light states reported above opens up many questions, some of which arise because the Hilbert spaces are infinite dimensional. Here are some irritants to further thought:

1. In the analysis above, use has been made of reduced states. These uses of reduced states involve no postulate of state reductions; instead they follow from simple applications of Bayes' rule to joint probabilities defined by the trace rule of quantum mechanics.[1, 9]
2. Consider a Hilbert space $\mathcal{H}$ as in Sect. 15.2 that contains state vectors defined by integrals $\int d\omega\, f(\omega)a^\dagger(\omega)|0\rangle$. The norm of $a^\dagger(\omega)|0\rangle$ is infinite, so that $a^\dagger(\omega)|0\rangle$ cannot be a vector of the Hilbert space $\mathcal{H}$;

nonetheless it can be seen as a vector in a larger vector space $\mathcal{V}$ equipped with any of a variety of norms or perhaps no norm at all, with $\mathcal{H} \subset \mathcal{V}$. How the structure of $\mathcal{V}$ impacts on the modeling of light is an open question.

3. For density operators ρ defined on a separable Hilbert space, the condition that $\text{Tr}(\rho) = 1$ by no means guarantees the existence of a finite von Neumann entropy $S(\rho) \stackrel{\text{def}}{=} -\text{Tr}(\rho \ln \rho)$. If transmitted signals are expressed by density operators of infinite von Neumann entropy, the effort to estimate information in terms of the Holevo bound[14] fails by producing ∞ minus ∞; on the other hand error probabilities for distinguishing two density operators make use of trace distances which are still well-defined and finite.
4. An infinite dimensional reduced density operator obtained by a partial trace on ρ_{BA} defined in Eq. 15.10 cannot be a constant times the identity matrix, as it would be in some well-known cases of finite-dimensional entangled states; in particular, this reduced density operator still depends on the frequency spectrum of ρ_{BA}.
5. If $g^*(\omega, \omega') = g(\omega', \omega)$, then there is a 1-photon, linearly polarized state of the form

$$\rho_g = \int d\omega d\omega' g(\omega, \omega') a^\dagger(\omega) \, |0\rangle\langle 0| \, a(\omega'). \tag{15.27}$$

Only if g can be written as a product $g_1(\omega)g^*_1(\omega')$ is this a pure state. As the in-state to an interferometer, modeled linearly, any mixed-state ρ_g exhibits the same second-order coherence as would a pure 1-photon linearly polarized state. Although the von Neumann entropy of any pure state is 0, depending on g, the von Neumann entropy of $S(\rho_g)$ can be any non-negative value; indeed g can be chosen to make $S(\rho_b)$ infinite.

6. Quantum cryptography and quantum computing have brought concepts from information theory, notably measures of entropy and information, into quantum optics. Jaynes[15] displays thermodynamic macrostates as classes of microstates, showing that entropy must be relative to variables viewed as under experimental control. Entropy is also relative to choices made in modeling. Another stimulus to clear thinking comes from Alfrèd Rényi[16]: "in proving limit theorems of probability theory by considering information, it is usually an advantage that one can choose between different measures." He guides us away from thinking of entropy as something physical or from asking whether one definition is better than another apart from its application to elucidating probability distributions in a particular context.
7. Gisin, Renner, and Wolf[17] prove that "a quantum state between two parties is entangled if and only if the classical random variables resulting from optimal measurements provide some mutual classical information between the parties." In this claim lies buried

an assumption of "unentangled detectors." A light detector can involve a probe particle with which the light to be detected interacts, and when two or more detectors are used jointly, their probe particles can be entangled, so that one can speak of entangled detectors.[1] The correlations typical of entangled light registered by unentangled detectors can be produced also by entangled detectors illuminated by unentangled light. For this reason we see that the correlations usually interpreted as experimental evidence of entangled light are a property of neither the light nor the detectors separately, but of their combination.

Within the confines of the mathematics of quantum-mechanical equations, there have been several occasions to use the word *photon*. In Ref. 5, I have said "n-photon state" to indicate a weighted integral of an n-fold product of creation operators acting on the vacuum. If one limits attention to operators $a^{\dagger}_1$ and $a^{\dagger}_2$, then the polarization-entangled state vector $|\psi_{BA,n}\rangle$ defined in Eq. 15.10 could be called an n-photon state; however when both $a^{\dagger}$ and $b^{\dagger}$ operators are viewed, one would be more apt to speak of *n bi-photons*. The possibilities for constructing light states in terms of integrals over polynomials in creation operators for various modes are far richer than the word *photon* can conveniently express. When we want precision, we had best skip the word and stick to the integrals.

We began by emphasizing the proven irreducible logical looseness in linking state vectors and operators to the probability distributions by which quantum mechanics makes contact with experiments.[1] Recognizing the categorical logical looseness of links between equations and experiments brings physicists opportunities for creativity, including formerly unsuspected freedoms to invent models,[18, 19] several of which have been exemplified above. Because of the same looseness, no quantum model can promise any "infallible doctrine"[20]; instead I offer the optics models above, like early maps of the new world, to stimulate the design and interpretation of experimental endeavors. Collaborations are welcome.

Acknowledgments

I thank Chandra Roychoudhuri for suggesting the topics covered in this chapter, and for encouraging the discussion of challenging open problems. I am much indebted to John Schlafer of BBN Technologies for posing the question that stimulated the work reported here. I thank Chip Elliott, David Pearson, Oleksiy Pikalo of BBN Technologies and Martin Jaspan, Gregg Jaeger and Alexander Sergienko of Boston University for discussions that helped put quantum cryptography in perspective. I thank Tai Wu of Harvard for showing me the importance of spatial variables in quantum cryptography and quantum computing.

This work was supported in part by the Air Force Research Laboratory and DARPA under Contract F30602-01-C-0170 with BBN Technologies.

Appendix A Review of Quantum Decision Theory

Let curved brackets around an upper and a lower list denote the joint probability that the random variables in the upper list take the values in the corresponding lower list. For $\binom{x}{a} \neq 0$, denote the probability of y conditioned on x by

$$f\left(\begin{matrix} y \\ b \end{matrix}\,\middle|\,\begin{matrix} x \\ a \end{matrix}\right) \stackrel{\text{def}}{=} \begin{pmatrix} x & y \\ a & b \end{pmatrix} \Big/ \begin{pmatrix} x \\ a \end{pmatrix}. \tag{A.1}$$

For a more complete discussion, including illustrations of the power of this notation, see Ref. 21.

Suppose B (for Bob) measures light prepared by A (for Alice); we model the light as density operator $\rho\,(i)$, and model the measurement with outcome j by the detection operator $M_B(j)$ of a positive operator-valued measure (POVM). Thus the conditional probability of B's outcome j given that A prepares $\rho\,(i)$ is

$$\left(\begin{matrix} B \\ j \end{matrix}\,\middle|\,\begin{matrix} A \\ i \end{matrix}\right) = \mathrm{Tr}[M_B(j)\rho(i)]. \tag{A.2}$$

This general quantum form sets up a decision problem,[13] as follows. Assume an *a-priori* probability $\binom{A}{i}$ for A preparing state $\rho\,(i)$. It follows that

$$\begin{pmatrix} B & A \\ j & i \end{pmatrix} = \mathrm{Tr}[M_B(j)\sigma(i)], \tag{A.3}$$

where $\sigma\,(i)$ is what might be called a "scaled" density operator, defined by

$$\sigma(i) \stackrel{\text{def}}{=} \begin{pmatrix} A \\ i \end{pmatrix} \rho(i). \tag{A.4}$$

Assume B makes a maximum-likelihood decision[13] of what state A prepared on the basis of outcome j. According to Bayes' rule, the posterior probability of $\rho\,(i)$ is

$$\left(\begin{matrix} A \\ i \end{matrix}\,\middle|\,\begin{matrix} B \\ j \end{matrix}\right) = \begin{pmatrix} B & A \\ j & i \end{pmatrix} \Big/ \sum_{i'} \begin{pmatrix} B & A \\ j & i' \end{pmatrix}. \tag{A.5}$$

Given any outcome j, B decides on a value of i that maximizes this posterior probability. The probability (averaged over outcomes) that B makes a correct decision is then

$$\text{pCorrect}_B = \sum_j \begin{pmatrix} B \\ j \end{pmatrix} \max_i \left(\begin{array}{c|c} A & B \\ i & j \end{array} \right) = \sum_j \max_i \begin{pmatrix} B & A \\ j & i \end{pmatrix}, \tag{A.6}$$

and the probability of B making an incorrect decision is

$$\text{pErr}_B = 1 - \text{pCorrect}_B. \tag{A.7}$$

Binary Decisions

Now specialize to the case that A chooses between two states, so $i \in \{0, 1\}$ and B's decision is between these two values. (The possible outcomes j can still range over more than two values.) Recalling that

$$(\forall\, x, y \geq 0) \min(x, y) = (x + y - |x - y|)/2, \tag{A.8}$$

we find for the case of just two values for i

$$\max_i \begin{pmatrix} B & A \\ j & i \end{pmatrix} = \frac{1}{2}\left\{ \begin{pmatrix} B & A \\ j & 0 \end{pmatrix} + \begin{pmatrix} B & A \\ j & 1 \end{pmatrix} + \left| \begin{pmatrix} B & A \\ j & 0 \end{pmatrix} - \begin{pmatrix} B & A \\ j & 1 \end{pmatrix} \right| \right\}, \tag{A.9}$$

which, with the usual rule relating joint to conditional probabilities, implies

$$\begin{aligned} \sum_j \max_i \begin{pmatrix} B & A \\ j & i \end{pmatrix} &= \frac{1}{2}\left\{ \begin{pmatrix} A \\ 0 \end{pmatrix} + \begin{pmatrix} A \\ 1 \end{pmatrix} + \sum_j \left| \begin{pmatrix} B & A \\ j & 0 \end{pmatrix} - \begin{pmatrix} B & A \\ j & 1 \end{pmatrix} \right| \right\} \\ &= \frac{1}{2}\left\{ \text{Tr}[\sigma(0) + \sigma(1)] + \sum_j |\,\text{Tr}[M_B(j)\sigma(0)] - \text{Tr}[M_B(j)\sigma(1)]\,| \right\} \\ &= \frac{1}{2}\left\{ \text{Tr}[\sigma(0) + \sigma(1)] + \sum_j |\,\text{Tr}[M_B(j)\,(\sigma(0) - \sigma(1))]\,| \right\}. \end{aligned} \tag{A.10}$$

To get an upper bound on pCorrect_B (and hence a lower bound on pErr_B), we notice that the construction in Ref. 14 works for operators that need not have unit trace. For any operator C, define $|C| = |C| = \sqrt{C^\dagger C}$, taking the positive square root. Now let C be $\sigma(0) - \sigma(1)$, which is self-adjoint. Assuming C is also compact,[22] there is a unitary operator U and some (possibly infinite) real diagonal matrix D such that $C = UDU^\dagger$. Let $D = D^{(+)} - D^{(-)}$, where $D^{(+)}$ and $D^{(-)}$ are positive semi-definite. Let $Q = UD^{(+)}U^{\dagger}$ and $S = UD^{(-)}U^{\dagger}$. Hence $C = Q - S$ with Q and S semi-positive and simultaneously diagonalizable. By construction, $D^{(+)}D^{(-)} = 0$, so we have $|C| = \sqrt{(Q-S)^2} = \sqrt{U(D^2)U^{+}} = Q + S$. It follows that

$$\begin{aligned}\text{pCorrect}_B &= \frac{1}{2}\left\{\text{Tr}[\sigma(0)+\sigma(1)] + \sum\nolimits_j |\text{Tr}[M_B(j)(\sigma(0)-\sigma(1))]|\right\}\\ &\le \frac{1}{2}\left\{\text{Tr}[\sigma(0)+\sigma(1)] + \sum\nolimits_j \text{Tr}[M_B(j)\,|\,\sigma(0)-\sigma(1)\,|\,]\right\} \qquad \text{(A.11)}\\ &\le \frac{1}{2}\left\{\binom{A}{0} + \binom{A}{1} + \text{Tr}\,|\,\sigma(0)-\sigma(1)\,|\right\}.\end{aligned}$$

In the case $\binom{A}{0} + \binom{A}{1} = \text{Tr}[\sigma(0)+\sigma(1)] = 1$, this implies the customary bound[13]

$$\text{pErr}_B \ge \frac{1}{2} - \frac{1}{2}\text{Tr}\,|\,\sigma(0)-\sigma(1)\,|. \qquad \text{(A.12)}$$

Furthermore, $M_B(0)$ can be the projection onto the space spanned by eigenvectors of Q, and $M_B(1)$ can be the projection onto the space spanned by eigenvectors of S. Because these two projections are mutually orthogonal, we have for this choice of M_B, $|\text{Tr}[M_B(j)(Q - S)]| = \text{Tr}[M_B(j)(Q + S)]$, so (as discussed by Helstrom) this choice of M_B achieves the optimum defined by the bound shown in Eq. A.12.

Equations A.12, A.2, A.6, and A.7 also imply for binary decisions the following:

Lemma: Given any two non-negative operators $\sigma(0)$ and $\sigma(1)$, all POVM's M_B satisfy

$$\sum\nolimits_j \text{Tr}[M_B(j)\sigma(j)] \le \frac{1}{2}\{\text{Tr}[\sigma(0)+\sigma(1)] + \text{Tr}\,|\,\sigma(0)-\sigma(1)\,|\,\}. \qquad \text{(A.13)}$$

Note that for any self-adjoint matrix C

$$\text{Tr}\,|\,C| = \sum\nolimits_j |\lambda_j|, \qquad \text{(A.14)}$$

where $\{\lambda_j\}$ is the set of the eigenvalues of C.

Appendix B Trace Distance

To deal with trace distances in the infinite-dimensional spaces called for in modeling frequency dependence of quantum states, we need to venture into separable Hilbert spaces, that is, Hilbert spaces requiring a denumerably infinite basis. Here I sketch the basics, most of which were worked out by von Neumann.[23]

Let $\mathcal{B}(\mathcal{H})$ denote the set of bounded operators on a separable Hilbert space $\mathcal{H}$. I call $A \in \mathcal{B}(\mathcal{H})$ *positive* if $(\forall\ |x\rangle \in \mathcal{H})\ \langle x|A|x\rangle \geq 0$ (this is what von Neumann calls *definite*). Let $\mathcal{B}^{+}(\mathcal{H}) \subset \mathcal{B}(\mathcal{H})$ denote the subset of positive bounded operators. Let $\{\psi_n\}_{n=1}^{\infty}$ be any orthonormal basis of $\mathcal{H}$. For a (finite-dimensional) square matrix, the trace is just the sum of the diagonal elements. We need traces of operators on infinite dimensional spaces; fortunately, however, all that we require is the special case dealt with in Chap. II, Sect. 11 of von Neumann's book,[23] namely traces of bounded operators on a separable Hilbert space of the form $A = BC$, where $B, C \in \mathcal{B}^{+}(\mathcal{H})$. (Of course one of these can be the identity operator.) Von Neumann shows how for operators BC one essentially gets away with thinking of the operators as limiting cases of a sequence of operators with finite-dimensional ranges (Dunford and Schwartz,[24] p. 515, proposition 32). If A is of this form, its trace is defined by

$$\mathrm{Tr}(A) \stackrel{\text{def}}{=} \sum_{n=1}^{\infty} \langle \psi_n | A | \psi_n \rangle. \tag{B.1}$$

The value of Tr(A), whether finite or infinite, is independent of the choice of basis,[23] and

$$\mathrm{Tr}(AB) = \mathrm{Tr}(BA). \tag{B.2}$$

For any bounded operator A, define $|A| = \sqrt{A^{\dagger}A}$; this $|A|$ is positive. For any $A, B \in \mathcal{B}(\mathcal{H})$ define the trace distance between them by

$$D(A, B) \stackrel{\text{def}}{=} \frac{1}{2}\mathrm{Tr}\,|A - B|. \tag{B.3}$$

Lemma: For $A, A_1, A_2, U \in \mathcal{B}(\mathcal{H})$ and U unitary, $U\sqrt{A^{\dagger}A}U^{\dagger} = \sqrt{U\ A^{\dagger}AU^{\dagger}}$ which implies

$$\mathrm{Tr}|A| = \mathrm{Tr}|UAU^{\dagger}|, \tag{B.4}$$

$$\mathrm{Tr}|A_1 - A_2| = \mathrm{Tr}|UA_1U^{\dagger} - UA_2U^{\dagger}|. \tag{B.5}$$

Lemma: For $A, B \in \mathcal{B}^{+}(\mathcal{H})$,

$$\mathrm{Tr}(BA) \geq 0. \tag{B.6}$$

PROOF: $\mathrm{Tr}(BA) = \mathrm{Tr}(B^{\dagger 1/2} A B^{1/2}) \geq 0$. □

Lemma: For $A, B \in \mathcal{B}^{+}(\mathcal{H})$,

$$\mathrm{Tr}(AB) \leq \mathrm{Tr}|AB|. \tag{B.7}$$

PROOF: Although not all bounded operators on an infinite dimensional Hilbert space have a polar decomposition, there is some partial isometry U such that $AB = U|AB| = U|AB|^{1/2}|AB|^{1/2}$ (Dunford and Schwartz,[24] p. 935). From the Cauchy-Schwarz inequality we then have

$$\begin{aligned}\mathrm{Tr}(AB) &= \mathrm{Tr}(U|AB|^{1/2}|AB|^{1/2}) \le \sqrt{\mathrm{Tr}(|AB|^{1/2}U^{\dagger}U|AB|^{1/2})\mathrm{Tr}(|AB|^{1/2}|AB|^{1/2})} \\ &= \sqrt{\mathrm{Tr}(U^{\dagger}U|AB|)\mathrm{Tr}|AB|} \le \mathrm{Tr}|\boldsymbol{AB}|.\end{aligned} \tag{B.8}$$

Because U is a partial isometry, $U^{\dagger}U$is a projection (Dunford and Schwartz,[24] p. 1248), which with Eq. B.8 implies the lemma. □

Lemma: For $A, B \in \mathcal{B}^{+}(\mathcal{H})$,

$$AB + BA = 0 \Rightarrow \mathrm{Tr}|A - B| = \mathrm{Tr}(A + B). \tag{B.9}$$

PROOF:

$$\begin{aligned}AB + BA = 0 \Rightarrow \mathrm{Tr}|A - B| &\overset{\text{def}}{=} \mathrm{Tr}\sqrt{A^2 - AB - BA + B^2} = \mathrm{Tr}\sqrt{A^2 + AB + BA + B^2} \\ &= \mathrm{Tr}|A + B| = \mathrm{Tr}(A + B).\end{aligned}$$

□

Lemma: For $\sigma_E \in \mathcal{B}^{+}(\mathcal{H}_E)$ and $\sigma_B(i) \in \mathcal{B}^{+}(\mathcal{H}_B)$,

$$\mathrm{Tr}_{EB}|\sigma_E \otimes \sigma_B(i) - \sigma_E \otimes \sigma_B(j)| = \mathrm{Tr}_E(\sigma_E)\mathrm{Tr}_B|\sigma_B(i) - \sigma_B(j)|. \tag{B.10}$$

B.1 Fidelity of Order ν

In the case Tr(A) = Tr(B), bounds for the trace distance $D(A,B)$ in terms of measures of fidelity are known.[14] I extend the definition to speak of *fidelity of order* ν, analogous to Rényi entropy of various orders, by

$$F_{\nu}(A, B) \overset{\text{def}}{=} \mathrm{Tr}[(A^{\nu/4}B^{\nu/2}A^{\nu/4})^{1/\nu}]. \tag{B.11}$$

What is usually called fidelity is F_2; however good use has been made[1] of $F_1(A, B) = \mathrm{Tr}(A^{1/4}B^{1/2}A^{1/4}) = \mathrm{Tr}(A^{1/2}B^{1/2})$. From lemma (B7) it follows that

$$F_1(A, B) = \mathrm{Tr}(A^{1/2}B^{1/2}) \le \mathrm{Tr}|B^{1/2}A^{1/2}| = \mathrm{Tr}\left(\sqrt{A^{1/2}BA^{1/2}}\right) = F_2(A, B). \tag{B.12}$$

Remark: It would be nice to find out the circumstances under which this generalizes to $\nu \le \mu \Rightarrow F_{\nu}(A, B) \le F_{\mu}(A, B)$.

Remark: Quantum decision theory employs trace distance not only for operators of unit trace, but for positive operators generally. It is trivial to extend the definition of fidelity by removing the requirement of unit trace for A and B; however, if this is done, fidelity of any order has an invariance under scale changes: for s real and positive and $A, B \in \mathcal{B}^+(\mathcal{H})$,

$$F_v(A, B) = F_v(sA, s^{-1}B), \tag{B.13}$$

with the implication that $F_v(sA, s^{-1}A) = \mathrm{Tr}|A|$, independent of s. In contrast, trace distance has no such invariance; rather we have $D(sA, s^{-1}A) = \frac{1}{2}(s - s^{-1})^2\mathrm{Tr}|A|$. Because of its invariance under this scale transformation, fidelity is of doubtful value in constructing distance measures over scaled density operators.

References

1. F. H. Madjid and J. M. Myers, "Matched detectors as definers of force," arXiv: quant-ph/0404113 v2, 2004; accepted for publication in *Annals of Physics.*
2. D. Kennedy, ed., "Editorial: Twilight for the enlightenment?," *Science* **308**, 165 (2005).
3. C. H. Bennett and G. Brassard, "Quantum cryptography: Public key distribution and coin tossing," in *Proc. IEEE International Conference on Computers, Systems and Signal Processing, Bangalore, India,* pp. 175–179, IEEE, New York, 1984.
4. J. M. Myers, T. T. Wu, and D. S. Pearson, "Entropy estimates for individual attacks on the BB84 protocol for quantum key distribution," *Proceedings of SPIE,* Vol. 5436, *Quantum Information and Computation II,* E. Donkor, A. R. Pirich, H. E. Brandt, eds., pp. 36–47, SPIE, Bellingham, WA, 2004.
5. J. M. Myers, "Framework for quantum modeling of fiber-optical networks, Part I," arXiv:quant-ph/041107 v2, 2004; "Framework for quantum modeling of fiber-optical networks, Part II," arXiv:quant-ph/041108 v2, 2004.
6. I am indebted to Chip Elliott of BBN Technologies for the distinction between attacks to which published proofs apply, which he has called "signature attacks," and a more vicious attack that modifies Bob's receiver, which he has called a "boundary attack."
7. B. A. Slutsky, R. Rao, P.-C. Sun, and Y. Fainman, "Security of quantum cryptography against individual attacks," *Phys. Rev. A* **57**, pp. 2383–2398, 1998.
8. B. A. Slutsky, R. Rao, P.-C. Sun, L. Tancevski, and S. Fainman, "Defense frontier analysis of quantum cryptographic systems," *Applied Optics* **37**, pp. 2869–2878, 1998.
9. J. M. Myers, "Conditional probabilities and density operators in quantum modeling," in preparation.
10. J. M. Myers, "Polarization-entangled light for quantum key distribution: how frequency spectrum and energy affect detection statistics," to be published in *Proceeding of SPIE,* Vol. 5815, *Quantum Information and Computation III,* E. Donkor, A. R. Pirich, H. E. Brandt, eds., SPIE, Bellingham, WA, 2005.
11. J. Lowry, BBN Technologies, "To crack a cryptographic system, find an abstraction and look under it" (discussion, 2001).
12. H. P. Yuen and J. H. Shapiro, "Optical communication with two-photon coherent states, Part I: Quantum-state propagation and quantum-noise reduction," *IEEE Trans. Info. Theory,* **IT-24**, pp. 657–668, 1978.

13. C. W. Helstrom, *Quantum Detection and Estimation Theory*, Academic Press, New York, 1976.
14. M. A. Nielsen and I. L. Chuang, *Quantum Computation and Quantum Information*, p. 403, Cambridge University Press, Cambridge, UK, 2000.
15. E. T. Jaynes, "The Gibbs paradox," in *Maximum Entropy and Bayesian Methods*, C. R. Smith, G. J. Erickson, and P. O. Neudorfer, eds., pp. 1–22, Kluwer, Dordrecht, 1992.
16. A. Rényi, "On measures of entropy and information," in *Proc. 4th Berkeley Symposium on Mathematical Statistics and Probability*, Vol. 1, pp. 547–561, University of California Press, Berkeley, 1961.
17. N. Gisin, R. Renner, S. Wolf, "Linking classical and quantum key agreement: Is there a classical analog to bound entanglement?," *Algorithmica* **34**, pp. 389–412, 2002.
18. J. M. Myers and F. H. Madjid, "A proof that measured data and equations of quantum mechanics can be linked only by guesswork," *Quantum Computation and Information*, S. J. Lomonaco, Jr. and H. E. Brandt, eds., *Contemporary Mathematics Series*, Vol. 305, pp. 221–244, American Mathematical Society, Providence, RI, 2002.
19. J. M. Myers and F. H. Madjid, "Gaps between equations and experiments in quantum cryptography," *J. Opt. B: Quantum Semiclass. Opt.* **4**, pp. S109–S116, 2002.
20. "We must especially fight against those ideologies which pronounce the infallibility of their doctrines and thus separate the world into irreconcilable camps." Max Born, quoted by Nancy Greenspan in "Max Born and the peace movement," *Physics World*, p. 38, April 2005.
21. F. H. Madjid and J. M. Myers, "Linkages between the calculable and the incalculable in quantum theory," *Annals of Physics* **221**, pp. 258–305, 1993.
22. G. J. Murphy, *C*-Algebras and Operator Theory*, Academic Press, New York, 1990.
23. J. von Neumann, *Mathematische Grundlagen der Quantenmechanik*, Springer, Berlin, 1932; translated with revisions by the author as *Mathematical Foundations of Quantum Mechanics*, Princeton University Press, Princeton, 1955.
24. N. Dunford and J. T. Schwartz, *Linear Operators, Parts I and II*, Interscience Publishers, New York, Part I, 1958, Part II, 1970.

16

Photon–The Minimum Dose of Electromagnetic Radiation

Tuomo Suntola
Suntola Consulting Ltd., Tampere University of Technology, Finland

CONTENTS

A radio engineer can hardly think about smaller amount of electromagnetic radiation than given by a single oscillation cycle of a unit charge in a dipole. When solved from Maxwell's equations for a dipole of one wavelength, the energy of the emitted radiation cycle obtains the form $E_\lambda = 2/3$ hf, where the Planck constant h can be expressed in terms of the unit charge, e, the vacuum permeability, μ_0, the velocity of light, c, and a numerical factor as $h = 1.1049 \cdot 2\pi^3 e^2 \mu_0 c = 6.62607 \cdot 10^{-34}$ [kgm²/s]. A point emitter like an atom can be regarded as a dipole in the fourth dimension. The length of such dipole is measured in the direction of the line element *cdt*, which in one oscillation cycle means the length of one wavelength. For a dipole in the fourth dimension, three space directions are in the normal plane which eliminates the factor 2/3 from the energy expression thus leading to Planck's equation $E_\lambda = hf$ for the radiation emitted by a single electron transition in an atom. The expression of the Planck constant obtained from Maxwell's equations leads to a purely numerical expression of the fine structure constant $\alpha = 1/(1.1049 \cdot 4\pi^3) \approx 1/137$ and shows that the Planck constant is directly proportional to the velocity of light. When applied to Balmer's formula, the linkage of the Planck constant to the velocity of light shows that the frequency of an atomic oscillator is directly proportional to the velocity of light. This implies that the velocity of light is observed as constant in local measurements. Such an interpretation makes it possible to convert relativistic spacetime with variable time coordinates into space with variable clock frequencies in universal time, and thus include relativistic phenomena in the framework of quantum mechanics.

16.1 Introduction

We are used to thinking that the emission of electromagnetic radiation described by Planck's equation is different from the emission of radiation from a dipole according to Maxwell's equations. Based on observations on black body radiation, the emission of electromagnetic radiation from a heated body, Max Planck in about 1900 concluded that the dose of electromagnetic radiation, a quantum, that can be emitted grows in a direct proportion to its frequency, expressed as $E = hf$. In this chapter, we will find out that emission of electromagnetic radiation from an electric dipole has basically the same property—once we solve for the energy of one cycle of radiation.

In explaining Philipp von Lenard's experiments on the photoelectric effect, Albert Einstein in 1905 applied an opposite aspect of Planck's postulate. To have electrons emitted from a solid surface, the energy quantum of incoming radiation shall exceed the work function needed in releasing an electron. Einstein's explanation was verified by the successful determination of Planck's constant from the photoelectric effect.

The works of Planck and Einstein inspired Niels Bohr to combine particle and wave properties of an electron in his model for hydrogen atom. In Bohr's model, discrete stationary energy states are characterized by standing waves with momentum equal to the momentum of electrons orbiting the nucleus in a classical Coulomb field.

Planck's postulate and the explanation of the photoelectric effect using the concept of the quantum led towards a dualistic view of electromagnetic radiation as a wavelike form of energy described in terms of Maxwell's equations and also as a flow of particles like quanta. Such dualistic view was strengthened through the analysis of Compton scattering of radiation based on the works of Arthur H. Compton and Peter Debye in the early 1920's. An important aspect was the momentum of a quantum which, as a zero rest mass particle in the framework of special relativity, could be identified as equal to the momentum of electromagnetic radiation according to Maxwell's equations, i.e. $E = c|p|$. A complementary view of the dualism between particles and waves was established through the work of Louis de Broglie who generalized the concept of the wavelength equivalence, the de Broglie wavelength $\lambda_{dB} = h/|p|$, of mass particles with momentum $\mathbf{p}$ in space, an idea implicitly included in the Bohr hydrogen atom about ten years earlier. Schrödinger's equation completed the framework of quantum mechanics in the late 1920's.

Key conclusions leading to quantum mechanics have been drawn from phenomena related to atoms and small particles. Emission of electromagnetic radiation from atoms as small point sources could not be quantitatively explained in the framework of Maxwell's equations. When an atomic source is described as an electric dipole emitting electromagnetic radiation, the displacement of the charge resulting in electric dipole momentum is considered as being of the order of atomic size, about 10^{-10} m, which is orders of magnitudes smaller than the wavelengths of radiation emitted. The situation, however, is radically changed if we consider a point source a dipole in the fourth dimension, in the direction of line element cdt, which in one oscillation cycle means the displacement of one wavelength—regardless of the emission frequency from the source.

When solved from Maxwell's equations, the energy of one cycle of electromagnetic radiation emitted from a dipole in the fourth dimension due to a single transition of a unit charge obtains the form of Planck's equation $E = hf$. Such a result gives the quantum a clear meaning as the energy of one cycle of electromagnetic radiation generated by a single electron transition in a point source.

Interpretation of a point source as a dipole in the fourth dimension suggests a fourth dimension of metric nature. Displacement of a point source by one wavelength in a cycle requires motion of space at velocity c in the metric fourth dimension. Such an interpretation is consistent with spherically closed space expanding in a zero energy balance of motion and gravitation in the direction of the 4-radius. A consequence of the conservation of the zero energy balance in interactions in space is that all velocities in space become related to the velocity of space in the fourth dimension, and all gravitational states in space become related to the gravitational state of spherically closed space.

16.2 Oscillating Electromagnetic Dipole

16.2.1 Electric Dipole in 3-Dimensional Space; the Standard Solution

Moving electric charges result in electromagnetic radiation through the buildup of changing electric and magnetic fields as described by Maxwell's equations. The electric and magnetic fields produced by an oscillating electric dipole at distance r $(r/z_0 > 2z_0/\lambda)$ can be expressed as

$$\vec{\varepsilon} = \frac{\Pi_0 \omega^2 \sin\theta}{4\pi\varepsilon_0 r c^2} \sin(kr - \omega t)\hat{\mathbf{r}}_\theta \tag{16.1}$$

and

$$\vec{\mathcal{B}} = \frac{1}{c}\varepsilon\,\hat{\mathbf{r}}_\varphi = \frac{\Pi_0 \omega^2 \sin\theta}{4\pi\varepsilon_0 r c^3} \sin(kr - \omega t)\hat{\mathbf{r}}_\varphi \tag{16.2}$$

where θ is the angle between the dipole and the distance vectors and

$$\Pi_0 = N e z_0 \tag{16.3}$$

is the peak value of the dipole momentum, where N is the number of unit charges, e, oscillating in a dipole of effective length z_0. Both field vectors, $\vec{\varepsilon}$ and $\vec{\mathcal{B}}$ and are perpendicular to the distance vector r. The Poynting vector, showing the direction of the energy flow, has the direction off $\hat{\mathbf{r}}$ (see Figure 16.1).

The energy density of radiation can be expressed as

$$\mathrm{E} = \varepsilon_0 \mathcal{E}^2 = \frac{\Pi_0^2 \mu_0 \omega^4 \sin^2\theta}{16\pi^2 r^2 c^2} \sin^2(kr - \omega t) \tag{16.4}$$

where $\mathcal{E}_0$ has been expressed in terms of μ_0 as $\mathcal{E}_0 = 1/\mu_0 c^2$. The average energy density of radiation is

$$\mathrm{E}_{\mathrm{ave}} = \frac{\mathrm{E}_0}{2\pi}\int_0^{2\pi} \sin^2(kr - \omega t)\, d(\omega t) = \frac{1}{2}\mathrm{E}_0 = \frac{\Pi_0^2 \mu_0 \omega^4}{32\pi^2 r^2 c^2}\sin^2\theta \tag{16.5}$$

The average power radiating through a sphere with radius r around the radiating dipole is

$$p = \left\langle \frac{dE}{dt} \right\rangle = \int_s c\mathrm{E}_{\mathrm{ave}}\, dS = \frac{\Pi_0^2 \mu_0 \omega^4}{32\pi^2 r^2 c}\int_s \sin^2\theta\, dS = \frac{\Pi_0^2 \mu_0 \omega^4}{12\pi c} \tag{16.6}$$

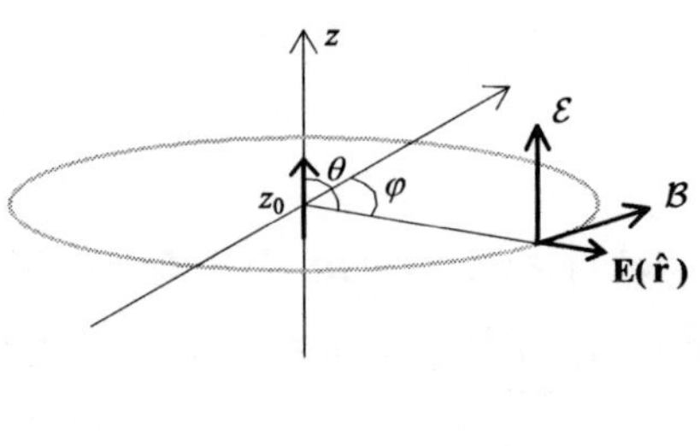

FIGURE 16.1
An electric dipole in the direction of the z-axis results in maximum radiation density in the normal plane of the dipole, θ = π/2.

By substituting equation 16.3 for Π_0 in equation 16.6, the energy flow in one cycle can be expressed as

$$E_\lambda = \frac{p}{f} = \frac{N^2 e^2 z_0^2 \mu_0 16\pi^4 f^4 \lambda}{12\pi c f} N^2 \left(\frac{z_0}{\pi}\right)^2 \frac{2}{3}\left(2\pi^3 e^2 \mu_0 c\right) f \tag{16.7}$$

In equation 16.7 the angular frequency ω has been converted to frequency $f = \omega/2\pi$, and the length of the dipole z_0 has been related to the wavelength $\lambda = c/f$.

Equation 16.7 means that the energy emitted by an electric dipole in a cycle is directly proportional to the frequency emitted. The factor 2/3 in equation 16.7 is the ratio of the average power emitted to all space directions to the maximum power emitted in the normal plane of the dipole. The factor $2\pi^3 e^2 \mu_0 c$, has the dimensions of momentum-length, like Planck's constant h, and has the numerical value of $2\pi^3 e^2 \mu_0 c_0$ is $5.997 \cdot 10^{-34} = h/1.1049$ [kgm²/s].

16.2.2 Point Source as an Electric Dipole in the Fourth Dimension

In one cycle of emission, a point source at rest in space moves a distance

$$z_4 = cdt = \frac{c}{f} = \lambda \tag{16.8}$$

in the fourth dimension characterized by line element i cdt in an imaginary direction perpendicular to space directions. Accordingly, emission to any space direction from a dipole in the fourth dimension appears like emission in the normal plane; the angle θ in equations 16.1 and 16.2 is constrained to the value $\pi/2$ for electric and magnetic fields in any space direction. This means that in the integrated energy of radiation of one cycle in equation 16.7, the factor 2/3 in the power density distribution is replaced by 1.

A quantum emitter, a hypothetical ideal dipole in the fourth dimension ($z_0 = z_4 = \lambda$), in which a single oscillation cycle of a unit charge ($N = 1$) results in the emission of one energy quantum in one cycle of radiation $E_0\lambda = hf$, can be expressed as

$$\begin{aligned} E_{0\lambda} &= \chi_\lambda (2\pi^3 e^2 \mu_0 c) f = \chi_\lambda \cdot 5.99695618 \cdot 10^{-34} \cdot f \\ &= hf = 6.626068765 \cdot 10^{-34} \cdot f \end{aligned} \tag{16.9}$$

The numerical values e, μ_0, h, and c equation 16.9 are based on CODATA 1998 recommended values. The constant χ_λ obtains the numerical value

$$\chi_\lambda \approx 1.104905316 \tag{16.10}$$

χ_λ combines the effects of the local geometry of space on the local velocity of light and a possible geometrical factor related to a dipole in the fourth dimension. Applying equation 16.9, Planck's constant can be expressed as

$$h = 2\pi^3 \chi_\lambda e^2 \mu_0 c \tag{16.11}$$

which expresses Planck's constant in terms of the dimensionless constant χ_λ the unit charge e, the vacuum permeability μ_o, and the velocity of light c. For a unified expression of energies we rewrite equation 16.9 as

$$E_{0\lambda} = hf = h_0 fc = \frac{h_0}{\lambda} c^2 = c\,|\,p\,| = m_{0\lambda} c^2 \tag{16.12}$$

where h_0 is defined as the intrinsic Planck's constant with dimensions of [kgm] instead of [kgm²/s] of the traditionally defined Planck's constant

$$h_0 \equiv \frac{h}{c} = \chi_\lambda \cdot 2\pi^3 e^2 \mu_0 = 2.210219 \cdot 10^{-42} [\text{kg} \cdot \text{m}] \tag{16.13}$$

and $m_{0\lambda}$ is the mass equivalence of a quantum of radiation

$$m_{0\lambda} = \frac{h_0}{\lambda} [\text{kg}] \tag{16.14}$$

Applying the intrinsic Planck's constant, the momentum of a quantum of radiation with wavelength λ can be expressed as

$$p_{0\lambda} = \frac{h_0}{\lambda} c = h_0 f = m_{0\lambda} c \tag{16.15}$$

Equation 16.14 relates the wavelength to the mass equivalence of a quantum of radiation

$$\lambda = \frac{h_0}{m_{0\lambda}} \tag{16.16}$$

As shown by equations 16.12 to 16.16, the intrinsic Planck's constant is related to the wavelength of radiation rather than to the momentum of radiation, which is how the traditional Planck's constant is related.

16.2.3 The Fine Structure Constant

Application of the intrinsic Planck's constant h_0 to the traditional definition of the fine structure constant α gives the expression of the fine structure constant in the form

$$\alpha \equiv \frac{e}{2h\varepsilon_0 c} = \frac{e^2 \mu_0 c}{2h} = \frac{e^2 \mu_0 c}{2h_0 c} = \frac{e^2 \mu_0}{2h_0} \tag{16.17}$$

which shows that the fine structure constant is not a function of the velocity of light. By applying equation 16.13 in equation 16.17, the fine structure constant obtains the form

$$\alpha = \frac{e^2\mu_0}{2 \cdot \chi_\lambda \cdot 2\pi^3 e^2 \mu_0} = \frac{1}{4\pi^3\chi_\lambda} \simeq 7.297352533 \cdot 10^{-3} \simeq \frac{1}{137} \tag{16.18}$$

which shows α as a purely mathematical, dimensionless constant without connections to any physical constants.

16.2.4 Unified Expression of Electromagnetic Energy

Equation 16.12 shows the energy of a cycle of electromagnetic radiation emitted by a single transition of a unit charge in a point source. The energy of a cycle of radiation emitted by a transition of N unit charges is

$$E_\lambda = c\,|\mathbf{p}_\lambda| = N^2 \frac{h_0}{\lambda} c^2 = m_\lambda c^2 \tag{16.19}$$

where m_λ is the mass equivalence of the a cycle of radiation. By applying the vacuum permeability μ_0 or the fine structure constant α, the Coulomb energy can be expressed as

$$E_{EM} = -\frac{q_1 q_2}{4\pi\varepsilon_0 r} = -N_1 N_2 \frac{e^2\mu_0}{4\pi r} c^2 = -N_1 N_2 \alpha \frac{h_0}{2\pi r} c^2 = -m_{EM} c^2 \tag{16.20}$$

where

$$m_{EM} = \left| \frac{q_1 q_2 \mu_0}{4\pi r} \right| \tag{16.21}$$

has the dimension of mass [kg] and is referred to as the mass equivalence of an electromagnetic energy object. As illustrated by equations 16.19 and 16.20, electromagnetic energy both as radiation and as Coulomb energy obtains a form identical to the expression of rest energy of a mass object.

16.2.5 Energy States of Hydrogen-Like Atoms

Due to the fundamental nature of the fine structure constant, it is illustrative to express the energy states of atoms in terms of the fine structure constant rather than in terms of Rydberg's constant R. The standard non-relativistic solution of energy states of electrons in a hydrogen-like atom is solved from Schrödinger's equation as

$$E_{z,n} = R \cdot hc \left(\frac{Z}{n}\right)^2 = \frac{m_e e^4}{8\varepsilon_0^2 h^2} \left(\frac{Z}{n}\right)^2 \tag{16.22}$$

where m_e is the mass of an electron, e is the unit charge of the electron, Z is the number of protons in the atom, and n is a positive integer. By applying the fine structure constant defined in equation 16.17, equation 16.22 can be expressed in the form

$$E_{z,n} = \frac{\alpha^2}{2}\left(\frac{Z}{n}\right)^2 m_e c^2 \tag{16.23}$$

where m_e is the rest mass electron (corrected with the effect of the nucleus mass M_N $[1/(1 + m_e/M_N)]$), and α is the fine structure constant defined in equation 16.17. The expression given in equation 16.23 is of special importance when drawing conclusions from the effects of the novel interpretation of a quantum on the rest energy of an electron and the energy states and characteristic emission frequencies of atoms.

The successful interpretation of a point source as a dipole in the fourth dimension suggests the interpretation of space as three dimensional environment moving at velocity c in a fourth dimension with metric nature. In such an interpretation the rest energy of mass appears as the energy of motion mass possesses due to the motion of space. Conservation of total energy in space means that all velocities in space become related to the velocity of space in the fourth dimension. As a further consequence, the local rest energy of mass appears a function of local motion and gravitation, which in equation 16.23 means that the energy states and the characteristic emission frequencies of atoms become functions of the local motion and gravitation. In fact, the effect of motion and gravitation on locally "available" rest energy converts Einsteinian spacetime with proper time and distance to dynamic space in absolute time and distance [1,2].

16.3 Space as Spherically Closed Surface of a 4-Sphere

16.3.1 Momentum of Mass Due to the Motion of Space in the Fourth Dimension

A fourth dimension of metric nature makes it possible to describe three-dimensional space as a closed "surface" of a 4-sphere expanding at velocity c in a zero-energy balance with the gravitation of the structure in the direction of the 4-radius as described in the Dynamic Universe approach [1,2]. In such a concept, mass has the meaning of the substance for the expression of energy rather than a form of energy. Mass at rest in space has momentum $p_4 = mc_4$ due to the motion of space in the fourth dimension, and like for radiation propagating at velocity $c = c_4$ in space, the energy of motion becomes equal to $E = c|\mathbf{p}_4|$ (see Figure 16.2).

The expansion velocity c_4 of space in the direction of the R_4 is determined by a zero energy balance between the energies of motion and gravitation of

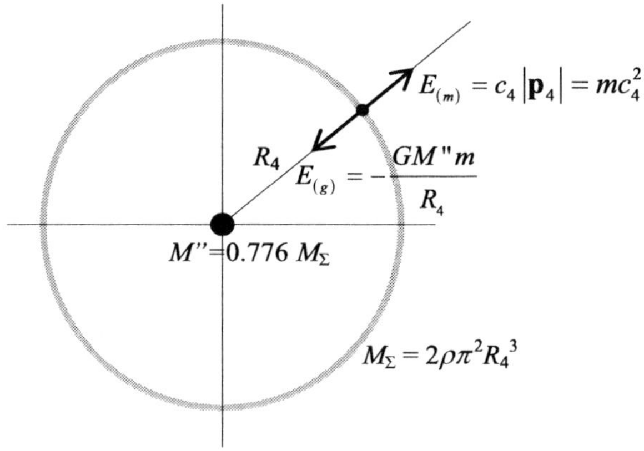

FIGURE 16.2
Space as a spherically closed structure. The barycenter of the structure is in the center of the 4-sphere. Integrated gravitational energy of mass *m* in spherically closed space can be expressed with the aid of the mass equivalence $M''= 0.776 - M_\Sigma$ of space, where M_Σ is the total mass in space.

the 4-sphere

$$c_4 = \pm\sqrt{\frac{GM''}{R_4}} = \pm\sqrt{\frac{GI_g M_\Sigma}{R_4}} \tag{16.24}$$

where G is the gravitational constant, c_4 the velocity in the direction of the radius R_4 of the 4-sphere, and $M'' = I_g \cdot M_\Sigma$ the mass equivalence of the total mass M_Σ in space. The factor $I_g = 0.776$ comes from the integration of the gravitational energy of a 4-sphere. Conservation of energy in interactions in space requires that the maximum velocity obtainable in space is equal to the expansion velocity c_4, which means that $c_0 = c_4$ is the velocity of light in hypothetical homogeneous space. *The velocity of light is not an independent physical constant but bound to the velocity of space in the direction of the 4-radius.*

16.3.2 The Effect of Local Gravitation and Motion on the Rest Energy of an Object

In the Dynamic Universe approach, the energy of mass due to the momentum in the direction of the 4-radius of space is $E_0 = c_0 |\mathbf{p}_4|$, which is the rest energy of mass at rest in hypothetical homogeneous space, the primary energy of mass in space. Conservation of the primary energy in interactions in space means that an increase of momentum in space is associated with a reduction of the momentum the mass object possesses in the fourth dimension. In a detailed analysis [1,2,5] the rest energy of mass object m in space can expressed as

$$E = c_0 mc \tag{16.25}$$

where c_0 is the velocity of light in hypothetical homogeneous space equal to the velocity of the expansion of space in the 4-radius of the structure, c is the local velocity of light which is reduced due to tilting of space close to local mass centers. Taking into account the system of n cascaded gravitational frames in space the local velocity of light can be expressed as

$$c = c_0 \prod_{i=l}^{n} (1 - \delta_i) \tag{16.26}$$

Mass m in equation 16.25 is the rest mass "vailable" in the n:th local energy frame

$$m = m_0 \prod_{i=1}^{n-1} \sqrt{1 - \beta_t^2} \tag{16.27}$$

where m_0 is the rest mass of the object at rest in hypothetical homogeneous space. Velocity $\beta_{n-1} (= v_{n-1}/c_{n-1})$ means the velocity on the n:th frame (as an energy object) in the $(n-1)$:the frame and δ_i is the gravitational factor of the object in the i:th frame

$$\delta_1 = \frac{GM_t}{r_t c^2} \tag{16.28}$$

16.3.3 Characteristic Emission and Absorption Frequencies and Wavelengths of Atoms

Application of equations 16.25, 16.26, and 16.27 in equation 16.23 allows the expression of the Balmer's formula for characteristic frequencies to be expressed as

$$f_{(n1,n2)} = \frac{\Delta E_{(n1,n2)}}{h_0 c_0} = Z^2 \left[\frac{1}{n_1^2} - \frac{1}{n_2^2} \right] \frac{\alpha^2}{2h_0} m_e c = f_{0(n1,n2)} \prod_{i=1}^{n} (1 - \delta_i) \sqrt{1 - \beta_i^2} \tag{16.29}$$

where $f_{0(n1,n2)}$ is the frequency of the transition for an atom at rest in hypothetical homogeneous space

$$f_{0(n1,n2)} = Z^2 \left[\frac{1}{n_1^2} - \frac{1}{n_2^2} \right] \frac{\alpha^2}{2h_0} m_{e(0)} c_0 \tag{16.30}$$

As shown by the second form of equation 16.29, the characteristic frequency is directly proportional to the local velocity of light, which means the velocity of light is observed as constant in local measurements with an atomic clock. The velocity of the expansion of space, $c_0 = c_4$, is a function of the time from singularity. Accordingly, the velocity of light and the frequency of atomic oscillators slow down equally with the expansion of space.

Balmer's formula for characteristic wavelengths obtains the form

$$\lambda_{(n1,n2)} = \frac{c}{f_{(n1,n2)}} = \frac{c_0}{f_{0(n1,n2)}} \frac{\prod_{i=1}^{n}(1-\delta_i)}{\prod_{i=1}^{n}(1-\delta_i)\sqrt{1-\beta_i^2}} = \frac{\lambda_{0(n1,n2)}}{\prod_{i=1}^{n}\sqrt{1-\beta_i^2}} \tag{16.31}$$

which shows that unlike the characteristic frequencies, the characteristic wavelengths of atoms are not a function of the velocity of light. By applying the Bohr radius $a_0(0)$, the characteristic wavelength of atoms can be expressed as

$$f_{0(n1,n2)} = Z^2\left[\frac{1}{n_1^2} - \frac{1}{n_1^2}\right]\frac{a^2}{2h_0} m_{e(0)}c_0 \tag{16.32}$$

which shows that the wavelength emitted is directly proportional to the Bohr radius of the atom. Equation 16.32 is just another form of Balmer's formula, which does not require any assumptions tied to the nature of the fourth dimension or the motion of space. Equation 16.32 also means that, like the dimensions of an atom, the characteristic emission and absorption wavelengths of an atom are unchanged in the course of the expansion of space.

When applied in a single frame equation 16.29 can be expressed as

$$f_{\delta,\beta} = f_{0,0}(1-\delta)\sqrt{1-\beta^2} \approx f_{0,0}\left(1-\delta-\frac{1}{2}\beta^2-\frac{1}{8}\beta^4+\frac{1}{2}\delta\beta^2\right) \tag{16.33}$$

which for the first order of β^2 and δ is the same as the corresponding equation derived in the general relativity theory for an oscillator moving in a gravitational frame

$$f_{\delta,\beta(GR)} = f_{0,0}\sqrt{1-2\delta-\beta^2} \approx f_{0,0}\left(1-\delta-\frac{1}{2}\beta^2-\frac{1}{8}\beta^4+\frac{1}{2}\delta\beta^2-\frac{1}{2}\delta^2\right) \tag{16.34}$$

In a constant gravitational potential characterized by gravitational factor δ_A, equation 16.33 obtains the form

$$f_\beta = f_{0,0}(1-\delta_A)\sqrt{1-\beta^2} = f_{\delta A,0}\sqrt{1-\beta^2} \tag{16.35}$$

which shows the effect of motion on the frequency. Equation 16.35 is formally identical to the corresponding result of special relativity. However, instead of relying on the concept of proper time and a velocity relative to an observer, equation 16.35 relies on the on the effect of the velocity on the characteristic frequency through the effect of a reduced rest energy of electrons in equation 16.29. The velocity in equation 16.35 means velocity relative to the state of rest in the local energy system where the velocity has been obtained; in an accelerator it means the state of a non-accelerated object.

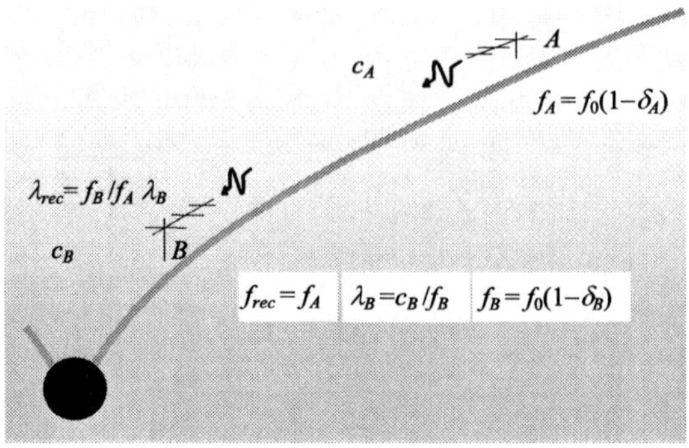

FIGURE 16.3
The velocity of light is lower close to a mass center, $c_B < c_A$, which results in a decrease of the wavelength of electromagnetic radiation transmitted from A to B. Accordingly, the signal received at B is blueshifted relative to the reference wavelength observed in radiation emitted by a similar transmitter in the δ_B-state. The frequency of the radiation, the number of quanta in a time interval, is unchanged.

16.3.4 Gravitational Shift of Electromagnetic Radiation

As shown by equations 16.29, 16.33, and 16.35 the frequencies of atomic oscillators are functions of the gravitational potential. As shown by equation 16.31 the wavelength of the radiation emitted by an atomic oscillator at different gravitational potentials is unchanged because of the equal changes in the frequency and the velocity of light.

The frequency of electromagnetic radiation passing from one gravitational potential to another, the number of cycles (or quanta) transmitted in a time interval is not subject to a change during the transmission. The wavelength of radiation sent from a different gravitational potential, however, is changed due to the difference in the velocity of light in different gravitational potentials (see Figure 16.3).

16.3.5 The Doppler Effect of Electromagnetic Radiation

Equation 16.30 allows the derivation of the Doppler effect of electromagnetic radiation by combining the effect of motion on the frequency and wavelength in equations 16.26 and 16.28 with a classical wave mechanical procedure. In a general form, the frequency transmitted from an oscillator A (δ_A,β_A) to a receiver (reference oscillator) $B(\delta_A,\beta_A)$ is expressed as

$$f_{A(B)} = f_B \frac{\prod_{j=k+1}^{n} (1-\delta_{B_j})\sqrt{1-\beta_{B_j}^2}\,(1-\beta_{jB(r)})}{\prod_{j=k+1}^{m} (1-\delta_{Ai})\sqrt{1-\beta_{Ai}^2}\,(1-\beta_{iA(r)})} \tag{16.36}$$

where $\beta_{iA(r)}$ is the component of the velocity of A in the direction of the distance vector r_{Ai}, and $\beta_{jB(r)}$ is the component of the velocity of B in the direction of the distance vector $\mathbf{r}_{Ai,Bj}$ in the i:th and j:th frame, respectively.

Conclusions

The solution given by Maxwell's equations for the energy of a single oscillation cycle of a unit charge in a dipole in the fourth dimension gives a natural interpretation to the nature of a quantum as the minimum dose of electromagnetic radiation. The interpretation of a point source as a dipole in the fourth dimension becomes obvious if we give the fourth dimension a metric meaning instead of considering it a time-like dimension of the Einsteinian spacetime. A fourth dimension of a metric nature makes it possible to describe three-dimensional space as a closed "surface" of a 4-sphere expanding at velocity c in a zero-energy balance with the gravitation of the structure in the direction of the 4-radius [1,2].

Spherically closed dynamic space converts Einsteinian spacetime in dynamic coordinates to dynamic space in absolute coordinates. The dynamic perspective to space became quite natural since the observations of Edwin Hubble which were not available in early 1900's when the spacetime concept was created. Also, many contemporary questions related to atomic clocks and GPS satellites are easier to tackle and understand on the basis of the dynamic approach studied in detail in the Dynamic Universe theory.

The Dynamic Universe theory actually introduces a paradigm shift comparable to that of Copernicus when he removed the center of universe from Earth to the Sun. In the present perspective, the universe is revealed to be a four dimensional entity which orders space to appear as the surface of a four dimensional sphere. This sphere, the three-dimensional space, is not held static by the famous cosmological constant, but it is expanding because of an overall zero energy balance between motion and gravitation. Conservation of the total energy in space also links local motion and gravitation to the rest energy of objects allowing the build-up of localized energy structures and material objects. The same pattern makes the ticking frequency of atomic clocks a function of the gravitational state and motion of the clock.

In addition to the nature of quantum as the minimum dose of electromagnetic radiation, Mach's principle, the nature of inertia, and the rest energy of matter, this comprehensive framework gives precise predictions to recent observations on the redshift and magnitude of distant supernova explosions without a need to postulate dark energy or accelerating expansion of space. It also explains the Euclidean appearance of distant space and the apparent discrepancy between the ages of oldest stars obtained by radioactive dating and the age of expanding space, which has remained a mystery.

Acknowledgments

The presentation of this paper in the symposium "The Nature of Light: What is a Photon?" in SPIE conference in San Diego 2005 has been supported by the Foundation of the Finnish Society of Electronics Engineers. The author expresses his gratefulness to the Foundation and the Society.

References

Tuomo Suntola, *Theoretical Basis of the Dynamic Universe*, ISBN 952-5502-1-04, 290 pages, Suntola Consulting Ltd., Helsinki, 2004.

Tuomo Suntola, *"Dynamic space converts relativity into absolute time and distance"*, Physical Interpretations of Relativity Theory IX (PIRT-IX), London, 3–9 September 2004.

Einstein, A., *Kosmologische Betrachtungen zur allgemeinen Relativitätstheorie*, Sitzungsberichte der Preussischen Akad. d. Wissenschaften (1917).

Feynman, R., Morinigo, W., Wagner, W., Feynman Lectures on Gravitation (during the academic year 1962–63), Addison-Wesley Publishing Company (1995), p. 10.

Tuomo Suntola, *"Observations support spherically closed dynamic space without dark energy"*, SPIE Conference "The Nature of light: What is a Photon? (SP200)", San Diego, 2005.

Tuomo Suntola and Robert Day, "Supernova observations fit Einstein-deSitter expansion in 4-sphere", arXiv/astro-ph/0412701 (2004).

17

Propagating Topological Singularities: Photons

R. M. Kiehn

Physics Department, University of Houston, Houston, Texas, USA

CONTENTS

Abstract

At its foundations, Maxwell's theory of electrodynamics, like thermodynamics, is a topological theory independent from geometric constraints of metric, scales, or gauge symmetries. One of the most interesting features of electromagnetism is its relationship to the transport of momentum and energy by means of photons. This article utilizes a topological perspective to discuss the features and concepts associated with photon, including spin, helicity and chirality.

Key words: photon, topological torsion, topological spin, polarization, helicity, propagating topological singularities.

17.1 Topological Perspective

At its foundations, Maxwell's electrodynamics is a topological theory independent from the geometric constraints of metric scales or gauge symmetries. The fundamental partial differential equations were shown to be metric free by Van Dantzig[14] in the 1930's. In the first half of the 20th century the dogma of quantum mechanics combined with relativity, led to the idea that electromagnetic radiation was composed of quanta or photons carrying integer spin (angular momentum), $L = n\hbar$, energy $E = n\hbar\omega$ momentum, $p = n\hbar/\lambda$. In 1932 Fock[3] demonstrated that the singular solutions to the Maxwell PDE's satisfied the eikonal equation, and gave formal realization of what constituted an electromagnetic signal (a definition well beyond the Einstein conjecture that "needle" radiation would follow geodesic paths).

The zero sets of the eikonal solutions represent propagating discontinuities (a topological defect formed by limit sets) in the field strengths. The finite propagation speeds were 4 fold degenerate (equal to C) in spaces constrained by the geometric symmetries of the Lorentz constitutive equations. In general, the propagation speeds of the singular solutions admit different speeds for different polarizations and for different directions of propagation; i.e., the 4 fold geometric degeneracy can be broken and the speed of light need not be the same in the outbound and inbound directions, say, in a rotating expanding plasma.[9] Equivalence classes of inertial frames of reference can be defined such that each observer in the equivalence class would agree that an electromagnetic signal was a propagating singularity. Fock demonstrated that the only linear transformations that preserved the signal discontinuity were the Lorentz transformations. It is this invariance of the field discontinuity that gives physical stature to the equivalence class of reference frames constructed with Lorentz transformations. However, it is now known, but not widely utilized in engineering practice, that the extended (conformal or Poincare) Lorentz transformations also preserve the concept of signal discontinuity. Moreover, the eikonal solutions can be identified with isotropic null vectors, defined by Cartan as spinors, which are not necessarily single valued with respect to extended Lorentz transformations.

The topological theory of classical electromagnetism is constructed in terms of two exterior differential systems, which have a correspondence with thermodynamics in that the first exterior differential system deals with thermodynamic intensities, and the second exterior differential system deals with thermodynamic quantities (or differential densities). The two exterior differential systems, $F - dA = 0$, and $J - dG = 0$, act as topological constraints on the variety of independent variables, say $\{x, y, z, t\}$. These two fundamental constraints lead algebraically to two other independent topological concepts of topological torsion, $A \hat{} F$, and topological spin, $A \hat{} G$, both of which are explicitly dependent upon the concept of potentials, $\{\mathbf{A}, \phi\}$. The exterior derivative of these 3-forms creates the two familiar Poincare deformation invariants as topological limit sets of an electromagnetic system, valid in the vacuum or plasma state. Non-zero values of the Poincare invariants are the source of topological change and irreversible phenomena in non-equilibrium thermodynamics. When the Poincare invariants vanish, the closed integrals of $A \hat{} F$ and $A \hat{} G$ exhibit topological invariant properties similar to the "quantized" chiral and spin properties of a photon. The "quantization" result is a topological result (independent from any microscopic or macroscopic constraint) related to the integers, and similar to the obvious fact that the number of holes in a surface is always an integer; 1.439 holes does not make sense. In the opinion of the author, the new 3-forms and their dynamics (which vanish in equilibrium electrodynamic systems) will lead to many new practical applications which will utilize non equilibrium thermodynamic properties of electromagnetic systems.

17.2 Down With Dogma

This chapter may startle the reader with what might appear to be a bit of heresy relative to the classical teachings of electromagnetism, which currently are presented dogmatically in terms of a geometrical perspective. The ultimate topic of discussion herein is the photon. The perspective of this chapter is based upon topology, not geometry.

The first somewhat heretical claim is: Maxwell's theory of electromagnetism is a topological theory, not a geometric theory, and can be deduced from logical principles.

Although admittedly "discovered" through a historical series of geometrically dominated or constrained experiments, and then summarized and augmented with an inspired guess by J. C. Maxwell, it should be recognized that the PDE's of electrodynamic theory can be deduced from mathematical logic, without the use of geometric constraints of metric, size and shape, or even experiment. For example, the sequence of logical steps which produce the Maxwell Faraday partial differential equations starts with:

1. An ordered set {1, 2, 3, 4 ...}, followed by
2. An ordered set of independent variables with neighborhoods,

$$\{\xi^1, \xi^2, \xi^3, \xi^4 \ldots; d\xi^1, d\xi^2, d\xi^3, d\xi^4 \ldots\}, \tag{17.1}$$

3. Upon which an ordered set of C2 functions $\{A_k(\xi^1, \xi^2, \xi^3, \xi^4 \ldots)\}$ is used to construct a C2 differentiable 1-form of Action, A. For electromagnetism, the coefficients of the 1-form play the role of the classic vector and scalar potentials:

$$A = A_k(\xi^j)d\xi^k. \tag{17.2}$$

4. An abstract topological neighborhood constraint is imposed in terms of an exterior differential system,

$$\text{Constraint of thermodynamic Intensities } F - dA = 0. \tag{17.3}$$

The 2-form $F = dA$ is required to be exact, which leads to the classic electromagnetic flux conservation law. The topological constraint implies that the domain of support of the 2-form F (in engineering language, the **E** and the **B** field intensities) cannot be compact without a boundary. In effect, it denies the existence of magnetic monopoles. Relative to even dimensional spaces, a 2-form of maximal rank generates a Symplectic manifold as the domain of support of the field intensities.

5. Exterior differentiation of the topological constraint, and use of the Poincare theorem on C2 differentiable functions, creates an ordered set of partial differential equations from the coefficients of the equations:

$$dF = ddA = 0. \tag{17.4}$$

The first four equations of this ordered set of of PDE's have the format of the Maxwell Faraday partial differential equations, which, by relabeling the partial derivatives[16] of the abstract coefficients, $A_k(\xi^j)$, are equivalent to the expressions,

$$\text{Maxwell Faraday :}$$

$$div\ \mathbf{B} = 0, \tag{17.5}$$

$$curl\ \mathbf{E} + \partial\mathbf{B}/\partial t = 0. \tag{17.6}$$

There are no additional terms, and no other field functions, no matter how many independent variables (≥4) are used in the construction of the abstract 1-form of action. If more than 4 independent variables (geometric dimensions) are used, the new "coordinates" add new PDE's that couple "new" field variables to the **E** and **B** field variables of the first four (Maxwell) equations, but do not alter the format of the first four PDE's—the Maxwell Faraday equations—in any way. The Maxwell Faraday equations are valid in a universal sense, nested in the totality of the ordered set of variables. No metric ideas were used in this logical "deduction" of the Maxwell Faraday PDEs. The concept of Faraday induction is universal for all thermodynamic systems that can be encoded by a 1 form of action, A. It may be startling, but true, that hydrodynamics and mechanics, as well as electromagnetics, when encoded in terms of a 1-form of action, are governed by the Faraday induction law.

From a thermodynamic point of view, the 2-form F, is related to thermodynamic intensities (objects which are homogeneous of degree 0, like temperature and pressure). However, the complete Maxwell system utilizes not only an exact 2-form, F, but also recognizes that there exists another thermodynamic set of conjugate variables, related to quantities or excitations (objects which are homogeneous of degree 1, like entropy and volume). In short, topological electromagnetism presumes that there exists a 2-form density G, which is closed but not exact. The non-exact 2-form G can have domains of support which are compact without boundary, while the exact 2-form F cannot. Exterior differentiation of G produces a 3-form of charge—current density, J, equivalent to a second topological constraint:

$$\text{Constraint of thermodynamic quantities (densities) } J - dG = 0. \quad (17.7)$$

This topological constraint leads to the Maxwell ampere PDE's, and as J is exact, leads to the conservation of charge. With appropriate relabeling, the Maxwell ampere equations are:

Maxwell ampere :

$$div\ \mathbf{D} = \rho \quad (17.8)$$

$$curl\ \mathbf{H} = \partial \mathbf{D}/\partial t + \mathbf{J}. \quad (17.9)$$

The guess of a $\partial\mathbf{D}/\partial t$ term introduced by Maxwell is automatic in the topological system.

Note that differential form densities, such as G and J, can be integrated without metric. The two systems of PDE's generated by exterior differentiation of the topological constraints are diffeomorphically invariant, meaning they are functionally well defined for all diffeomorphically equivalent coordinate systems, be they Galilean, Lorentz, spherical, or anything else if the mapping functions are homeomorphically equivalent and differentiable. However, the differential form constraints are not constrained to diffeomorphic (tensor) equivalences. The topological differential form statements, and therefore Maxwell PDE's, are well defined (via the pullback substitutions) with respect to submersions from higher dimensional spaces (think fiber bundles) which are not invertible, but are differentiable. The bottom line is that Lorentz (diffeomorphic) invariance of the PDE's is trivial, as they are tensor equations. So the what makes the Lorentz equations so dogmatically important? The answer resides with the fact that the *singular solutions* of the PDE's, *not the equations*, have a linear equivalence class generated by only the Lorentz transformations.

The second somewhat heretical claim is: An electromagnetic signal is a propagating discontinuity (a propagating topological defect), not a sinusoidal wave!

Actually this idea was developed by V. Fock about 1932, where he demonstrated (following Hadamard's ideas[4] of characteristics) that the hyperbolic PDE's of Maxwell admitted singular solutions upon which the field intensities were not uniquely defined. These singular point sets can admit zero field intensities on one side and finite non zero field intensities on the other side of the singular solution submanifold. The singular point sets are not stationary and represent propagating discontinuities, with a speed $C = \pm 1/\sqrt{\varepsilon\mu}$ in simple cases. The equivalence class of reference systems which are linearly related and preserve the singular solutions have a common fact: the singularities propagate at a finite constant and invariant speed, C. The singular set was defined by Fock in terms of a solution, ϕ, to the eikonal equation, which is a non-linear first order quadratic PDE equal to zero:

$$\{\pm (\partial\phi/\partial x)^2 \pm (\partial\phi/\partial y)^2 \pm (\partial\phi/\partial z)^2 \mp (\partial\phi/c\partial t)^2\} = 0. \tag{17.10}$$

The zero set of the singular solution set defines an implicit hypersurface in space time. (Note that Majorana, Weyl and Dirac spinors are related to the differences in signs and there solution representations.)

More importantly, Fock demonstrated that the only *linear* transformation of coordinates that preserved the propagating field discontinuity was the Lorentz transformation. That is, if two observers were related by a Lorentz transformation, then if the first observer claimed to see a propagating discontinuity (signal), then so would the (Lorentz related) second observer claim to see a propagating discontinuity (signal) and each would say the propagation speed was the same constant C. The importance of the Lorentz transformation is that it defines an equivalence class of ("inertial") systems (for observers that use electromagnetic means of measurement) that preserve the idea of a propagating discontinuity (signal). The Maxwell PDE's are well defined with respect to **all** diffeomorphic observers, but the singular eikonal solutions at constant speed C are well defined only with respect to the linear Lorentz equivalent observers.

It should be noted that Fock also demonstrated that there was a non-linear transformation that also preserved the concept of a propagating discontinuity. It is the fractional projective (Moebius) transformation. The speed of discontinuity propagation is not a constant, and can range from zero to infinity. Hence for Moebius related observers, the speed of a signal is not the constant value C of the Lorentz equivalent class. Such situations are also related to the conformal group. This mathematical result of Fock has yet to be exploited in practical electromagnetism.

The connection of the Fock–eikonal idea to the Einstein–null geodesic idea is that both are quadratic forms of a Minkowski signature. However, the Fock concept makes a direct connection to electromagnetic theory, while the Einstein concept does not.

$$\text{Null geodesic}: (đs)^2 = (dx)^2 + (dy)^2 + (dz)^2 - (dt)^2 \Rightarrow 0, \tag{17.11}$$

$$\text{Eikonal}: (\partial\phi/\partial x)^2 + (\partial\phi/\partial y)^2 + (\partial\phi/\partial z)^2 - (\partial\phi/c\partial t)^2 \Rightarrow 0. \tag{17.12}$$

Note that the square of the line element is not the square of an exact differential form; $đs$ can have path dependent values.

The eikonal solutions are not necessarily solutions to the wave equation. However, if an eikonal solution is also a solution to the wave equation, then any function of the eikonal solution is also a solution to the wave equation. The classic example is given by the (linear) phase function,

$$\phi = kz \pm \omega t, \tag{17.13}$$

which satisfies both the eikonal equation and the wave equation, if the constants satisfy the equation,

$$\omega/k = \pm c. \tag{17.14}$$

An important concept is that if an eikonal solution, ϕ, is also a solution to the wave equation, then any function of the eikonal, $F(\phi)$ is also a solution to the wave equation. In 1914, in a small monograph entitled "Electrical and Optical Wave Motion," H. Bateman introduced a number of interesting solutions to Maxwell's equations that emulate propagating singular strings (not plane waves). Bateman is perhaps more famous for his work on the equations that describe the decay chains of radioactive species.

However, as pointed out by Whittaker,[17] it was Bateman who determined in 1910 that the Maxwell equations were invariant with respect to the conformal group, a much wider group than the Lorentz transformations. Bateman in 1910 also recognized the relationship of his work to the tensor calculus of Ricci and Levi-Civita, several years before the Einstein development of general relativity. Bateman[1] discusses various forms of transformations which lead to forming one wave function from another, including the Moebius transformation. He even describes methods for constructing a wave function from a solution to the diffusion equation. Bateman mentions that Stokes and Wiechert thought of x-rays as "pulses traveling through the aether, the energy being confined within a thin shell" (of discontinuities). However, there are solutions to the eikonal equation that are not solutions to the wave equation. This difference distinguishes a "signal" from a "wave".

The third somewhat heretical claim is: The concept of spinor solutions to Maxwell's equations is a topological idea that does not depend upon microphysical scales.

The impact of quantum mechanics, starting with Planck's concept of the "quantized" oscillator energy enabling the thermodynamic deduction of the blackbody radiation distribution law, the Einstein model for explaining the photoelectric effect, the Bohr atom description of the emission of light carrying off integer units of angular momentum and energy, the Compton analysis of the distribution peaks in the scattering of electromagnetic radiation by electrons, the deBroglie conjecture that energy and momentum were related to a "wave" analysis involving Planck's constant, frequency and reciprocal wavelength, and Dirac's description of the relativistic hydrogen atom, all have led to the idea that the "bundle" of energy and momentum now known as the Photon has a deep relationship to microphysics, and would appear to be associated with what Cartan called spinors. The philosophical problem is that these bundles of energy and momentum, these photons, can have extent and coherent interactions that are many orders of magnitude greater than the dimensions of the atoms and molecules, from which they supposedly originate. A fundamental question is how do the quantal properties of the photon emerge from a topological perspective?

First consider the concept of spinors. Without the use of micro scales, the idea of spinor solutions to Maxwell's electrodynamics comes from the topological perspective that the 2-form of field excitations, $F = dA$, can be represented by an anti-symmetric matrix. Then, depending on the rank of the matrix $[F]$ (in say 4D) the eigenvectors either have zero eigenvalues, or complex eigenvalues. In every case, if **e** is an eigenvector with eigenvalue γ such that

$$[\mathbb{F}]\circ|\mathbf{e}\rangle = \gamma|\mathbf{e}\rangle, \tag{17.15}$$

then,

$$\langle\mathbf{e}|\circ[\mathbb{F}]\circ|\mathbf{e}\rangle = \gamma\langle\mathbf{e}|\circ\langle\mathbf{e}|. \tag{17.16}$$

Due to antisymmetry of $[\mathbb{F}]$, it follows that

$$\langle\mathbf{e}|\circ[\mathbb{F}]\circ|\mathbf{e}\rangle = 0. \tag{17.17}$$

Hence, it must be true that

$$\gamma\langle\mathbf{e}|\circ|\mathbf{e}\rangle = 0. \tag{17.18}$$

For division algebras there are two choices: either $\gamma = 0$, or $\langle\mathbf{e}|\circ|\mathbf{e}\rangle = 0$. The implication is that for non zero eigenvalues γ, the quadratic form must vanish:

$$\langle\mathbf{e}|\circ|\mathbf{e}\rangle = (e^1)^2 + (e^2)^2 + (e^3)^2 + (e^4)^2 = 0. \tag{17.19}$$

(This concept can be extended to a diagonal unit matrix of any signature.) The null quadratic form is equal to the sum of squares of the components, if the eigenvalue is not zero. Either the eigenvector has zero components, or the eigenvector of the antisymmetric matrix is a complex vector which has been defined as a "null isotropic vector" in the theory of differential geometry. The null isotropic eigenvector direction fields are similar to vectors, but have complex components and non-zero complex eigenvalues. Such null isotropic vectors define spinors.[2] They have metric properties in the sense of a quadratic form (that has zero value), but not the unique affine properties (see p. 3,[2]) of tensors. Spinors generate harmonic forms and also are related to conjugate pairs of minimal surfaces. The bottom line is that spinors are normal consequences of antisymmetric matrices, and as topological artifacts are not restricted to physical microscopic or quantum constraints. According to the topological thermodynamic arguments, they should appear at all scales. Note that the 1-form of thermodynamic work, $W = i(V)dA$,[15] can be expanded in terms of a basis of spinors, and the extremal field, if it exists.

As an example, consider the 1-form of action and its associated Pfaff sequence given by the expressions

$$A = ydx - xdy + sdz - zds, \tag{17.20}$$

$$F = dA = 2dy\hat{}dx + 2ds\hat{}dz, \tag{17.21}$$

$$A\hat{}F = 2\{xdy\hat{}dz\hat{}ds - ydx\hat{}dz\hat{}ds + zdx\hat{}dy\hat{}ds - sdx\hat{}dy\hat{}dz, \tag{17.22}$$

$$F\hat{}F = 8dx\hat{}dy\hat{}dz\hat{}ds, \tag{17.23}$$

Note that the 4×4 antisymmetric matrix is of the form

$$[\mathbb{F}] = \begin{bmatrix} 0 & 1 & 0 & 0 \\ -1 & 0 & 0 & 0 \\ 0 & 0 & 0 & 1 \\ 0 & 0 & -1 & 0 \end{bmatrix}, \tag{17.24}$$

with eigenvalues and eigenvectors,

$$\text{Eigenvectors} = [0, 0, 1, i], [1, i, 0, 0] \text{ with eigenvalue} = i, \tag{17.25}$$

$$\text{Eigenvectors} = [0, 0, 1, -i], [1, -i, 0, 0] \text{ with eigenvalue} = -i, \tag{17.26}$$

Each eigenvector is null isotropic such that the sum of squares of the coefficients is zero. This example is a simple example generated by the 1-form, A, whose coefficients form the adjoint field to the three exact differentials generated by the Hopf map (a submersion from 4D to 3D).

The fundamental idea is that spinors are the natural format of propagating singularities generated from the eikonal equation. Topologically then, photons are represented by spinors that generate propagating discontinuities. It should also be noted that spinors are natural generators of conjugate pairs of minimal surfaces.

The fourth somewhat heretical claim is: The concept of photon quantization is a topological idea that does not depend upon microphysical scales.

From the topological formulation given above, in terms of exterior differential forms, $\{A, F, G, J\}$ the question arises as to how discrete (quantum) features of the photon enter into the topological theory. From thermodynamic arguments, if the Maxwell equations are uniquely integrable, then the maximum topological dimension of the 1-form of action is 2. That is, there exist two functions on the geometrical domain of 4D which generate all of the differential topology associated with the field intensities. Such is the domain of an isolated, or equilibrium, thermodynamic system. Exterior differential 3-forms do not exist on domains of isolated topology; the topological structure consists of a single connected component. On non-equilibrium domains, the topological dimension can be 3 or 4. Such domains support exterior differential 3-forms and 4-forms on multiple components of the topological structure. The question is how do you formulate the possible multiple component topological structures of non-equilibrium electrodynamic system? The answer is in terms of closed, but not exact, exterior differential 3-forms which are homogeneous of degree zero.

By inspection, from the set of exterior forms $\{A, F, G, J\}$ it is possible to construct two important 3-forms, that are related to 4 component "currents" on a 4D domain of $\{x, y, z, t\}$. The 3-forms are written in terms of engineering variables in the following equations. The objects are zero in isolated equilibrium systems. They are (topological) artifacts of non-equilibrium electromagnetic systems:

$$\text{Topological torsion} = A\hat{}F \text{ units h/e} \tag{17.27}$$

$$\mathbf{T}_4 = [\mathbf{E} \times \mathbf{A} + \mathbf{B}\phi, \mathbf{A} \circ \mathbf{B}] \tag{17.28}$$

$$\text{Topological spin} = A\hat{}G, \text{ units h} \tag{17.29}$$

$$\mathbf{S}_4 = [\mathbf{A} \times \mathbf{H} + \mathbf{D}\phi, \mathbf{A} \circ \mathbf{D}]. \tag{17.30}$$

These topological objects are universally defined for non equilibrium electromagnetic systems, yet their dynamics and properties have been little utilized. These 3-forms can have non-zero exterior differentials (which are exact exterior differential 4-forms) related to the historical Poincare invariants of the electromagnetic field:

$$\text{Poincare II } d(A\hat{}F) = F\hat{}F = 2(\mathbf{E} \circ \mathbf{B})dx\hat{}\, dy\hat{}\, dz\hat{}\, dt \tag{17.31}$$

$$\text{Poincare I } d(A\hat{}G) = F\hat{}G - A\hat{}J = \{\mathbf{B} \circ \mathbf{H} - \mathbf{D} \circ \mathbf{E}\} - \{\mathbf{A} \circ \mathbf{J} - \rho\phi\}. \tag{17.32}$$

The closed integrals of these 4-forms are topological properties that are evolutionary invariants of all processes that can be represented (to within a factor) by vector fields, $\mathbf{V}_4$:

$$\text{Poincare II invariant} \quad L_{(\rho \mathrm{V}_4)} \oint_{4D} F\hat{}F = 0 \tag{17.33}$$

$$\text{Poincare I invariant} \quad L_{(\rho \mathrm{V}_4)} \oint_{4D} \{F\hat{}G - A\hat{}J\} = 0 \tag{17.34}$$

Even more importantly, when and where the exterior derivatives of each 3-form vanish, then by deRham's topological theorems, the closed cyclic integrals of each 3-form will have values that have rational integer ratios; i.e., the closed cyclic integrals are integers times some universal constant. The cyclic integrals are "quantized" relative to the physical constant, h/e, for topological torsion, and to the physical constant, h, for topological spin. These concepts have not made any use of geometric ideas of size and shape, yet yield "quantum" numbers. The do not depend upon geometric scales, nor any explicit use of quantum theory.

In terms of topological thermodynamics, the manifolds upon which $d(A\hat{}F) = F\hat{}F = 0$ are non-equilibrium domains of Pfaff topological dimension 3. These submanifolds of space time can emerge (as if by a condensation process) from dissipative thermodynamic systems of Pfaff topological dimension 4 ($d(A\hat{}F) = F\hat{}F \neq 0$). Further note that the ratios of these two topological quantum numbers yields the Hall impedance, $Z_{Hall} = h/e^2$ (to within a rational fraction), indicating the fact that the emergence of multiple component topological systems can have topological coherence.[8]

The fifth somewhat heretical claim is: Long lived propagating states can occur in non equilibrium electrodynamic systems, and the photon is an example of such a soliton.

The non-equilibrium electrodynamic system consists of systems where the Pfaff topological dimension is greater than 2. For a 4D space time set of independent variables, the possibilities are that the domain of interest is of Pfaff dimension 3 or Pfaff dimension 4. Pfaff dimension 3 domains can emerge from Pfaff topological dimension 4 domains by means of thermodynamic irreversible processes. What is remarkable is that thermodynamic domains of Pfaff topological dimension 3 admit evolutionary processes that can be described by a unique extremal Hamiltonian field. Such submanifold domains then can evolve as soliton structures maintaining a topological coherence and a long life time. The submanifold structures of Pfaff topological dimension 3 do not depend upon geometric scales, yet they are precisely the domains required such that the 3-forms of topological torsion and topological spin have zero divergence. They are sets that have the properties required for the "quantized" topological properties of spin quanta and flux quanta.

Such unique Hamiltonian fields exist for all odd Pfaff topological dimensional systems greater than 2. Such manifolds belong to the class of contact manifolds. All even Pfaff topological manifolds belong to the class of symplectic manifolds, and do NOT admit such extremal Hamiltonian processes. In fact, it appears that the class of thermodynamically irreversible processes is an artifact of Pfaff topological dimension 4. The important idea is that non equilibrium electromagnetic systems involve the 3-forms of topological torsion, $A\hat{}F$, and topological Spin, $A\hat{}G$, whose closed homogeneous forms furnish the quantum numbers associated with photons.[10,13]

17.3 Can Photons Detect Vacuum Chirality?

From the disciplines of astronomy, general relativity, and quantum mechanics comes an increased interest in possible chiral phenomena that could be associated with the vacuum state. Yet the classic literature of electromagnetism does not seem to address such a chiral effect. The conventional Lorentz vacuum state for classical electromagnetism is defined in terms of solutions to the Maxwell Faraday equations for the intensities, **E** and **B**, and the Maxwell ampere equations for the excitations, **D** and **H**, which produce no charge densities or current densities, and satisfy the constitutive equations of constraint, $\mathbf{D} = \varepsilon_0\mathbf{E}$ and $\mathbf{H} = \mathbf{B}/\mu_0$. Such solutions for the field intensities satisfy not only both Maxwell equations, but also the vector wave equation with a propagation speed of $c = 1/\sqrt{\varepsilon_0\mu_0}$. The permittivity, ε_0, and the permeability, μ_0, of the Lorentz vacuum domain are presumed to be isotropic and homogeneous constants.

It is remarkable that a chiral constitutive relation of the form $\mathbf{D} = \varepsilon_0\mathbf{E} + [\gamma] \circ \mathbf{B}$ and $\mathbf{H} = \mathbf{B}/\mu_0 - [\gamma^\dagger]\circ \mathbf{E}$ will also satisfy both Maxwell equations, without generating real charge densities and real current densities. The assumption of a simple complex scalar form for chiral constitutive matrix, $[\gamma] = (g + i\gamma)$, leads to two general cases. In one case, the only detectable difference between the chiral vacuum and the Lorentz vacuum is to be found in the value for radiation impedance, Z, a value which depends on the chiral coefficients g and γ, as well as the ratio $\sqrt{\mu_0/\varepsilon_0}$, through the determinant of the constitutive matrix. In the other case, the propagation phase velocities of left handed and right handed helical waves can be slightly different leading to a reactive impedance contribution to the classic radiation impedance of the Lorentz vacuum.

The Lorentz vacuum will be defined as the case where $\gamma = 0$, $\gamma^\dagger = 0$, and the chiral vacuum will be defined as the case when $\gamma \neq 0$, $\gamma^\dagger \neq 0$.

Substitution of the Lorentz vacuum constraints

$$\mathbf{D} = \varepsilon_0\mathbf{E} \; \mathbf{H} = \mathbf{B}/\mu_0. \tag{17.35}$$

into the Maxwell ampere equation yields

$$\text{grad div } \mathbf{E} - \text{curl curl } \mathbf{E} - \varepsilon\mu\partial^2\mathbf{E}/\partial t^2 \tag{17.36}$$

In other words a necessary condition for the Lorentz vacuum is that the fields satisfy the vector wave equation (with *div* $\mathbf{E} = 0$).

Following Bateman, form the inner 3D productof the Maxwell faraday equation with $\mathbf{H} = \mathbf{B}/\mu$, and the inner product of the source free Maxwell ampere equation with **E**. Use the constitutive definitions for the Lorentz vacuum where $\mathbf{H} = \mathbf{B}/\mu$ and $\mathbf{D} = \varepsilon\mathbf{E}$. Subtract the second resultant from the first, (assuming $\gamma = 0$), to produce the famous Poynting equation,

$$\text{div } (\mathbf{E} \times \mathbf{H}) + \mathbf{H} \circ \partial\mathbf{B}/\partial t + \mathbf{E} \circ \partial\mathbf{D}/\partial t \Rightarrow \tag{17.37}$$

$$\text{div } (\mathbf{E} \times \mathbf{H}) + \partial(1/2\mathbf{B}^2/\mu + 1/2\varepsilon\mathbf{E}^2)/\partial t = 0. \tag{17.38}$$

The result is an equation of continuity in terms of the field variables. By comparison to a "fluid", this "equation of continuity" yields a field energy density, ρ_e, and an energy current density, $\rho_e\mathbf{v}$, given by the expressions:

$$\rho_e c^2\mathbf{v} = (\mathbf{E} \times \mathbf{H}) = (\mathbf{D} \times \mathbf{B})c^2 \quad \text{and} \quad \rho_e c^2 = (1/2\mathbf{B}^2/\mu + 1/2\varepsilon\mathbf{E}^2). \tag{17.39}$$

It is important to note that the energy flux, $(\mathbf{E} \times \mathbf{H})$, and the momentum flux, $(\mathbf{D} \times \mathbf{B})$, are in the same direction and propagate with the same speed.

It should be remembered that these equations can be complex. The energy current density and the energy density can be formed from complex numbers. Bateman finds the extraordinary result, equivalent to the expression,

$$\rho_e^2 (1/\mu\varepsilon - \mathbf{v} \circ \mathbf{v}) = \rho_e^2 (c^2 - \mathbf{v} \circ \mathbf{v}) \tag{17.40}$$

$$\equiv (1/c^2)\{[(1/2)(\mathbf{D} \circ \mathbf{E}) - (1/2)(\mathbf{B} \circ \mathbf{H})]^2 \tag{17.41}$$

$$+ (\mathbf{E} \circ \mathbf{B}/Z_{freespace})^2\}. \tag{17.42}$$

under the assumption that $\varepsilon\mu c^2 = 1$. The factor (μ/ε) is the square of the radiation impedance of free space, $Z_{freespace} = \sqrt{\mu/\varepsilon}$. It is apparent that the first term on the right is the first Poincare (conformal) invariant equivalent to the Lagrange energy density of the field (the difference between the deformation and the kinetic energy densities). The second term is the second Poincare invariant of the field, and is to be associated with topological parity and thermodynamic irreversibility.[11] Bateman remarks that "the rate at which energy flows through the field is less that the velocity of light", unless the two Poincare invariants on the RHS vanish. The importance of the null Poincare invariants becomes obvious, as they furnish the requirement that the field energy propagates with the speed of light. It is important to remember that these equations can involve complex vector fields.

In general, for the Lorentz vacuum, the energy density of the field is defined as

$$\text{Ham} = (1/2)(\mathbf{D} \circ \mathbf{E}) + (1/2)(\mathbf{B} \circ \mathbf{H}) = 1/2\mathbf{B}^2/\mu + 1/2\varepsilon\, \mathbf{E}^2 \tag{17.43}$$

while the field Lagrangian is defined classically as

$$\text{Lag} = (1/2)(\mathbf{D} \circ \mathbf{E}) - (1/2)(\mathbf{B} \circ \mathbf{H}) = 1/2\varepsilon\, \mathbf{E}^2 - 1/2\mathbf{B}^2/\mu. \tag{17.44}$$

The development above describes classic results valid for a Lorentz vacuum, but now the question arises as to how these results change for a chiral vacuum, defined as a vacuum for which the constitutive matrices represented by $[\gamma]$ are not zero, but for which there are no real charge densities or current densities. The objective of this article is to assume that $[\gamma]$ is a complex domain constant, not zero, and then to determine what are the consequences of such an assumption. Such an assumption, which if applicable to the vacuum, would imply that the chiral vacuum, and therefore the universe itself, may not have a center of symmetry. The chiral adjective is appropriate, for a pure imaginary $[\gamma]$ replicates certain features of media which are optically active. The classic example of an optically active media is a solution of right handed helical molecules, such as sugar, in water. The phenomena has practical use in the wine industry and has been used to permit the grower to determine the sugar content of his grapes. (This is the basis of the word *brix* often found on French wine labels.)

Once a constitutive matrix is assumed it is possible to compute the characteristics of the combined Maxwell Faraday and Maxwell ampere partial differential system. These surfaces, independent from any gauge assumptions, define point sets upon which the solutions to the partial differential system are not unique. The characteristic point sets, in general, form non-stationary Kummer–Fresnel quartic surfaces, of which the constitutive equations of the chiral vacuum generate a special case.[16] The theory for such surfaces has been worked out in detail, and the references below contain links to Maple programs that will generate such surfaces for arbitrary constitutive equations. There is an added importance to the recognition that the characteristic surfaces are Kummer surfaces, for then a connection between classical electromagnetism and Clifford algebras can be made, with the possibility that classical solutions to Maxwell's equations can involve spinors. Examples of such quaternionic solutions that indicate that the phase velocity of propagation in the inbound and outbound directions are not the same have been published.[9]

Along these lines, it is of interest to note that, in 1914, Bateman[1] realized that a complex 3-dimensional vector, $\mathbf{M} = \mathbf{B} \pm i\sqrt{\varepsilon\mu}\mathbf{E}$ could be used to express both the Maxwell Faraday and the Maxwell ampere equations for the Lorentz vacuum as one combined set of complex vector equations. Bateman determined that it is possible to find a conjugate pair of solutions $\mathbf{M}$ and $\mathbf{M}'$ that satisfy the complex equation

$$\mathbf{M} \circ \mathbf{M}' = 0. \tag{17.45}$$

Each solution satisfies the equation

$$\mathbf{M} \circ \mathbf{M} = (\mathbf{B}^2 - \varepsilon\mu\mathbf{E}^2) \pm 2i(\sqrt{\varepsilon\mu}\mathbf{E} \circ \mathbf{B}) = \mathbf{I}_1 \pm 2i\mathbf{I}_2, \tag{17.46}$$

where $\mathbf{I}_1$ and $\mathbf{I}_2$ are the Poincare conformal invariants of the field, $\mathbf{M}$.

If the complex solution vector satisfies the complex equation of constraint,

$$\mathbf{M} \circ \mathbf{M} = (\mathbf{B}^2 - \varepsilon\mu\mathbf{E}^2) + 2i(\sqrt{\varepsilon\mu}\mathbf{E} \circ \mathbf{B}) = 0, \tag{17.47}$$

then such a vector not only satisfies both the Maxwell Faraday and the Maxwell ampere (source free) equations for a Lorentz vacuum, but also propagates the field energy with the speed of light. Such solutions were defined by Bateman as self conjugate solutions. (Translate to self dual solutions in modern day language.) The self dual equation of constraint also leads to the Clifford algebras, and therefore indicates that the Bateman solutions can have spinor representations, as well as complex number representations.

The Bateman self conjugate condition requires that the (complex) magnetic energy density be the same as the (complex) electric energy density, and the (complex) electric field be orthogonal to the (complex) magnetic field, $\mathbf{E} \circ \mathbf{B} = 0$. Both of these Poincare conformal invariants must be zero to satisfy the Bateman self duality condition. It is the self dual solutions, these self conjugate solutions, that satisfy the eikonal expression, and therefore, as Bateman points out, can represent propagating electromagnetic discontinuities.[5] The Poincare invariants are additive, such that it is conceivable to construct a self-conjugate solution from two or more non-self conjugate solutions, each of which has different Poincare invariants, but which are equal to zero under addition.

Bateman apparently did not notice that the complex constraint equation of self duality on **M** is precisely the conditions that the complex position vector generated by **M** defines a minimal surface.[6] Moreover, Bateman did not notice that most of his results are to be obtained also for a chiral vacuum.

17.3.1 Details of Chiral Vacuum

Use the (complex) chiral vacuum constitutive equations in the format of Post,[7]

$$\mathbf{D} = \varepsilon_0 \mathbf{E} + [\gamma] \circ \mathbf{B} \qquad \mathbf{H} = -[\gamma^{\dagger}] \circ \mathbf{E} + \mathbf{B}/\mu_0\,, \tag{17.48}$$

along with the Maxwell Faraday equations and the Maxwell ampere equations, and replicate the steps of the preceding section. For simplicity, assume that the matrix

$$[\gamma] = (g + \sqrt{-1}\,\gamma)\sqrt{\mu/\varepsilon}\,[1] \tag{17.49}$$

and

$$[\gamma^{\dagger}] = (\alpha \cdot g - \sqrt{-1}\,\beta \cdot \gamma)\sqrt{\mu/\varepsilon}\,[1] \tag{17.50}$$

where $\alpha, \beta = \pm 1$. Note that if $\alpha = +1$, $\beta = +1$, then $[\gamma^{\dagger}]$ is the hermitean conjugate of $[\gamma]$. If $\alpha = 1$, $\beta = -1$, then the imaginary part of $[\gamma]$ is anti-hermitean. The Fresnel–Kummer wave surface equation for the characteristic of the Maxwell equations may be written as the polynomial,

$$\{R^4 + 1 - [2 - g^2(1-\alpha)^2 + \gamma^2(1+\beta)^2]R^2\} - i2\{g\gamma(1-\alpha)(1+\beta)\} = 0, \tag{17.51}$$

where $R^2 = n_x^2 + n_y^2 + n_z^2 = \mathbf{n} \circ \mathbf{n}$ represents the norm of the projectivized wave vector (index of refraction vector), $\mathbf{n} = \mathbf{k}/\omega$. Solutions of the characteristic polynomial yield the phase velocities of propagation in terms of the magnitude of the reciprocal index of refraction vector, **n**. The phase velocity solutions are isotropic and homogeneous constants, determined by the root of the characteristic polynomial. The phase velocity is complex unless $\alpha = +1$, or the numeric factors are zero, e.g., $g = 0$ or $\gamma = 0$. For this reason, the case of $\alpha = -1$ is ignored in this article.

If the hermitean conjugate constraints are used, $\alpha = 1$, and $\beta = 1$, then the phase velocity is determined from the formula for the (homogeneous, isotropic) index of refraction,

$$n = \pm\gamma \pm \sqrt{\gamma^2 + 1}. \tag{17.52}$$

For finite γ any g, there is a time-like dispersion of two helical waves. These chiral waves have phase velocities a bit greater and a bit less that the velocity of light $c = \sqrt{1/\varepsilon\mu}$, as determined by the chiral factor γ, and these phase velocities are independent of the chiral factor g.

If the constraints $\alpha = 1$, and $\beta = -1$ are used, then the phase velocities are those of the Lorentz vacuum, ($n = 1$), for any value of chiral factors, g and/or γ. The fundamental result is that the chiral vacuum and the Lorentz vacuum are almost indistinguishable.

For the case $\alpha = 1$, and $\beta = 1$, the determinant of the constitutive matrix is real and equal to

$$\det[\text{Constitutive}] = -(\varepsilon/\mu + g^2 + \gamma^2)^3, \tag{17.53}$$

a value which is proportional to the reciprocal of the free space impedance cubed. For $\gamma = 0$, the only difference between the chiral vacuum and the Lorentz vacuum would be in the value of the free space impedance, $Z = \sqrt{1/(\varepsilon/u + g^2)}$. If $\gamma \neq 0$, then there could exist a slight dispersion (in time) between left handed and right handed polarization states.

For the case $\alpha = 1$, and $\beta = -1$, the determinant of the constitutive tensor is more complicated. The determinant has complex values (implying dissipation) unless either $\gamma = 0$, or $g = 0$. In each non-dissipative case,

$$Z = \sqrt{1/(\varepsilon/u + g^2)} \text{ for } \gamma = 0, \quad Z = \sqrt{1/(\varepsilon/u - \gamma^2)} \text{ for } g = 0, \quad n = 1. \tag{17.54}$$

Reality constraints imply that all cases of interest to this article are such that $\alpha = 1$. Substitution of the constitutive equations into the Maxwell ampere equation yields

$$\mathbf{J} = curl\,\mathbf{H} - \partial\mathbf{D}/\partial t = \{curl\mathbf{B} - \varepsilon\mu\partial\mathbf{E}/\partial t\}/\mu \tag{17.55}$$

$$+g(-curl\,E - \partial B/\partial t) + \sqrt{-1}\,\gamma(\beta \cdot curl\,E - \partial B/\partial t) \tag{17.56}$$

$$\rho = div\mathbf{D} = \varepsilon div\mathbf{E} + (g + \sqrt{-1}\,\gamma)(div\,\mathbf{B}) \tag{17.57}$$

The point of this exercise is to note that in virtue of the Maxwell Faraday equation, the chiral vacuum constitutive relations produce no real charge currents or charge densities if $\beta = -1$, independent of the choice of chiral coefficients. The field intensities satisfy the vector wave equation with phase velocities that are those of the Lorentz vacuum.

If $\beta = +1$ then only an imaginary current density is created for non-zero γ. It is then possible to compute the reactive power, $\mathbf{J} \circ \mathbf{E}$, and therefor a reactive impedance that depends upon γ. (It is tempting to identify the chiral coefficient with the reciprocal Hall impedance, $\gamma = e^2/h$). The field intensities then satisfy a wave equation with a phase velocity that depends upon γ.

In no case do the chiral vacuum constitutive equations yield a free charge density, if $div\mathbf{E} = 0$ and $div\mathbf{B} = 0$. This result is valid if the field intensities are derived from a set of potentials. A second point is that the chiral factors of the type, g, do not have any effect on the Lorentz vacuum except to modify the radiation impedance, Z.

Similar substitutions of the chiral constitutive equations lead to the Poynting equation in the form:

$$\begin{aligned}\mathrm{div}(\mathbf{E}3\mathbf{H}) + \mathbf{H} \circ \partial\mathbf{B}/\partial t + \mathbf{E} \circ \partial\mathbf{D}/\partial t &= \mathrm{div}(\mathbf{E}3\mathbf{H}) + \partial(1/2\mathbf{B}^2/\mu + 1/2\varepsilon\mathbf{E}^2)/\partial t \\ &= \{(\alpha - 1)g - \sqrt{-1}(\beta + 1)\gamma\}\mathbf{E} \circ \partial\mathbf{B}/\partial t.\end{aligned} \tag{17.58}$$

If the RHS of the equation above vanishes, then the Poynting theorem of equation 17.39 is retrieved without change in form. For the choice $\alpha = +1$, $\beta = -1$, again there are no differences between the chiral vacuum and the Lorentz vacuum, for any value of the chiral factors. For the choice $\alpha = +1$, $\beta = +1$, the equation implies a chiral (imaginary or reactive) component to the Poynting equation, related to the time-like dispersion of the left handed and right handed helical waves. This term vanishes for $\gamma = 0$, and is independent from g.

The next step is to evaluate the expressions for the total field Hamiltonian energy density and the Lagrange density of the chiral vacuum. The expression for the Hamiltonian energy density becomes

$$\begin{aligned}\mathrm{Ham} &= (1/2)(\mathbf{D} \circ \mathbf{E}) + (1/2)(\mathbf{B} \circ \mathbf{H}) \\ &= 1/2\mathbf{B}^2/\mu + 1/2\varepsilon\mathbf{E}^2 + \{(\alpha - 1)g + \sqrt{-1}(\beta + 1)\gamma\}\mathbf{E} \circ \mathbf{B}/2\end{aligned} \tag{17.59}$$

while the field Lagrangian is becomes:

$$\begin{aligned}\mathrm{Lag} &= (1/2)(\mathbf{D} \circ \mathbf{E}) - (1/2)(\mathbf{B} \circ \mathbf{H}) \\ &= 1/2\varepsilon\mathbf{E}^2 - 1/2\mathbf{B}^2/\mu + \{(\alpha + 1)g + \sqrt{-1}(1 - \beta)\gamma\}\mathbf{E} \circ \mathbf{B}/2\end{aligned} \tag{17.60}$$

These results indicate that there are slight modifications to the energy density formulas, modifications that are dependent upon the second Poincare invariant. However, for systems where the field intensities are deducible from a 1-form of potentials, and the 1-form is of Pfaff dimension 3 or less, then $\mathbf{E} \circ \mathbf{B}$ vanishes, and all computations of Hamiltonian or Lagrangian energy densities are identical for the Lorentz vacuum, or for the chiral vacuum. It is only for cases where the 1-form of potentials is of Pfaff dimension 4, such that

$\mathbf{E} \circ \mathbf{B} \neq 0$ that the chiral factors can make a difference in the expressions for Hamiltonian or Lagrangian energy density.

Again study $\alpha = 1$. Then the choice $\beta = -1$, implies that the Hamiltonian energy density is the same as the Lorentz vacuum, but the Lagrangian depends upon the chiral factors. The choice $\beta = +1$, implies that the Lagrangian depends upon the chiral factor g and the Hamiltonian depends upon the chiral factor γ. All chiral effeects on the energy densities disappear if $F \hat{} F = -2 (\mathbf{E} \circ \mathbf{B}) dx \hat{} dy \hat{} dz \hat{} dt = 0$.

These are rather startling results for they demonstrate that the Lorentz vacuum and the chiral vacuum can be formally indistinguishable, except for the impedance of free space (which is related to the determinant of the constitutive tensor and therefore to the chiral coefficients).

Summary

From a topological and thermodynamic perspective of the electromagnetic field, there appears to be a common thread among eikonal solutions, spinors, propagating topological discontinuities or defects, minimal surfaces, and topological quantization. All of these properties suggest that the common topological thread is that which is usually perceived as the photon. A topological perspective of electromagnetism not only include features attributed to the photon, but also points out that non equilibrium thermodynamic concepts can be formulated to produce interesting experiments and practical devices. For example, the fact the irreversible dissipation occurs when the field intensities have a collinear component ($\mathbf{E} \circ \mathbf{B} \neq 0$) could be used to influence condensation. Stable long lived states in a plasma should be designed about the constraints that ($\mathbf{E} \circ \mathbf{B} = 0$) which yield non equilibrium dynamical systems described by Hamiltonian processes. Each of these ideas involve the concepts of topological torsion and topological spin, and hence the quantal properties of the photon.

References

1. H. Bateman, *Electrical and Optical Wave Motion,* Dover, New York, 1955.
2. E. Cartan, *The Theory of Spinors,* Dover, New York, p. 39, 1966.
3. V. Fock, *Space, Time and Gravitation,* Pergamon-Macmillan, New York, 1964.
4. J. Hadamard, *Lectures on Cauchy's Problem,* Dover, New York, p. 51, 1951.
5. R. K. Luneburg, *Mathematical Theory of Optics,* University of California Press, Berkeley, 1964.
6. R. Osserman, *A Survey of Minimal Surfaces,* Dover, New York, 1986.
7. E. J. Post, *Formal Structure of Electromagnetics,* Dover, New York, 1997.
8. R. M. Kiehn, Are there three kinds of superconductivity? *Int. J. Modern Phys,* 15, 1779, 1991.

9. R. M. Kiehn, G. P. Kiehn, and B. Roberds, Parity and time-reversal symmetry breaking, singular solutions and Fresnel surfaces, *Phys. Rev A* 43, 5165, 1991.
10. R. M. Kiehn, Topological evolution of classical electromagnetic fields and the photon, in, *Photon and Poincaré Group,* V. Dvoeglazov (ed.), Nova., Commack, New York, p. 246–262, 1999.
11. R. M. Kiehn, "Topological torsion and topological spin as coherent structures in plasmas", 2001, arXiv physics/0102001.
12. R. M. Kiehn, "Curvature and torsion of implicit hypersurfaces and the origin of charge", *Annales de la Foundation Louis de Broglie,* 27, 411, 2002.
13. R. M. Kiehn, The photon spin and other topological features of classical electromagnetism, in *Gravitation and Cosmology: From the Hubble Radius to the Planck Scale,* Amoroso, R., et al., eds., Kluwer, Dordrecht, pp. 197–206, 2002.
14. D. Van Dantzig. *Proc. Cambridge Philos. Soc.* 30, 421, 1934; also Electromagnetism independent of metrical geometry", *Proc. Kon. Ned. Akad. v. Wet.,* 37, 521–531, 644–652, 825–836, 1934.
15. R. M. Kiehn, "Non Equilibrium Thermodynamics", vol 1, 2004, on line, Lulu. com.
16. R. M. Kiehn, "Plasmas and Non equilibrium Electrodynamics". vol 4, 2004, online, Lulu.com.
17. E. T. Whittaker, *Analytical Dynamics,* Dover, New York, 1944.

18

The Photon: A Virtual Reality

David L. Andrews

Nanostructures and Photomolecular Systems, School of Chemical Sciences, University of East Anglia, Norwich NR4 7TJ, United Kingdom

CONTENTS

Abstract

It has been observed that every photon is, in a sense, virtual—being emitted and then sooner or later absorbed. As the motif of a quantum radiation state, the photon shares these characteristics of any virtual state: that it is not directly observable; and that it can signify only one of a number of indeterminable intermediates, between matter states that are directly measurable. Nonetheless, other traits of real and virtual behavior are usually quite clearly differentiable. How "real", then, is the photon? To address this and related questions it is helpful to look in detail at the quantum description of light emission and absorption. A straightforward analysis of the dynamic electric field, based on quantum electrodynamics, reveals not only the entanglement of energy transfer mechanisms usually regarded as "radiative" and "radiationless"; it also gives significant physical insights into several other electromagnetic topics. These include: the propagating and non-propagating character in electromagnetic fields; near-zone and wave-zone effects; transverse and longitudinal character; the effects of retardation, manifestations of quantum uncertainty and issues of photon spin. As a result it is possible to gain a clearer perspective on when, or whether, the terms "real" and "virtual" are helpful descriptors of the photon.

Key words: virtual photon, photonics, quantum electrodynamics, resonance energy transfer, retardation, photon spin.

18.1 Introduction

It is no longer so straightforward to explain what is meant by a "photon".[1] Although the term belongs to a concept first formulated a hundred years ago, this book eloquently bears witness to the present truth of this concise understatement. In recent literature, there is further disconcerting evidence in the number adjectival qualifiers that can be found attached to the term, as for example in "superluminal",[2] "electric",[3] "magnetic",[4] "ballistic",[5] "transverse",[6] and "longitudinal",[7] photons. "Real" and "virtual" photons are the subject of the present discourse. Based on the elementary definition that a virtual photon is one not directly observed, it has been correctly commented that every photon is, in a sense, virtual—being emitted and then sooner or later absorbed.[8] As the defining motif of a quantum radiation state, the photon exhibits the characteristic indeterminacy of any quantum virtual state, signifying its role as intermediary between states of matter that are directly measurable.

Nonetheless, it is usually considered that traits of virtual behavior are distinctive and unambiguous. To address the question of what it means to categorize a photon as "real" or "virtual" in an optical context, this chapter revisits the detailed quantum description of a photon history comprising creation and propagation. The photophysics exemplifies an interplay of quantum theory, electromagnetism and the principles of retardation; analysis based on quantum electrodynamics (QED) not only confronts key issues of photon character; it also elucidates a number of related matters such as the entanglement of "radiative" and "radiationless" mechanisms for energy transfer, two distinct senses of photon transversality, and photon spin issues.

18.2 QED Formulation

The photon has a character that, *inter alia*, reflects the electromagnetic gauge. In the Coulomb gauge the radiation field is ascribed an unequivocally transverse character,[9] in the sense that its electric and magnetic fields are orthogonally disposed with respect to the wave-vector. As will be shown, this transversality condition of electromagnetic fields is not necessarily transferable to a disposition with respect to the interpreted direction of electromagnetic energy transduction. To engage in a detailed study of these features it is appropriate to fully develop the theory of energy transfer within the framework of quantum electrodynamics, which treats both fields and matter on the same quantum basis. The system Hamiltonian comprises unperturbed operators for the radiation and for two material components, a source/donor A and a detector/acceptor B differentiated by a label ξ, and also two corresponding light-matter interaction terms;

$$H = H_{\text{rad}} + \sum_{\xi=A,B} H_{\text{centre}}(\xi) + \sum_{\xi=A,B} H_{\text{int}}(\xi). \tag{18.1}$$

The first two components of equation 18.1 determine a basis in terms of which states of the system can be described, i.e., a direct product of eigenstates of the radiation field Hamiltonian and the Hamiltonian operators for the two components of matter. The third, radiation field-matter interaction, summation term can be expressed either in minimal coupling form (expressed in terms of coupling with the vector potential of the radiation field) or the generally more familiar multipolar formulation directly cast in terms of electric and magnetic fields.

These two options lead to identical results for real processes, that is those subject to overall energy conservation;[10,11] for convenience the following theory is to be developed in multipolar form. [*Note*, in its complete form the multipolar interaction Hamiltonian can itself be partitioned as: (i) a linear coupling of the electric polarization field (accommodating all electric multipoles) with the transverse electric field of the radiation; (ii) a linear coupling of the magnetization field (all magnetic multipoles) with the magnetic radiation field; (iii) a quadratic coupling of the diamagnetization field with the magnetic radiation field. It may be observed that, although the following analysis focuses on electric polarization coupling, the same principles concerning the identity and transversality characteristics of real and virtual photons apply to each and every multipolar term.[12]] In equation 18.1, the absence of any terms with $\xi' \neq \xi$ signifies that the transduction of energy between A and B is not effected by direct instantaneous (longitudinal) interactions, but only through coupling with the quantum radiation field—a feature that is in marked contrast to most classical descriptions. In the lowest order, electric-dipole term in the multipole expansion, each $H_{\text{int}}(\xi)$ operator is given by;

$$H_{\text{int}}(\xi) = -\sum_{\xi} \mu(\xi) \cdot \mathbf{e}^{\perp}(\mathbf{R}_{\xi}). \tag{18.2}$$

where the electric-dipole moment operator, $\mu(\xi)$, operates on matter states and the transverse electric field operator, $\mathbf{e}^{\perp}(\mathbf{R}_{\xi})$ on radiation states. The latter operator is expressible in a plane-wave mode expansion summed over all wave-vectors, $\mathbf{p}$, and polarizations, λ;

$$\mathbf{e}^{\perp}(\mathbf{R}_{\zeta}) = \mathrm{i}\sum_{\mathbf{p},\lambda} \left(\frac{\hbar c p}{2\varepsilon_0 V}\right)^{1/2} [\mathbf{e}^{(\lambda)}(\mathbf{p}) a^{(\lambda)}(\mathbf{p}) e^{\mathrm{i}(\mathbf{p}\cdot\mathbf{R}_{\zeta})} - \bar{\mathbf{e}}^{(\lambda)}(\mathbf{p}) a^{\dagger(\lambda)}(\mathbf{p}) e^{-\mathrm{i}(\mathbf{p}\cdot\mathbf{R}_{\zeta})}]. \tag{18.3}$$

Here $\mathbf{e}^{(\lambda)}(\mathbf{p})$ is the polarization unit vector (plane or circular, but always orthogonal to $\mathbf{p}$) and $\bar{\mathbf{e}}^{(\lambda)}(\mathbf{p})$ is its complex conjugate; V is an arbitrary quantization volume and $a^{\dagger(\lambda)}(\mathbf{p})$, $a^{(\lambda)}(\mathbf{p})$ respectively are photon creation and annihilation operators for the mode $(\mathbf{p},\lambda)$. Accordingly, each action of H_{int} signifies photon creation or annihilation.

Consider an energy transfer process for which the initial state $|i\rangle$ of the system may be written $|A^{\alpha}; B^{0}; 0\rangle$ and the final state $|f\rangle$ as $|A^{0}; B^{\beta}; 0\rangle$. Here the superscript 0 signifies the ground energy level, with α and β denoting the appropriate excited levels for the source and detector, respectively. Overall conservation of energy demands that $E^{A}{}_{\alpha 0} \equiv E^{A}{}_{\alpha} - E^{A}{}_{0} = E^{B}{}_{\beta 0} \equiv E^{B}{}_{\beta} - E^{B}{}_{0} \equiv \hbar ck$ where the last equality serves to introduce a convenient metric k. Energy transfer is mediated by coupling to the vacuum radiation field, invoking (a minimum of) one $a^{\dagger(\lambda)}(\mathbf{p})$ and also one $a^{(\lambda)}(\mathbf{p})$ operator, whose two distinct time-orderings correspond to: (*a*) the creation of a virtual photon at A and its subsequent annihilation at B; (*b*) *vice-versa*. Both pathways have to be considered, in order to take account of the non-energy conserving route allowed by the Uncertainty Principle at very short times; the virtual photon can be understood as "borrowing" energy from the vacuum, consistent with an energy uncertainty $\hbar/t$, where t is the photon time-of-flight—here determined by the displacement of the detector from the source. This principle also indicates a temporary relaxation of exact energy conservation in the isolated photon creation and annihilation events. When the whole system enters its final state, i.e. after the virtual photon is annihilated, energy conservation is restored. With two virtual photon-matter interactions and $H_{\text{int}}(\xi)$ acting as a perturbation, the quantum amplitude, $M_{fi}^{\text{e-e}}$, for energy transfer is calculated from the second term of an expansion in time-dependent perturbation theory;

$$M_{fi}^{\text{e-e}} = \frac{\langle f|H_{\text{int}}|r_a\rangle\langle r_a|H_{\text{int}}|i\rangle}{(E_i - E_{r_a})} + \frac{\langle f|H_{\text{int}}|r_b\rangle\langle r_b|H_{\text{int}}|i\rangle}{(E_i - E_{r_b})}. \tag{18.4}$$

The ensuing calculation leads into some relatively straightforward vector analysis and contour integration; the major didactic issues and also some of the mathematical intricacies have both been the subject of recent reviews.[13,14] Using the convention of summation over repeated Cartesian indices, the result for the transfer quantum amplitude emerges as follows:

$$M_{fi}^{\text{e-e}} = \mu_i^{0\alpha(A)} V_{ij}(k,\mathbf{R}) \mu_j^{\beta 0(B)}, \tag{18.5}$$

Here $\mathbf{R} = \mathbf{R_B} - \mathbf{R_A}$ is the source-detector displacement vector, the source transition dipole moment is $\mu^{0\alpha(A)} \equiv \langle A^0|\mu^{(A)}|A^{\alpha}\rangle$, and for the detector $\mu^{\beta 0(B)} \equiv \langle B^{\beta}|\mu^{(B)}|B^{0}\rangle$; also $V_{ij}(k, \mathbf{R})$ is the retarded resonance electric dipole—electric dipole coupling tensor, expressible as;

$$V_{ij}(k,\mathbf{R}) = \frac{e^{ikR}}{4\pi\varepsilon_0 R^3}\{(\delta_{ij} - 3\hat{R}_i\hat{R}_j) - (ikR)(\delta_{ij} - 3\hat{R}_i\hat{R}_j) - (kR)^2(\delta_{ij} - \hat{R}_i\hat{R}_j)\}. \tag{18.6}$$

18.3 Retarded Electric Fields and Photon Transversality

The quantum amplitude equation 18.5 can legitimately be interpreted as the dynamic dipolar interaction of the detector with a retarded electric field $\mathbf{e}_{\mathrm{R}}(\mathrm{B})$, generated by the source. From equation 18.5 it follows that this field has Cartesian components given by;

$$e_{\mathrm{R}j}(\mathrm{B}) = -\mu_i^{0\alpha(\mathrm{A})} V_{ij}(k,\mathbf{R}). \tag{18.7}$$

Notwithstanding its quantum electrodynamical derivation outlined above, the result has an identical form[15,16] to that which, when cast in SI units, emerges from classical retarded electrodynamics;[17]

$$\mathbf{e}_{\mathrm{R}} = k^2(\hat{\mathbf{R}} \times \mu^{0\alpha}) \times \hat{\mathbf{R}} \frac{e^{ikR}}{4\pi\varepsilon_0 R} + [3\hat{\mathbf{R}}(\hat{\mathbf{R}} \cdot \mu^{0\alpha}) - \mu^{0\alpha}] \left(\frac{1}{4\pi\varepsilon_0 R^3} - \frac{ik}{4\pi\varepsilon_0 R^2} \right) e^{ikR}. \tag{18.8}$$

Previous analyses have mostly focused on the striking variation in range-dependence exhibited within the results. Both in equations 18.6 and 18.8 the first term, proportional to R^{-3}, is dominant in the short-range or near-zone region ($kR << 1$), whereas the third term, proportional to R^{-1}, dominates in the long-range or wave-zone ($kR >> 1$). Consequently short-range energy transfer is characterized by a (Fermi Rule) *rate* that runs with R^{-6}, familiarly known as "radiationless" (*Förster*) resonance energy transfer,[18] whereas the long-range transfer rate carries the R^{-2} dependence that is best known as the *inverse square law.*

These two cases are asymptotic limits of a completely general rate law illustrated in Fig. 18.1. The Uncertainty Principle again affords a simple way of understanding the exhibited behavior. In terms of a transit time, t, for the energy transfer we have; $\hbar^{-1}\,\Delta E\,\Delta t \equiv c\Delta k\,\Delta t \equiv \Delta k\,\Delta R \sim 1$. It is because energy is transferred that the propagating electric field does not display the same inverse power dependence on the separation R for all times. For energy transfer over very short times, associated with short-range transfer distances $kR << 1$, the energy cannot be localized in either A or B and the result essentially reflects the R^{-3} form of a static dipolar field. However at distances where $kR >> 1$, corresponding to relatively large times, the propagating character of the energy becomes more evident, and leads to the characteristic radiative R^{-1} behavior.

Despite the fact that the virtual photon formulation leading to equation 18.6 is cast in terms of electromagnetic fields that are purely transverse with respect to the photon propagation direction $\hat{\mathbf{p}}$, the field equation 18.8 contains elements that are manifestly non-transverse against $\hat{\mathbf{R}}$. To exhibit this explicitly, the given expression can be decomposed into terms that are

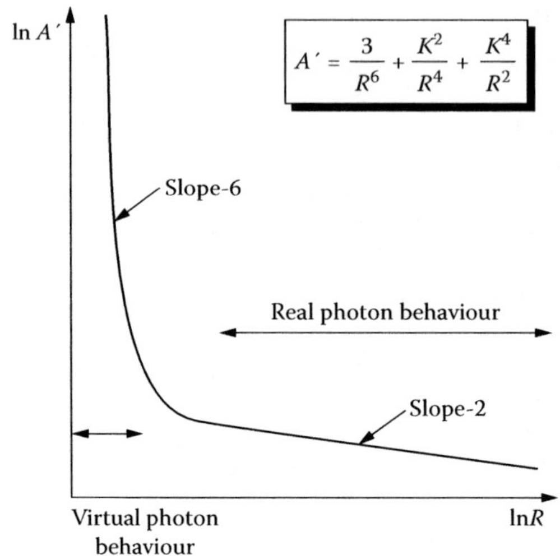

FIGURE 18.1
Logarithmic plot of the rate of dipole–dipole energy transduction against distance, with short- and long-range asymptotes. The formula for the dimensionless function A' (insert) determines the rate for an isotropically oriented system; for details see ref. 19.

transverse (⊥) and longitudinal (||) with respect to $\hat{\mathbf{R}}$;

$$\mathbf{e}_{\perp} = \frac{e^{ikR}}{4\pi\varepsilon_0 R^3}[\hat{\mathbf{R}}(\hat{\mathbf{R}}\cdot\mu^{0\alpha}) - \mu^{0\alpha}](1 - ikR - k^2R^2); \tag{18.9}$$

$$\mathbf{e}_{||} = \frac{e^{ikR}}{2\pi\varepsilon_0 R^3}\hat{\mathbf{R}}(\hat{\mathbf{R}}\cdot\mu^{0\alpha})(1 - ikR); \tag{18.10}$$

One immediate conclusion to be drawn from the prominence of the longitudinal component in the short-range region is the fact that photons with **p** not parallel to $\hat{\mathbf{R}}$ are involved in the energy transfer—which is consistent with the position-momentum Uncertainty Principle. By contrast the absence of an overall R^{-1} term in equation 18.10, compared to equation 18.9, signifies that the component of the field that is longitudinal with respect to $\hat{\mathbf{R}}$ is not sustained in the wave-zone $kR >> 1$ (equivalently $R \gg ƛ$, where $ƛ = 2\pi/k$ designates the wavelength regime of the energy being transferred). Physically, this relates to the fact that with increasing distance the propagating field loses its near-field character and is increasingly dominated by its transverse component, conforming ever more closely to what is expected of "real" photon transmission.

18.4 Quantum Pathways

It is of passing interest to note the results of a recent analysis which, for the first time, allowed the identification of contributions to the propagating field equation 18.8 separated on another basis, reflecting terms arising through either one of the two alternative quantum pathways discussed in Sect. 2. These signify (*a*) the physically intuitive propagation of a virtual photon from A to B; (*b*) the counterintuitive case of virtual photon propagation from B to A. In the short-range, $kR \ll 1$, both such contributions to the field unequivocally exhibit R^{-3} dependence; both play a significant role in the mechanism for energy transfer, as is once again consistent with quantum mechanical uncertainty. However in the long-range (which features only terms transverse to $\hat{\mathbf{R}}$), contributions of type (*a*) carry an R^{-1} radiative dependence, whereas those arising from type (*b*) unexpectedly fall off as R^{-4}. Although it was anticipated that the "reverse propagation" terms would dwindle in importance compared to type (*a*), as distance increases and the photon acquires an increasingly real character, it was not previously recognized that the rate of diminution actually increases with distance.[20]

18.5 Spin and Photon Angular Momentum

While a number of issues associated with the interplay of transversality and angular momentum have been explored in the general context of spontaneous emission,[21] the developing technology of *spintronics*[22] invites a consideration of energy transduction between quantum dots. In determining the transverse field produced by an electric dipole spin transition, it transpires that noteworthy features arise in the case of a source whose transition moment is spin-aligned with respect to $\hat{\mathbf{R}}$, i.e., whose complex transition moments lie in a plane orthogonal to the transfer direction and therefore expressible as:

$$\mu_{(\pm)}^{0\alpha} = \frac{\mu}{\sqrt{2}}[\hat{\mathbf{i}} \pm i\hat{\mathbf{j}}]. \tag{18.11}$$

Here, the corresponding result for the electric field, from equation 18.9, is:

$$\mathbf{e}_{\perp(\pm)} = -\frac{\mu e^{ikR}}{4\pi\varepsilon_0 R^3}\left[\frac{1}{\sqrt{2}}(\hat{\mathbf{i}} \pm i\hat{\mathbf{j}})\right](1 - ikR - k^2R^2). \tag{18.12}$$

As is readily shown, the complex vector in equation 18.12 that is designated by the terms in square brackets corresponds to a circularly polarized photon of left/right helicity, signifying retention of ±1 units of *spin* angular momentum.[23] This feature has the potential for considerable importance in connection with energy migration down a column of quantum dots, oriented in a common direction.[24] Even though, in the technically most significant near-zone region, the coupling cannot be ascribed to real photon propagation—and the power law on distance also changes between near-zone and far-zone displacements—the fundamental symmetry properties are the same in each regime and angular momentum is therefore conserved.

Finally, it is of interest to make an observation prompted by the rise to prominence of the technology of twisted laser beams—beams with a helical wavefront that convey what has become termed *orbital* angular momentum.[25] The connotations of the term "photon" in such a context have been the subject of much recent work, particularly in connection with Laguerre-Gaussian modes, and it has been shown that the photons in such beams convey multiples of the usual spin, the integer multiplier corresponding to the topological charge. Intriguingly, there have also been recent cases of non-integer vortex production.[26] Here, there is an obvious issue to be addressed concerning a rapprochement with the bosonic character of quantized radiation states; the validity of the photon concept in the case of such beams therefore remains to be established. In processes where photon emission and absorption are together encapsulated within a theory of energy transduction, it is legitimate to use any complete basis set for the photon of *de facto* virtual character and there is nothing to be gained (or lost) by employing vortex modes.

Conclusion

Based on a consideration of the "life" of a photon as it propagates from its source of creation towards the site of its annihilation at a detector, a case can be made that every such photon in principle exhibits both virtual and "real" traits. In the short-range limit significant retardation is absent and the virtual nature of the photon in a sense justifies the widely adopted term "radiationless" as a descriptor of the energy transfer. The effect of increasing transfer distance is to diminish the virtual character of the coupling; the energy transfer exhibits an increasingly "radiative", propagating behaviour—though a partly virtual character always remains; the coupling photons are never *fully* real. Thus the radiative and radiationless mechanisms for energy transduction, traditionally viewed as separate, are accommodated within a single theoretical construct, and it is significant that they never compete.

Further analysis reveals hitherto unsuspected features in the asymptotic behaviour of the quantum pathways for resonance energy transfer. The results formally vindicate the accommodation of both source-creator and detector-creator pathways in the near-zone, and the domination of the source-creator pathway in the wave-zone. Physically, this behavior is consistent with a rapid diminution in significance of the pathway in which the virtual photon propagates from the detector "back" to the source, consistent with a diminishing virtual character for the coupling photon. Finally, a consideration of the angular momentum aspects of the photon field shows that the possibility for retention of angular momentum, associated with circular photon polarizations, can apply even in the near-zone. The result offers new possibilities for implementation in spintronic devices.

Acknowledgment

The author would like to thank Mohamed Babiker and David Bradshaw for useful discussions.

References

1. R. Loudon, *The Quantum Theory of Light*, 3rd ed. (University Press, Oxford, 2000) p. 2.
2. F. Cardone and R. Mignani, "A unified view to Cologne and Florence experiments on superluminal photon propagation", *Phys. Lett. A* **306,** pp. 265–270, 2003.
3. Y. Koma, M. Koma, E. M. Ilgenfritz, T. Suzuki and M. I. Polikarpov, "Duality of gauge field singularities and the structure of the flux tube in Abelian-projected SU(2) gauge theory and the dual Abelian Higgs model", *Phys. Rev. D* **68**, 094018, 2003.
4. R. S. Lakes, "Experimental test of magnetic photons", *Phys. Lett. A* **329**, pp. 298–300, 2004.
5. I. Prochazka, K. Hamal and B. Sopko, "Recent achievements in single photon detectors and their applications", *J. Mod. Opt.* **51**, pp. 1289–1313, 2004.
6. V. V. Karasiev, E. V. Ludena and O. A. Shukruto, "Relativistic Dirac-Fock exchange and Breit interaction energy functionals based on the local-density approximation and the self-consistent multiplicative constant method", *Phys. Rev. A* **69**, 052509, 2004.
7. V. R. Pandharipande, M. W. Paris and I. Sick, "Virtual photon asymmetry for confined, interacting Dirac particles with spin symmetry", *Phys. Rev. C* **71**, 022201, 2005.
8. F. Halzen and A. D. Martin, *Quarks and Leptons* (Wiley, New York, 1984) p. 140.
9. D. P. Craig and T. Thirunamachandran, *Molecular Quantum Electrodynamics* (Dover, Mineola, New York, 1998) p. 10.
10. E. A. Power and T. Thirunamachandran, "Energy of a dipole near a plane metal-surface", *Am. J. Phys.* **50**, pp. 757–757, 1982.
11. E. A. Power and T. Thirunamachandran, "Quantum electrodynamics with non-relativistic sources. 3. Intermolecular interactions", *Phys. Rev. A* **28**, pp. 2671–2675, 1983.
12. A. Salam, "A general formula for the rate of resonant transfer of energy between two electric multipole moments of arbitrary order using molecular quantum electrodynamics", *J. Chem. Phys.* **122**, 044112, 2005.

13. D. L. Andrews and D. S. Bradshaw, "Virtual photons, dipole fields and energy transfer: a quantum electrodynamical approach", *Eur. J. Phys.* **25**, pp. 845–858, 2004.
14. G. J. Daniels, R. D. Jenkins, D. S. Bradshaw, and D. L. Andrews, "Resonance energy transfer: The unified theory revisited", *J. Chem. Phys.* **119**, pp. 2264–2274, 2003.
15. E. A. Power and T. Thirunamachandran, "Quantum electrodynamics with non-relativistic sources. 2. Maxwell fields in the vicinity of a molecule", *Phys. Rev. A* **28**, pp. 2663–2670, 1983.
16. A. Salam, "Resonant transfer of excitation between two molecules using Maxwell fields", *J. Chem. Phys.* **122**, 044113, 2005.
17. J. D. Jackson, *Classical Electrodynamics,* 3rd ed. (Wiley, New York, 1998) p. 411.
18. T. Förster, "Experimentelle und theoretische Untersuchungen des zwischenmolekularen Übergangs von Elektronenanregungsenergie", *Z. Naturforsch.* **4a**, pp. 321–327, 1949.
19. D. L. Andrews, "A unified theory of radiative and radiationless molecular-energy transfer", *Chem. Phys.* **135**, pp. 195–201, 1989.
20. R. D. Jenkins, G. J. Daniels and D. L. Andrews, "Quantum pathways for resonance energy transfer", *J. Chem. Phys.* **120**, pp. 11442–11448, 2004.
21. S. Franke and S. M. Barnett, "Angular momentum in spontaneous emission", *J. Phys. B: At. Mol. Opt. Phys.* **29**, pp. 2141–2150 (1996).
22. S. A. Wolf, D. D. Awschalom, R. A. Buhrman, J. M. Daughton, S. von Molnár, M. L. Roukes, A. Y. Chtchelkanova, and D. M. Treger, "Spintronics: A spin-based electronics vision for the future", *Science* **294**, pp. 1488–1495, 2001.
23. L. Mandel and E. Wolf, *Optical Coherence and Quantum Optics* (University Press, Cambridge, 1995) p. 490.
24. G. D. Scholes, D. L. Andrews, V. M. Huxter, J. Kim, and C. Y. Wong, "Transmission of quantum dot exciton spin states via resonance energy transfer", *Proc. SPIE* 5929 (86–93, 2005).
25. L. Allen, M. J. Padgett and M. Babiker, "The orbital angular momentum of light", *Prog. Opt.* **39**, 291–372, 1999.
26. J. Leach, E. Yao, and M. J. Padgett, "Observation of the vortex structure of a non-integer vortex beam", *New J. Phys.* **6**, 71 (2004).

19

The Photon and its Measurability

Edward Henry Dowdye, Jr.

CONTENTS

Abstract

Abstractly, the photon is looked at in Euclidean Space Geometry, this time strictly under the electrodynamics of Galilean Transformations of Velocities $c' = c \pm v$, where the velocity c refers to that velocity with which the photon is emitted from its moving primary source which moves with velocity v relative to the laboratory frame. A non-interfering hypothetical observer, not of the real world, would note from the laboratory frame that the interference free photon moves with velocity c'. Since any measurement by a real world observer involves interference, the window, lens or mirror of the observers measuring apparatus gives rise to a secondary photon that is in term re-emitted with the very same velocity c relative to its secondary source, namely, the window, lens or mirror of the observers measuring apparatus. This chapter will demonstrate that the problems in modern physics, involving both electromagnetism and gravitation, have their pure classical solutions under the electrodynamics of Galilean Transformations of Velocities, while abiding strictly by the rules of Galilean Transformations and employing the classical assumptions of the rectilinear behavior of both the photon and the graviton in Euclidean Space.

Key words: Euclidean space, Galilean transformation, rectilinear, primary emission, secondary emission, extinction, extinction shift.

19.1 Introduction

Emission theorists such as Sir Isaac Newton (1642–1727), Pierre Simon de Laplace (1749–1827), Jean-Baptiste Biot (1774–1862), Sir David Brewster (1781–1868) and Walter Ritz (1878–1909) never completed the very important fundamentals, the pure classical ideas based on the correct principles of optics that seemed to be on the correct path then! Many emission theories have come and gone in the past century. This principle of the non-measurability of the interference-free photon by the interfering observer, denoted here as the *extinction shift principle* is an emission theory, but unlike earlier emission theories, a clear distinction is made between that which can be measured and that which can only be calculated.

The *undisturbed, not measurable nature* of the photon is considered. It is seen immediately that no requirements of a medium or ether is necessary to explain the apparent phenomenon of the photon in this emission theory. No distortions of the standard coordinate system of space and time are required to formulate the explanations of the significant fractions of the velocity of light phenomena. The mathematical illustrations require only the correct use of Galilean transformations of velocities applied to the emissions and re-emissions of photons and the exchange of gravitons in Euclidean space geometry alone. This time only the Galilean transformations of velocities along with the principle of the rectilinear motion of the photon and of the graviton successfully accomplish the mathematically equivalence of relativity using pure classical tools.

19.2 Measurability of Photon

A purely classical treatment of the transit-time effects of electromagnetism and gravitation, using solely Galilean transformations of velocities $c' = c \pm v$ in Euclidean space, leads directly to exact solutions of the important set of problems responsible for the success and fame of both general and special relativity. [1]. In this emission theory, the Galilean transformations are applied to the undisturbed "free" propagating waves of a theoretically ideal vacuum. An ideal vacuum may be defined as that space which is void of interference, thus permitting an *undisturbed* motion of a *primary wave*, whose motion is exactly the velocity c relative to its most direct source that is moving with the velocity v relative to the reference frame. The inter-atomic space of a solid or deep interstellar space may approach such an ideal vacuum. It is mathematically illustrated that the solutions require absolutely no assumptions of a medium-dependent velocity or a *luminiferous ether.* The theoretical assumptions of distortion of space and dilation of time are unnecessary and are not considered at all in Euclidean space. The mathematical illustrations predict that a direct measurement or observation on a primary photon or wave by an interfering observer is impossible with contemporary technical means and methods.[1]

It follows that, as a consequence of Galilean transformations of velocities applied to *undisturbed waves* in Euclidean space, neither the *primary wavelength* nor the *velocity* of the primary photon or wave packet is measurable! The primary wave will be extinguished by all attempts to measure it and will be replaced by a re-emitted secondary wave. Under the correctly applied Galilean transformations in Euclidean space, it follows also that only the frequency of the secondary wave, propagating with velocity c in the frame of reference of the interference, is observed. An extinction or annihilation of the most primary wave emitted from a moving source actually takes place. The extinguished primary wave is replaced by a secondary wave as a consequence of direct interference by any attempts to measure it, and is re-emitted by the secondary source with an extinction-shifted wavelength. The secondary source here is a window, a lens or a mirror of the measuring apparatus. For this reason this effect is designated the *extinction shift principle.*

As a direct consequence of these emission effects, a resting observer measures a transverse relative time shift, mathematically equivalent to the time dilation of relativity. Similarly, it is easily shown that the wave equations are invariant under the electrodynamics of Galilean transformations in Euclidean space. Applying the very same rules of this emission theory to Galilean transformations of velocities of gravitation, important problems of general relativity are solved. The very same principal axioms of the *extinction shift principle* used for applying the Galilean transformation of velocities, this time to the emission and re-emission (exchange) of the gravitons in Euclidean space, were used to calculate the perihelion rotation effect of the planet Mercury, the PSR1913+16 binary neutron pulsar star system, the so-called solar light-bending effect and the gravitational redshift effect. The principle leads directly to the derivations of the equations of general relativity, but for pure classical reasons only. The solutions mathematically illustrate that the motion of both the photon and the graviton describe a rectilinear path, a fundamental principle of optics that has been practically forgotten in modern physics. [3] It is mathematically demonstrated that this very same emission theory is applicable to both gravitation and electromagnetism.

In this principle the *undisturbed nature* of a not-yet-measured or interference-free primary wave and the obvious consequence of the measurement of a primary wave, are considered. The mathematical illustrations imply that the *undisturbed wavelength* of a primary wave remains unchanged, and is independent of reference frames! Its velocity of motion is exactly c, relative

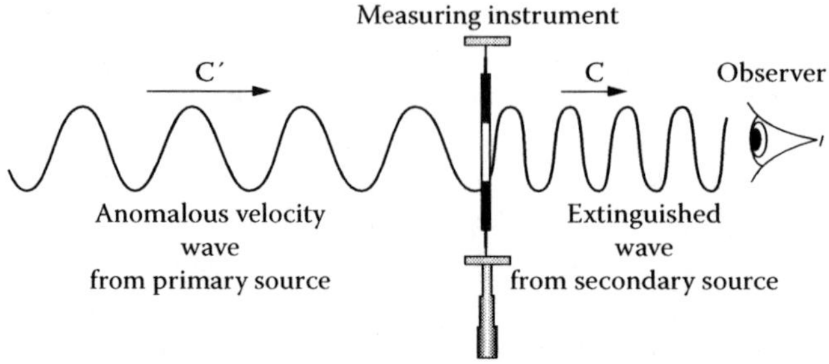

FIGURE 19.1
An interfering observer attempts to measure a previously undisturbed primary wave emitted from a primary source of a different frame of reference other than that of the observer.

to its most primary source alone. A most significant finding of this emission theory is that neither the wavelength nor the velocity of a primary undisturbed wave or photon is measurable. As a direct consequence of this principle, any knowledge of the velocity of motion of a single photon is also denied to all ordinary observers in the real material world. Knowledge of the velocity of a photon, a wave packet or a wavelet would require more than one direct measurement of at least two distinct positions and the corresponding times of detection at those positions. Since the very first detection of a photon requires direct interference with it, the *undisturbed* flight of the most primary photon is interrupted upon measurement. It is *extinction shifted* as depicted in Figure 19.1.

The primary photon or wave is extinguished, as is illustrated, by the measuring apparatus and its **true** wavelength is thereby *extinction shifted*! A naive observer would claim incorrectly that the velocity of the wave is always c. It is for this very reason that the experimental efforts of this past century were incorrectly interpreted as having observed a constancy in the velocity of light. The experiments were simply misinterpreted. The successful derivation of the equations of Relativity using assumptions of this *extinction shift principle* is in itself a direct mathematical physics proof that the phenomena taking place in the laboratories of nature are purely classical ones, describable only in the framework of *Euclidean space geometry.*

19.2.1 Constancy of Velocity of Light

The parameter c is the velocity of light constant, which has been measured very accurately to be about 299,792,458 meters per second in vacuum. There are additionally many issues pertaining to whether this constant has had different values at earlier times and/or in different regions of the universe. But the constant c is not the issue here at all. The real issue here is the constancy of the velocity of light in all frames of reference! The primary question remains: Does the *true velocity* of electromagnetic waves and gravitation in a given frame of reference depend on the motion of its *primary source* of a different frame of reference? Does the *Galilean transformation of velocities*

$$c' = c \pm v \tag{19.1}$$

apply to both electromagnetism and gravitation? The question is whether this equation, named after the famous Italian scientist and mathematician, Galileo Galilei (1564–1642), [2] is applicable to the physics of the photon and the graviton. Galileo, perhaps the most famous early astronomer, is considered one of the founders of modern science; far ahead of his time in many ways. The above questions have been answered in the affirmative by the mathematical proof of this emission theory. The mathematical illustrations and proofs, along with the cited observational evidence, show that the velocity of light is **not** constant in all frames of reference.

19.2.2 Rectilinear Motions of Photons and Gravitons

The rectilinear path of the photons and gravitons [3] is a fundamental basis of this emission theory. As a direct consequence of Galilean transformations in Euclidean space, the principle of emission and re-emission suggests that any undisturbed photon or graviton simply cannot change its path. It cannot deviate as long as its path is undisturbed. A primary photon moving along an undisturbed path will give rise to a secondary photon at the point of interference, thereby terminating the undisturbed path. The undisturbed phenomenon of rectilinear motion is hitherto not treated in modern physics texts.

As opposed to any light-bending effect or a warped space, as assumed in relativity, alternatively, altering the path of re-emitted photons is accomplished via electrodynamics of reemission in Euclidean space, as a direct consequence of relative phase and conservation of energy. The path of the new photon is characteristic of the interfering medium. [3] The primary photon upon **extinction** or interference no longer exists. In any refracting medium, the photon is subjected to processes of re-emission, i.e., from primary to secondary, from secondary to tertiary, on out to many n-ary re-emissions, each segment denoting infinitesimally short rectilinear (straight-line) paths along which the re-emitted photon or exchanged graviton moves.

19.2.3 Definition of Extinction Shift

As opposed to a Doppler shift, a re-emission at the point of interference of a primary not-yet-interfered-with photon or wave takes place. In Figure 19.2, an undisturbed primary wave moves independent of reference frames, from **primary** to **secondary** source frame, until which time it is re-emitted (extinction shifted) upon interference at the window. As illustrated, from left to right, the primary wave emitted from an approaching source on the left has the primary undisturbed wavelength of λ_{c+v} with velocity $c + v$ relative to the depicted fixed interference. The primary wave is extinguished at the point of interference, immediately re-emitting a new secondary wave

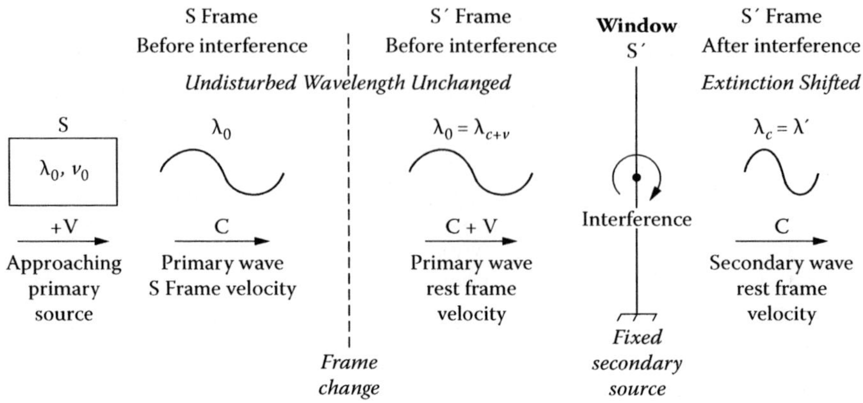

FIGURE 19.2
Reference-frame independent primary wave is re-emitted as a secondary wave whose wavelength is consequently extinction shifted.

with an extinction shifted wavelength of λ_c at the velocity c relative to the interfering secondary source (the fixed interference), and with the relative frequency of the primary wave as would be noted in the frame of reference of the interference.

For the case of the **approaching source** as depicted above, the new re-emitted secondary wave will have a shorter wavelength of

$$\lambda_c < \lambda_{c+v} \tag{19.2}$$

and will move with the velocity c relative to the point of interference, the new secondary source. The primary and secondary waves have exactly the same frequency v as would be noted in the frame of reference of the interference, i.e., the velocity-to-wavelength ratio of the primary wave equals the velocity-to-wavelength ratio of the secondary wave. For the approaching primary source always:

$$\left.\frac{c+v}{\lambda_{c+v}}\right|_{\substack{Before\\Interference}} = v = \left.\frac{c}{\lambda_c}\right|_{\substack{After\\Interference}} \tag{19.3}$$

For the case of the **receding source**, the new re-emitted secondary wave will have a longer wavelength of

$$\lambda_c > \lambda_{c-v} \tag{19.4}$$

and will move with the velocity c relative to the point of interference, the new secondary source. For the receding primary source always

$$\left.\frac{c-v}{\lambda_{c-v}}\right|_{\substack{Before\\Interference}} = v = \left.\frac{c}{\lambda_c}\right|_{\substack{After\\Interference}} \tag{19.5}$$

The point that has been missed in previous emission theories is that the ordinary real world observer can measure neither the *undisturbed velocity* $c \pm v$ nor the *undisturbed wavelength* $\lambda_{c\pm v}$ of the *primary undisturbed* wave. The measuring instrument can only discern the frequency ν of interference as is perceived in the frame of reference of the interference. Thus, any observer in the frame of reference of the interference would count the same number of waves passing a fixed point per unit time before and after the interference. Hence, the number of primary waves entering the interference equals the number of secondary waves leaving the interference.

As a consequence of the mathematical illustrations [1], it is thereby demonstrated that any wavelength of a primary undisturbed wave cannot be Doppler shifted, but rather re-emitted as an *extinction shifted* secondary wave, requiring absolutely no relativistic corrections whatsoever. And there is no direct observation or measurement on the primary wave! Solving equation 19.3 for λ_c we have, for the above illustrated approaching source

$$\lambda_c = \lambda_{c+v}\left(1+\frac{v}{c}\right)^{-1} \tag{19.6}$$

Solving equation 19.5 for a receding source (if the source were to move in the opposite direction) we have:

$$\lambda_c = \lambda_{c-v}\left(1-\frac{v}{c}\right)^{-1} \tag{19.7}$$

Thus, any primary wave along with its previously undisturbed wavelength is extinguished at the interference and replaced with a new secondary wave with a shifted, i.e., extinction-shifted wavelength, moving with velocity c in the frame of reference of the interference. It follows that any observation on the primary by the real-world observer is strictly denied. Expanding equations 19.6 and 19.7, one gets second order and higher order terms, the mathematically equivalence of the relativistically corrected Doppler shift. [3] [10] It is also important note that, unlike earlier emission theories, the principal axioms of the extinction shift principle make a clear distinction between the **measurable** and the **calculable**. [1]

19.3 Mathematical Illustrations

19.3.1 Invariance of Wave Equation

$$\frac{\partial^2\Phi}{\partial x^2}+\frac{\partial^2\Phi}{\partial y^2}+\frac{\partial^2\Phi}{\partial z^2}-\frac{1}{c^2}\frac{\partial^2\Phi}{\partial t^2}=0 \tag{19.8}$$

The invariance of the wave equation is mathematically illustrated under direct application of Galilean transformations of velocities using the principal axioms of the extinction shift principle. [1] The rules of emissions and re-emissions in Euclidean space geometry are strictly adhered to. Assume:

1. All undisturbed primary waves, i.e., $\Phi = \Phi_0 \sin 2\pi(\nu t + \frac{1}{\lambda}x)$ are emitted at velocity *c* relative to their most primary sources and upon any interference are then re-emitted at the same velocity c in the frame of reference of the interference. The undisturbed primary wave propagates with velocity *c* in all frames of reference other than that of the most primary source. The re-emitted secondary wave $\Phi' = \Phi'_0 \sin 2\pi(\nu' t' + \frac{1}{\lambda'}x')$ noted with relative frequency ν' and extinction shifted wavelength λ', propagates with velocity *c* relative to its secondary source.
2. The undisturbed (not measurable) wavelength λ, void of interference, remains unchanged in all frames of reference.
3. The laws governing emission and re-emission do **not** change with the frame of reference.

As a consequence of these rules, the apparent equations of motion, due to measurement or extinction of the primary wave, will be the same for all observers, regardless of the frame of reference, since the velocity of the re-emitted wave is always exactly *c* in the frame of reference of the interference only; a velocity of $c' \neq c$ in all others frames of reference. Only the observed frequency and the extinction shifted wavelength will depend on the frame of reference. From the principal axioms of the extinction shift principle (see Appendix IV of Reference [1]), all interfering observers will measure a frequency and a wavelength, the product of which is always *c*. In the frame of reference of the primary source, the velocity of the wave is $\nu\,\lambda = c$ relative to the primary source only, For any approaching source, the observable is always

$$\nu'\lambda' = \left[\nu\left(1+\frac{\upsilon}{c}\right)\right]\left[\lambda\left(1+\frac{\upsilon}{c}\right)^{-1}\right] = \nu\lambda = c. \tag{19.9}$$

For any receding source, the observable is always.

$$\nu'\lambda' = \left[\nu\left(1-\frac{\upsilon}{c}\right)\right]\left[\lambda\left(1-\frac{\upsilon}{c}\right)^{-1}\right] = \nu\lambda = c. \tag{19.10}$$

A hypothetical, non-interfering observer, however, would note that the velocity of an undisturbed wave moving, say along the x direction, would depend on the reference frame, strictly obeying Galilean transformations of velocities and that the undisturbed wavelength, not measurable by any interfering observer, would remain unchanged.

The hypothetical observer, who abides strictly by the principal axioms of the extinction shift principle, while correctly applying these rules to Galilean transformation in Euclidean space geometry, would correctly predict that all interfering observers would always note $\nu'\lambda' = \nu\lambda = c$. By differentiating the equation $\Phi' = \Phi'_0 \sin 2\pi\,(\nu' t' + \frac{1}{\lambda'} x')$ twice after t' and x', the interfering observer arrives at

$$\frac{\partial^2 \Phi'}{\partial t'^2} = -\Phi'(2\pi)^2 \nu'^2 = \nu'^2 \lambda'^2 \frac{\partial^2 \Phi'}{\partial x'^2} \tag{19.11}$$

Thus, the interfering observer, regardless of his frame of reference, derives the very same wave equation

$$\frac{\partial^2 \Phi'}{\partial x'^2} + \frac{\partial^2 \Phi'}{\partial y'^2} + \frac{\partial^2 \Phi'}{\partial z'^2} - \frac{1}{c^2}\frac{\partial^2 \Phi'}{\partial t'^2} = 0, \tag{19.12}$$

for quantities differing only in ν' and λ', but not in $\nu'\lambda' = \nu\lambda = c$.

The wave equation is found to be totally invariant under Galilean transformations of velocities, using the correctly formulated principle axioms of the extinction shift principle applied to emissions and re-emissions in Euclidean space geometry.

19.3.2 Transverse Relative Time Shift

Let a source move with constant velocity v in a direction transversely relative to a stationary observer as indicated in Figure 19.3. Assume the source has a lifetime of τ_0 seconds and emits two bursts of signals, an initial one at birth ($t = 0$) and a final one at death ($t = 3$). The resting observer is placed at a distance D from the nearest point on the path of the moving source. Let the initial burst serve as time reference and be emitted such that it is received at the observer's measuring apparatus when the source is positioned such that a line extended from the observer to the source is at right angle to the path of the source (dotted line). It is herewith mathematically illustrated that the difference in the times of arrival of the initial and final waves is actually $\tau' > \tau_0$; a transverse relative time shift, the inverse of a transverse relative frequency shift.

As a consequence of Galilean transformations and the rectilinear path of all constituent parts of a wave front, a simultaneous detection by a single observer of both the initial and the final signal burst is not possible! The initial wave front will arrive at the speed $c' = \sqrt{c^2 + v^2}$ for the distance $\sqrt{D^2 + D^2 \frac{v^2}{c^2}}$ and have the radius $D = \tau_0 c$. The final wave is emitted at distance $\tau_0 v$ past the point of emission of the initial wave. The final wave front is received at the observer delayed by τ' seconds, during which time the center of the spherical wave front moves the distance $\tau' v$ past the ($t = 3$) point to the ($t = 9$) point, while its radius increases to the length of $\tau' c$. It follows from geometry that

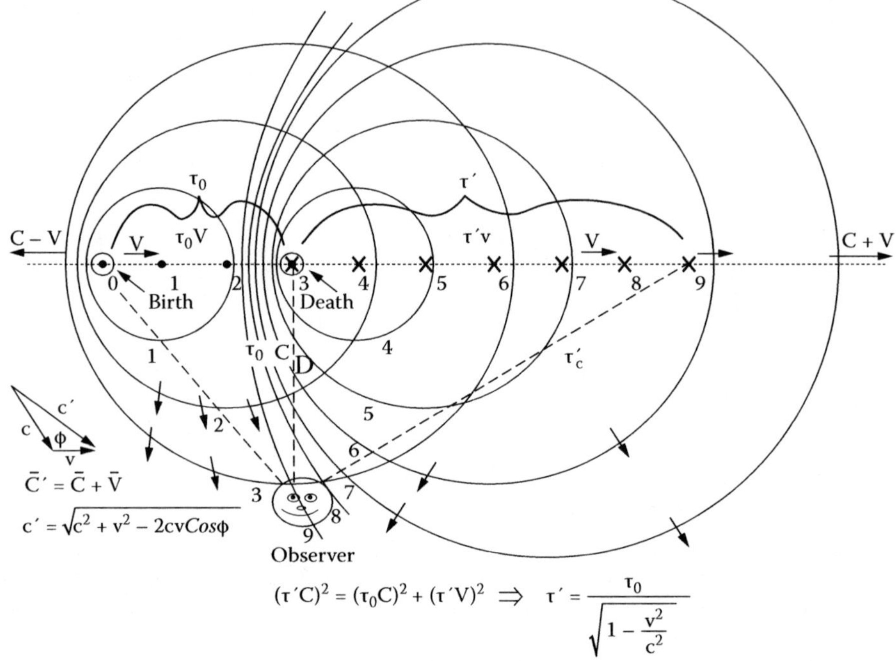

FIGURE 19.3
Transverse relative time shift as opposed to time dilation assumed by relativity.

$(\tau' c)^2 = (\tau_0 c)^2 + (\tau' v)^2$. Solving for τ' we get

$$\tau' = \frac{\tau_0}{\sqrt{1-\frac{v^2}{c^2}}} \tag{19.13}$$

Thus, a particle of lifetime τ_0 and velocity v will appear to any fixed observer to move the distance $\tau' v$ in time τ'.

This effect is therefore a transverse relative time shift, not a time dilation. [1] It should be noted that a procedure similar to that above derives a velocity dependent effective mass

$$m_{eff} = \frac{m_0}{\sqrt{1-\frac{v^2}{c^2}}} \tag{19.14}$$

which is the mathematical equivalent of the relativistic mass.

19.3.3 Perihelion Rotation Effect

We shall calculate the planet Mercury perihelion rotation effect using solely Galilean transformations of velocities applied to the transit time effect that is due to the exchange of gravitons between the mass bodies, each moving with a given velocity relative to the gravitational field set up by the other. The table lists important astrophysical parameters necessary for this calculation.

A one-way transit time effect for a gravitational interaction between mass particles separated by a distance r may be given as $\tau_{rec} = \frac{r}{c_{rec}}$ when they recede from one another. For an approaching case, the transit time may be given by $\tau_{app} = \frac{r}{c_{app}}$. Based on the table, the receding mass particles see a Galilean transformed velocity of the gravitational field set up by the other mass of velocity $c'_{rec} = c(1 - \frac{v^2}{c^2} + 2\frac{v}{c}\cos\phi)^{1/2}$. This translates to an effective distance of $r_{rec} = c\tau_{rec} = r(1 - \frac{v^2}{c^2} + 2\frac{v}{c}\cos\phi)^{-1/2}$. The mean orbital velocity of the planet Mercury is $v_{Mercury} = 48.96 Km/sec$ and $\frac{v_{Mercury}}{c} = 1.632 \cdot 10^{-4}$. From the Table, for the calculation of c' we equate $2vc' \cos\phi \approx 2vc \cos\phi$ and thus $2\frac{v}{c'} \cos\phi \approx 2\frac{v}{c} \cos\phi$, since $\frac{c'}{c} \approx 1$. We will see later on that the terms in $\frac{v}{c} \cos\phi$ will cancel due to sign change and the practical symmetry of the elliptical orbit!

Herewith, for both the receding and approaching cases, the angle ϕ is only slightly greater than $\frac{\pi}{2}$ radians, causing the value of $\cos\phi$ to take on negative values. For the receding case, the effective path for gravitational influence is therefore

$$r_{rec} \approx r\left(1 + \frac{1}{2}\frac{v^2}{c^2} - \frac{v}{c}\cos\phi\right) \quad \text{where} \quad r_{rec} > r. \tag{19.15}$$

TABLE 19.1

Effective Path Length, Resulting Effective Force, and Velocity Transformations from the Geometry (Figure 19.4) for Receding and Approaching Cases. Important Astrophysical Parameters used to Calculate Perihelion Rotation Effect under Galilean Transformations of Velocities in Euclidean Space are Listed

Velocity Dependent Parameters (Receding and Approaching Cases)	Astrophysical and Prbital Parameters (for Planet Mercury)
r_{rec} = effective length (receding)	$GM = 1.3271544 \cdot 10^{2}0 m^3/s^2$
r_{app} = effective length (approaching)	$a = 57.9 \cdot 10^9 m$
$F_{rec} = \frac{GM_m}{r_{rec}^2}$	$e = 0.205633$
$F_{app} = \frac{GM_m}{r_{app}^2}$	$r = a(1 - e^2)/(1 + e\cos\nu)$
$C'_{rec} = \sqrt{C^2 - \upsilon^2 + 2\upsilon C'\cos\phi}$ $\approx \sqrt{C^2 - \upsilon^2 + 2\upsilon C\cos\phi}$	$\omega = \sqrt{\frac{GM}{r^3}}$
$C'_{rec} = C\left(1 - \frac{\upsilon^2}{C^2} + 2\frac{\upsilon}{C}\cos\phi\right)^{\frac{1}{2}}$	$\upsilon = \sqrt{\frac{GM}{r}}$
$C'_{app} = \sqrt{C^2 + \upsilon^2 + 2\upsilon C\cos\phi}$	
$C'_{rec} = C\left(1 + \frac{\upsilon^2}{C^2} - 2\frac{\upsilon}{C}\cos\phi\right)^{\frac{1}{2}}$	$\frac{\upsilon^2}{C^2} = 2.63 \cdot 10^{-8}$

Similarly, for the approaching case, the effective path for gravitational influence is

$$r_{app} \approx r\left(1-\frac{1}{2}\frac{v^2}{c^2}+\frac{v}{c}\cos\phi\right) \quad \text{where} \quad r_{app} < r. \tag{19.16}$$

From the table of orbital parameters, the angular velocity $\omega = \sqrt{\frac{GM}{r^3}}$ can be modified to reflect the receding case, giving

$$\omega_{rec} = \sqrt{\frac{GM}{r_{rec}{}^3}} = \sqrt{\frac{GM}{r^3}}\left(1-\frac{v^2}{c^2}+2\frac{v}{c}\cos\phi\right)^{\frac{3}{4}}. \tag{19.17}$$

Similarly, for the approaching case,

$$\omega_{app} = \sqrt{\frac{GM}{r_{app}{}^3}} = \sqrt{\frac{GM}{r^3}}\left(1+\frac{v^2}{c^2}-2\frac{v}{c}\cos\phi\right)^{\frac{3}{4}}. \tag{19.18}$$

Since the means orbiting velocity of the planet Mercury is such that $\frac{v}{c} << 1$, then we have $\omega_{rec} \approx \sqrt{\frac{GM}{r^3}}[1-\frac{3}{4}\frac{v^2}{c^2}+\frac{3}{2}\frac{v}{c}\cos\phi]$ for the receding case.

Expressing the angular velocity ω as a function of velocity v, we have

$$\frac{d}{dv}\omega_{rec} = \sqrt{\frac{GM}{r^3}}\left[-\frac{3}{2}\frac{v}{c^2}+\frac{3}{2}\frac{1}{c}\cos\phi\right]. \tag{19.19}$$

wherefrom $\Delta\omega_{rec} = \omega(-\frac{3}{2}\frac{v}{c^2}+\frac{3}{2}\frac{1}{c}\cos\phi)\Delta v$ where $\omega = \sqrt{\frac{GM}{r^3}}$.

$$\Delta\omega_{rec}\Big|_{\Delta v=+v} = \omega\left(-\frac{3}{2}\frac{(+v)^2}{c^2}+\frac{3}{2}\frac{(+v)}{c}\cos\phi\right) \tag{19.20}$$

Similarly, for the approaching case,

$$\Delta\omega_{app}\Big|\Delta v = +v = \omega\left(+\frac{3}{2}\frac{(-v)^2}{c^2}-\frac{3}{2}\frac{(-v)}{c}\cos\phi\right) \tag{19.21}$$

A net change in angular velocity of the planet Mercury for a complete orbit may be given as $\Delta\omega = \Delta\omega_{rec} - \Delta\omega_{app}$ which results in a function of a second order in $\frac{v}{c}$ only!

$$\Delta\omega = \omega\left[\left(-\frac{3v^2}{2c^2}\right)-\left(+\frac{3v^2}{2c^2}\right)\right] \tag{19.22}$$

We note immediately that, under Galilean transformations of velocities, the first order terms in $\frac{v}{c}\cos\phi$ cancel as a consequence of sign changes during

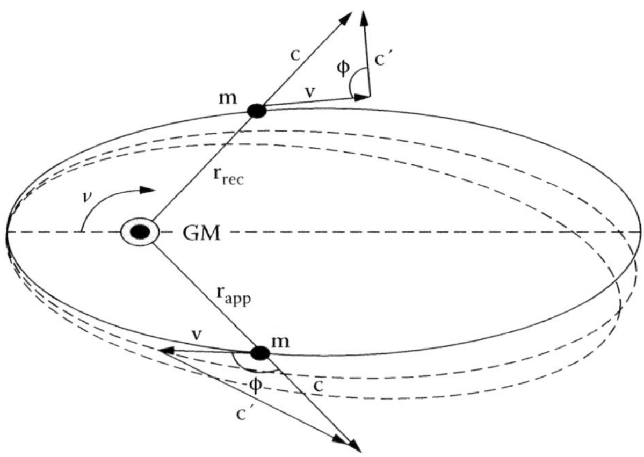

FIGURE 19.4
Direct application of extinction shift principle calculates the perihelion rotation effect due a transit time effect for an exchange of gravitons between orbiting mass bodies according to Galilean transformations in Euclidean space. The velocity c' is that of the gravitons of the field of M relative to m in the depicted sections of its elliptical orbit. For simplicity of problem solution, the angle ϕ is chosen to separate the velocity vectors for v and c' for the receding case, v and c for the approaching case.

the approach and receding portions of the orbit! From Figure 19.4 the resultant velocity c' of the gravitational field of gravitons from the solar mass M, as seen from the Mercury mass m, has practically the same angle ϕ history relative to the velocity v of Mercury's orbit, since $\frac{v}{c} << 1$. Hence, practically the same angular history for ϕ is swept for $0 < \nu < \pi$ and $\pi < \nu < 2\pi$. The second order terms in $\frac{v}{c}$ accumulate, as is expected, since a required net energy for the planetary orbit must be zero. Thus, the perihelion must shift! The net change in the angular velocity is calculated using the table of astrophysical and orbital parameters for Mercury and simply by setting $v = \frac{\pi}{2}$ as follows:

$$\Delta\omega = \omega \cdot 3\frac{v^2}{c^2} = \frac{3\omega GM}{a(1-e^2)c^2}; \quad \Delta\omega = 7.04814 \cdot 10^{-14}\ \text{rad/sec} \qquad (19.23)$$

Expressing this result in radians per period, we have

$$\Delta\omega\frac{2\pi}{\omega} = \frac{6\pi GM}{a(1-e^2)c^2} = 5.019568 \cdot 10^{-7}\ \text{rad/period} = 42.988\ \text{arcsec/century} \quad (19.24)$$

This result verifies that gravitation as well as light behaves strictly according to Galilean transformation of velocities in Euclidean space with the very same velocity c, relative to the primary source only, as that of the velocity of light. This principle of the graviton exchange has a direct analogy to the principles of the emission and re-emission of the photon according to the principal axioms of the extinction shift principle, the same principle used to arrives at the perihelion rotation results of the PSR1913+16 binary neutron pulsar star system, calculated in detail in Reference [1], arriving at the precise numerical result obtained by relativity, first published in Reference [11], a result claimed by relativity in 1990 for its fame and validation.

19.4 Observational Evidence

The past century of experiments in optics along with convincing observational evidence lend support the above demonstrated emission theory, summarized in detail in Reference [1]. The principal axioms that serve as the rules for applying the Galilean transformations were mathematically illustrated on the past century of velocity of light experiments, to include the Beckmann and Mandics Lloyd mirror experiment [4] in 1965 and the Babcock and Bergman rotating mirror experiment [5] repeated by Beckmann and Mandics [6] in 1964. Rotz [7], and James and Sternberg [8] performed variations of this experiment. One of the most important experiments was performed by Albert A. Michelson [9] and involved two mirrors rotated about a common center inside of an optical loop. The principal axioms serve as the maps to help explain the experimental outcomes, the details of which are given in Reference [1] in the Appendix pp. 23A–32A. Additionally, along with the planet Mercury perihelion rotation effect calculated here, the calculation of the perihelion rotation effect of the PSR1913+16 binary pulsar system, first calculated by Taylor et al. (1978) using general relativity and published in Reference [11], is mathematically illustrated with the very same technique of this emission theory and published in detail in Reference [1].

Summary and Conclusions

The obvious consequence of the measurement and the undisturbed nature of the not physically measurable phenomenon are herein considered. Significant findings include:

1. The non-measurability of the wavelength or velocity of a primary wave or a primary photon from a frame of reference other than that of the most primary source.
2. The primary, not-yet-interfered-with undisturbed wavelength remains unchanged and is independent of reference frames.
3. The extinction shift principle correctly predicts the outcome of important astrophysical phenomena taking place in the laboratories of nature for both electromagnetism and gravitation by applying

the very same rules of Galilean transformations of velocities in Euclidean space alone.

4. These pure classical treatments lead directly to the solutions of famous problems responsible for the success and fame of general and special relativity. [1]

Appendix: Principal Axioms of Extinction Shift Principle

There are various combinations of light paths that need to be considered for theoretically interpreting the results pertaining to electromagnetic emissions. The experiment always pertains to a source primary emitter, an interference, one or more secondary sources of emission or re-emitters and an observer or a detector. The principal axioms pertain to the various combinations of the state of the source, the interference and the observer and the direct application of the Galilean transformations to derive the observed frequencies, wavelengths and velocities in Euclidean space. For instance, one experiment may involve a fixed source, a fixed interfering window and a moving observer. Another experiment may involve a moving source, a fixed interfering window and a moving observer. Still another experiment may involve a fixed source, a moving interfering window and a fixed observer, and so on.

Similarly, for the case of gravitation, a given primary mass particle may be considered as the source of a primary field that perturbs a secondary mass particle that is the direct source of a secondary field. The secondary field set up by this secondary mass conveys indirect information on the primary mass particle via its secondary field to yet a third tertiary mass or some sensor mass under influence of the fields.

The same Galilean transformation of velocities applied to gravitation to solve problems in astrophysics and correctly predict the outcomes of nullified experiments in optics provides grounds for the correctness of this extinction shift principle! See Appendix IV of Reference [1].

References

[1] Dowdye, E., *Discourses & Mathematical Illustrations pertaining to the Extinction Shift Principle under the Electrodynamics of Galilean Transformations,* Craig Color Printing Corp, Chadds Ford, PA, (2001).

[2] Redondi, Pietro, Galileo: *Heretic* (*Galileo Eretico*), Princeton University Press, Princeton, New Jersey (1987).

[3] Born, M. and Wolf, E., *Principles of Optics,* Pergamon Press, London, **71, 100–104** (1975).

[4] Beckmann, Petr and Mandics, Peter, Test of the Constancy of the Velocity of Electromagnetic Radiation in High Vacuum, *Radio Sci. J. Res.* NBS/USNC/URSI 69D, No.4, **623–628** (1965).

[5] Babcock, G. C. and Bergman, T. G., Determination of the Constancy of the Speed of Light, *J. Opt. Soc. Am.* **54, 147–151** (1964).

[6] Beckmann, Petr and Mandics, Peter, Experiment on the Constancy of the Velocity of Electromagnetic Radiation, *Radio Sci. J. Res.* NBS/USNC/URSI 68D, No. 12, **1265–1268** (1964).

[7] Rotz, Fred B., New Test of the Velocity of Light Postulate, *Phys. Lett.*, 7, **252–254** (1963).

[8] James, J. F. and Sternberg, R. S., Change in Velocity of Light Emitted by a Moving Source, *Nature* 197, **1192** (1963).

[9] Michelson, A. A., Effect of Reflection from a Moving Mirror on the Velocity of Light, *Astrophys. J.*, 37, **190–193** (1913).

[10] Jackson, J. D., *Classical Electrodynamics,* John Wiley & Sons, New York, **512–515** (1975).

[11] Hulse, R. A. and Taylor, J. H., *Science* 250, 770 (1990); *Astrophys. J.* 195, L51 (1975).

20

Phase Coherence in Multiple Scattering: Weak and Intense Monochromatic Light Wave Propagating in Cold Strontium Cloud

David Wilkowski, Yannick Bidel, Thierry Chanelière, Robin Kaiser, Bruce Klappauf, and Christian Miniatura

Institut non linéaire de Nice, UMR, Valbonne F-06560, France

CONTENTS

Abstract

For large bulk disordered media, light transport is generally successfully described by a diffusion process. This picture assumes that any interference is washed out under configuration average. However, it is now known that, under certain circumstances, some interference effects survive the disorder average and in turn lead to wave localizations effects. In this chapter, we investigate coherence of a monochromatic laser light propagating in an optically thick sample of laser-cooled strontium atoms. For this purpose, we use the coherent backscattering effect as an interferometric tool. At low laser probe beam intensities, phase coherence is fully preserved and the interference contrast is maximal. At higher intensities, saturation effects start to set in and the interference contrast is reduced.

20.1 Introduction

One of the fascinating properties of photons, like all quantum objects, is interference. In the well-known two-slits experiment, the photon experiences the two slits at the same time, *i.e.*, takes two different paths to reach the detector. The detection probability P is obtained from the superposition principle which states that $P = |A_1 \exp(i\varphi_1) + A_2 \exp(i\varphi_2)|^2$ where $A_n \exp(i\varphi_n)$ is the quantum mechanical amplitude to go through slit n while the other is closed. Interference effects are then encoded in the phase-difference $\varphi = \varphi_2 - \varphi_1$. As a result, depending on the φ value, the two paths may interfere constructively or destructively and correspondingly lead to an increased or decreased detection probability and thus to interference fringes. These quantum interferences are very sensitive to any phase-breaking mechanisms destroying coherence.

The same principles apply for monochromatic light shining and being scattered off an optically thick disordered sample. For a given configuration of scatterers, the scattered light exhibit a well-known speckle pattern. This pattern originates from the coherent superposition of all possible quantum amplitudes $A_p \exp(i\varphi_p)$ associated to each possible scattering path p inside the medium. The detection probability is thus now $P = |\Sigma_p A_p \exp(i\varphi_p)|^2$. Averaging now over all possible scatterers configurations, one may think that all interference terms of the form $\Sigma_{p\neq q} A_p A_q \exp i(\varphi_p - \varphi_q)$ will be washed out. This is true *unless* paths p and q are *geometrically the same* but *travelled in opposite directions*. We then say that we face pairs of *reversed* paths. In this case, disorder average cannot break the *two-wave* interference associated to these pairs of scattering paths. This is the basic surviving interference effect at the heart of the coherent back scattering (CBS) phenomenon. Collecting light retro-reflected off the sample, the average detection signal exhibits a narrow angular cone around exact backscattering. The angular width of the CBS cone typically scales as $(k\ell)^{-1}$ where k is the light wave vector and ℓ the light scattering mean free path inside the sample. This CBS cone is a hall mark of interference effects in multiple scattering [1–3] even if other interference effects which survive disorder-average also exist: weak localization effects (interference corrections to the Boltzmann diffusion constant), universal conductance fluctuations [2,4,5], *etc.*

Technically speaking, the semi-classical picture developed so far to explain interference effects in multiple scattering is valid in the (weak localization) regime $k\ell >> 1$. When disorder is so strong that the onset $k\ell \approx 1$ is reached (Ioffe-Regel criterion), then a disorder-induced "metal-insulator" transition occurs. Optical states in the bulk are exponentially localized and transport is inhibited. This is the celebrated Anderson scenario (strong localization phenomenon) valid for any kind of linear waves [6, 7]. For infra-red optical light, only one experimental observation [8] has been reported so far but further investigation is still required due to possible residual absorption [9, 10].

The CBS effect can be used as a powerful interferometric tool to study possible phase-breaking mechanisms at work while the light wave interacts with a random medium. As a (multi) two-wave interferometer, the CBS interferometer shares many similarities with other, more common, two-wave interferometers like the two-slit set-up. However it has also some particular and unusual properties: it is an automatically self-aligned, zero path-length set-up. This interferometer is thus very robust. The CBS enhancement factor α, defined as the ratio between the intensity collected at exact backscattering and the intensity collected far off exact backscattering, is a measure of the degree of coherence of light leaving the medium. When coherence is fully preserved, α takes its maximal value in the so-called parallel polarization channels and is exactly 2 for spherically symmetric scatterers in the helicity-preserving polarization channel $h||h$. In this case, any phase-breaking mechanism inducing a coherence loss between the interfering multiple scattering reversed paths is expected to yield an enhancement factor smaller than 2 in the $h||h$ channel.

This makes the CBS interferometer an unique tool to study decoherence effects in multiple scattering. As an example, we may cite the decoherence induced by a Zeeman-degenerate internal structure (for experiment see [11,12] and [13] for theory) and also the corresponding surprising restoration of interference by applying an external magnetic field [14].

In this chapter we present a collection of CBS experiments done with a cold strontium (^{88}Sr) atomic cloud as an optically thick disordered medium. Using an atomic gas to investigate wave transport phenomena offers substantial advantages with respect to classical Mie or Rayleigh scatterers. First, as point-dipole scatterers, the maximum light scattering cross-section $\sigma = 6\pi/k^2$ is far larger than the square of the geometrical size itself. Scattering is thus very efficient. Second, as atoms are extremely resonants scatterers, a slight detuning of the incoming light with respect to the internal atomic resonance (by few linewidths Γ) can change by several orders of magnitude the light scattering efficiency. Third, atoms of a given species are perfect monodisperse scatterers. The major drawback is the large transport time $\tau^* \gtrsim \Gamma^{-1}$ (≈ 5 *ns* for strontium) [15] which imposes a Doppler broadening much smaller than Γ [16]. Cooling atoms in a Magneto-Optical Trap (MOT) circumvents this difficulty.

The strontium MOT and its main characteristics will be given in section II. Then in section III, we will describe the results obtained in two different regimes: the elastic scattering regime obtained at low laser intensities and the inelastic scattering regime obtained at high laser intensities. In the first case, coherence is fully preserved while in the second case it is altered due to vacuum-induced dipole fluctuations.

20.2 Cold Atomic Cloud

20.2.1 MOT Set-Up

The cold strontium cloud is produced in a magneto-optical trap (MOT) set-up. The cooling transition is the optical dipole transition line $^1S_0 - {}^1P_1$ at $\lambda =$ 461 *nm*. This transition thus connects a $J_g = 0$ ground state to a $J_e = 1$ excited state. The excited-state natural linewidth is $\Gamma/2\pi = 32$ *MHz* and the corresponding saturation intensity is $I_s = 42.5$ mW/cm^2.

First an effusive strontium beam is extracted from an oven operating at 500°C. Then a 27 *cm* long Zeeman slower reduces the strontium longitudinal velocity within the velocity capture range of the MOT, *i.e.*, ~ *50 m/s* (Fig. 20.1a). The Zeeman slower, MOT, and probe laser beams all operate at 461 *nm* and are generated from the same frequency-doubled source detailed in [17]. Briefly, a single-mode grating stabilized diode laser and a tapered amplifier are used in a master-slave configuration to produce 500 *mW* of light at 922 *nm*. This infrared light is then frequency-doubled in a semi-monolithic standing-wave cavity with an intracavity $KNbO_3$ nonlinear crystal. The cavity is maintained at resonance with the infrared light thanks to a feedback loop. The second harmonic exits the cavity through a dichroic mirror providing 150 *mW* of tunable single-mode light, which is then frequency locked on the 461 *nm* strontium line in a heat pipe (Fig. 20.1b). We use acousto-optic modulators for subsequent amplitude and frequency variations. The MOT is made of six independent trapping beams. Each beam is carrying an intensity of 5.2 mW/cm^2 and each beam waist is 8 *mm*. The trapping beams are red-detuned by $\delta = -\Gamma$ with respect to the atomic resonance line. Two anti-Helmoltz coils generate a 70 *G/cm* magnetic field gradient to trap the atoms.

20.2.2 MOT Parameters

20.2.2.1 Trapped Population

The 461 *nm* transition used for cooling is not a closed transition. Hence, atoms in the $^1P_1$ state can radiatively decay to the $^1D_2$ state and then to the triplet $^3P_1$ and $^3P_2$ states (see Fig. 20.2). Atoms ending in the long-lived $^3P_2$ state are then lost. The maximum optical pumping loss rate (obtained at large laser intensities) is 1300 s^{-1} whereas the loading atomic flux in our experiment is about 10^9 s^{-1}. Hence pumping losses can reduce the number of trapped atoms down to typically 10^6 atoms but do not prevent by themselves the formation

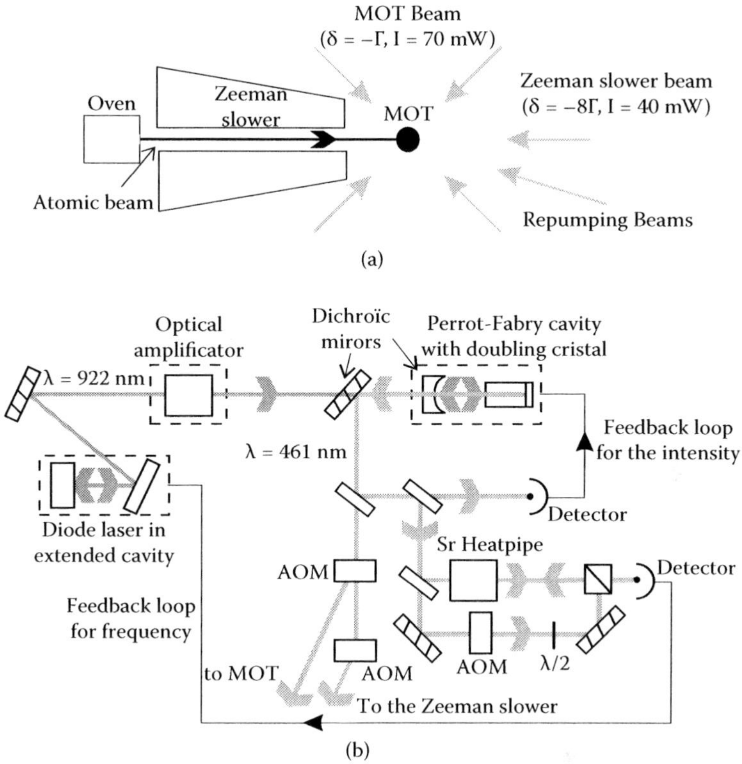

FIGURE 20.1
MOT (a) and Laser (b) set-ups. For more details see text.

of the cold atomic cloud. In Fig. 20.3, the MOT lifetime is shown as a function of the saturation parameter s

$$s = \frac{s_0}{1 + 4\delta^2/\Gamma^2} \tag{20.1}$$

where $s_0 = I/I_s$ is the on-resonance saturation parameter (I being the total MOT laser intensity) and δ is the laser detuning. The plain curve is obtained by considering optical pumping as the only loss mechanism with no adjustable parameters. We see that the overall experimental behavior is well reproduced by this simple model. We think that the small mismatch may come from systematic errors in the measurements of the laser intensity or detuning.

In principle, atoms should be efficiently shielded from these optical pumping losses by adding two additional lasers on resonance with the $^3P_2 \rightarrow {}^3S_1$ line at 707 *nm* and with the $^3P_2 \rightarrow {}^3S_1$ line at 679 *nm*. Using only the 707 *nm* laser, atoms are pumped to the $^3P_0$ metastable state. The relative maximum

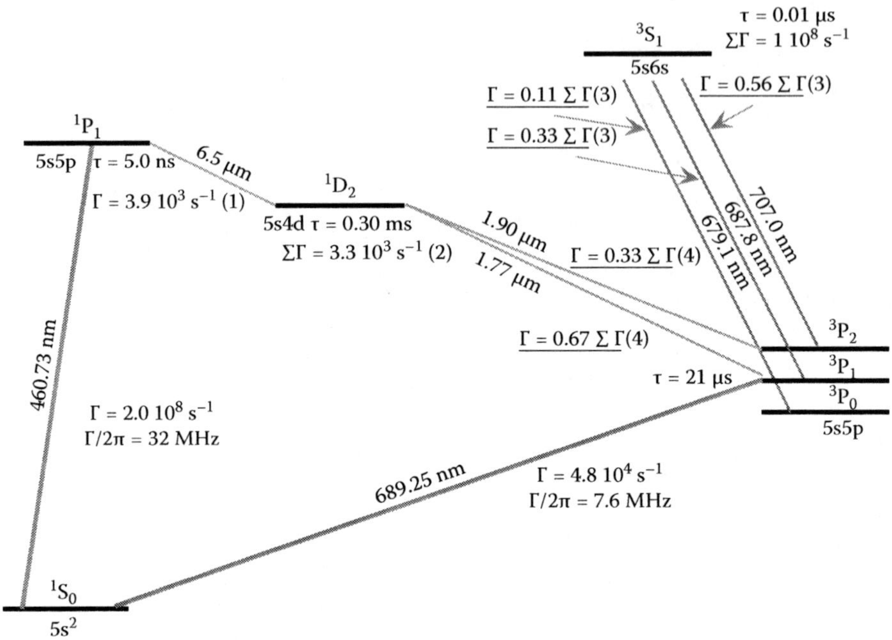

FIGURE 20.2
Energy diagram of ^{88}Sr atom.

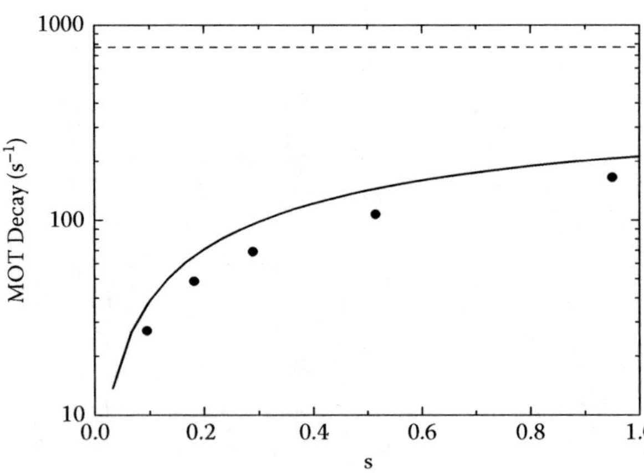

FIGURE 20.3
Measured MOT decay rate as a function of the saturation parameter s of the MOT beams (full circles) and its comparison to a theoretical model based on optical pumping losses (plain line). The dashed line corresponds to the maximum decay rate obtained at $s >> 1$.The experiment was done at $\delta = -1.4\Gamma$.

gain G of the trapped population should then be:

$$G = \frac{\Gamma_{^3S_1 \to {}^3P_1}}{\Gamma_{^3S_1 \to {}^3P_0}} \tag{20.2}$$

It corresponds to the ratio between the inverse decay probability of the $^3S_1$ state to the $^3P_0$ and the inverse decay probability of the $^3S_1$ state to the $^3P_1$ state. From Fig. 20.2, we get $G = 3$. Hence the maximum gain of the trapped population should be $1 + G = 4$. However the measured gain is only around 2.5, *i.e.* lower than the expected value. We think that this discrepancy is due to the MOT magnetic field gradient which expels the atoms pumped in the anti-trapping Zeeman states of the $^1D_2$ and $^3P_2$ levels.

By using the two pumping lasers, the number of trapped atoms is increased up to about $N \approx 10^8$. In this configuration, the MOT lifetime is essentially dominated by inelastic cold collisions, residual optical pumping and hot collisions with the uncooled strontium atoms of the atomic beam. Operating at low laser intensity, the number of trapped atoms is substantially decreased and the hot collision loss channel becomes the dominant one. In this case the MOT lifetime is found to be 0.5 s.

20.2.2.2 Size and Density

The MOT size is obtained by fluorescence imaging on a CCD camera. The MOT shape is roughly Gaussian with a *rms* radius of a fraction of *mm*. In Fig. 20.4a, the MOT *rms* volume V is plotted as a function of the number N

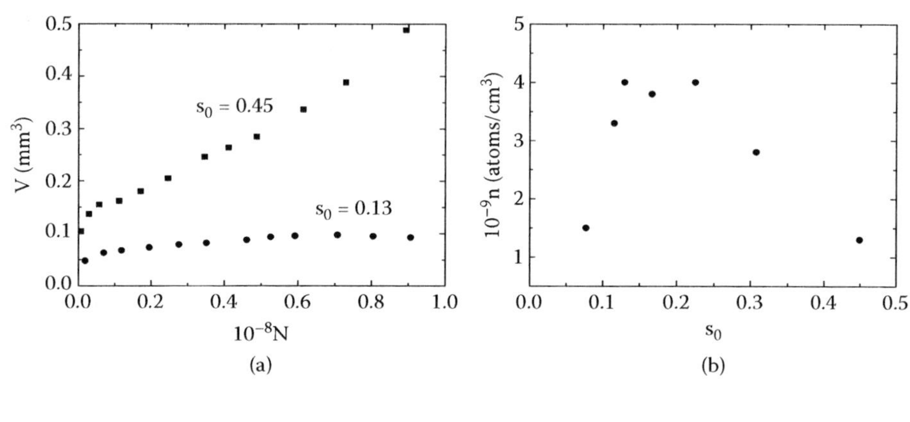

FIGURE 20.4
(a) *rms* MOT volume V as a function of the trapped population N for two different on-resonance MOT saturation parameters. When $s_0 = 0.13$, V is independent of N while it increases with N when $s_0 = 0.45$. (b) MOT spatial density *versus* the on-resonance MOT saturation parameter s_0. For s_0 small, N is reduced and accordingly n. For s_0 large, V increases faster than N and n is decreased. The maximum density is roughly $n \approx 4 \times 10^9$ atoms/cm^3. Both experiments were performed at $\delta = -0.9\Gamma$.

of trapped atoms. We have observed two different behaviors: at low laser intensities, the MOT volume is roughly independent of N while it increases with N at higher laser intensities.

This phenomenon is well known [18] and comes from multiple scattering of light in the cold cloud. Indeed if a scattered photon is re-scattered again in the MOT before leaving it, it induces a repulsive force between atoms and the MOT cloud inflates. If the scattering event is elastic, the repulsive force is compensated by the attractive force due to shadowing (trapping beam attenuations). At higher intensities (see section III C), the scattering becomes mostly inelastic, and the repulsive force dominates. The volume of the MOT cloud is then determined by balancing the trapping and the repulsive forces. This situation is well evidenced by the plot at $s_0 = 0.45$ in Fig. 20.4a.

Knowing the MOT volume V and the number N of atoms, we can deduce the MOT density $n = N/V$. In Fig. 20.4b, the MOT density n is plotted as a function of the on-resonance saturation parameter s_0. At low s_0, n is reduced because the trapping force is small and N decreases. At higher s_0, n is reduced because the multiple scattering repulsive force sets in and V is increased faster than N. The maximal density is about $n \approx 4 \times 10^9$ atoms/cm^3 and is obtained at $s_0 \approx 0.15$.

20.2.2.3 Optical Thickness

An important parameter for localization experiments is the optical thickness b of the cold cloud. This quantity is defined by the exponential attenuation of a light beam propagating through the MOT (Lambert-Beer law):

$$I_t = I_0 e^{-b} \tag{20.3}$$

where I_t is the transmitted intensity and I_0 the initial intensity. Noting by L the MOT *rms* diameter, then $b = L/\ell_{ex}$ where ℓ_{ex} is known as the extinction length. When the only attenuation mechanism is depletion by scattering, the extinction length reduces to the scattering mean free path ℓ:

$$\ell = \frac{1}{n\sigma} \tag{20.4}$$

where n is the atomic density and σ the light scattering cross-section

$$\sigma = \frac{\sigma_0}{1 + 4\delta/\Gamma^2}; \quad \sigma_0 = \frac{3\lambda^2}{2\pi} \tag{20.5}$$

Multiple scattering is said to set in when $b \gtrsim 1$. For b >> 1, light transport in the bulk is successfully described by a diffusion process. In our case, the optical depth at resonance is b ≈ 3: it is enough to evidence multiple scattering effects but not enough to reach the diffusive transport regime.

The light scattering mean free path is $\ell = L/b \approx 0.4\ mm$, giving $k\ell \approx 10^4$ ($k = 2\pi/\lambda$ is the incoming light wavevector). Our MOT cloud is thus far from achieving the Anderson localization threshold $k\ell \approx 1$ where strong localization of light is expected to occur. However, as we will see in the next sections, even in this weak localization regime where $k\ell >> 1$, interference effects influencing transport in multiple scattering can be evidenced.

20.2.2.4 Temperature and Phase-Space Density

Since the optical cooling dipole transition involves a $J_g = 0$ ground state, only Doppler cooling is present. Hence the lowest expected temperature is about 0.5 *mK* much higher than standard temperatures in MOTs operating with alkaline atoms where Sisyphus-type mechanisms are also present.

We have however measured here temperatures larger than the Doppler predictions. As an example, Fig. 20.5 shows the measured velocity dispersion σ_v in a 1D optical molasses as a function of the on-resonance saturation parameter s_0 (for more details see [19]). The dotted curve corresponds to the Doppler theory prediction and is completely off the experimental data. In fact, Doppler cooling proves very sensitive to heating induced by transverse spatial intensity fluctuations. The plain curve in Fig. 20.5 is the result of a Monte-Carlo simulation taking into account these intensity fluctuations. As one can see, a perfect agreement with experimental data is then recovered.

In the MOT, the typical measured temperature is about 5 *mK* ($\sigma_v \approx 1\ m/s$) Even if the situation is more complex here than in pure 1D molasses, we

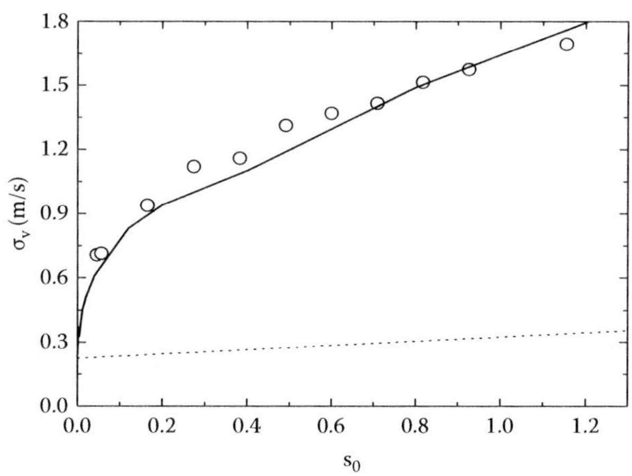

FIGURE 20.5
Measured velocity dispersion σ_v as a function of the on-resonance saturation parameter s_0 at $\delta = -\Gamma/2$. The experimental data (open circles) are compared to the bare Doppler prediction (dotted line) and to the Monte-Carlo simulation (solid line) taking into account transverse spatial intensity fluctuations. While the Doppler theory is completely off, very good agreement is found with our theoretical model (see text).

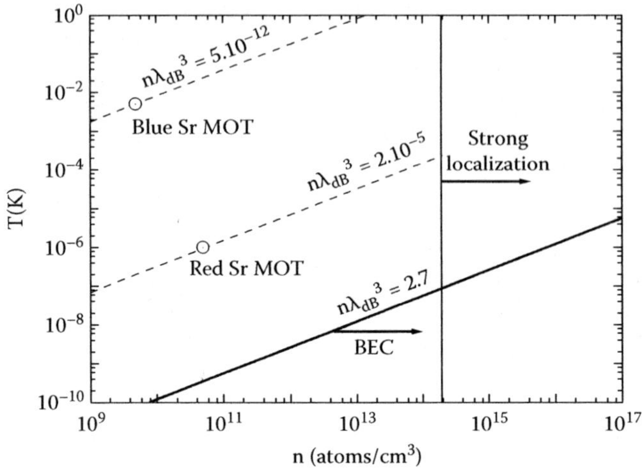

FIGURE 20.6
Strong localization and Bose-Einstein thresholds in the temperature and density plane in Log-Log units. The BEC threshold occurs at a phase-space density $n\lambda_d^3B \approx 2.7$ which gives T ~ $n^{-2/3}$. The strong localization threshold occurs at $k\ell \approx 1$ which fixes, at resonance, the density onset for localization at $n^* \approx k^3/6\pi$ giving $n^* \approx 2 \times 10^{14}$ atoms/cm^3 at $\lambda = 461$ *nm*. For the strontium MOT operating at $\lambda = 461$ *nm*, the phase-space density is $n\lambda_d^3B \approx 5.10^{-12}$. By cooling strontium atoms with the spin-forbidden transition at $\lambda = 689$ *nm*, the phase-space density can be increased by a factor about 10^7 while the spatial density is increased by a factor about 10. For $\lambda = 689$ *nm*, the density onset for strong localization is now $n \approx 4 \times 10^{13}$ atoms/cm^3.

think that heating due to spatial intensity fluctuations still exist and explain the high temperatures found in our experiment [19] but also in other earth-alkaline MOTs [20–23]. Note however that we still have $k\sigma_v/\Gamma \approx 5\% << 1$ so that Doppler broadening is completely negligible.

Figure 20.6 displays the Strong Localization and the Bose-Einstein thresholds in the temperature and density plane. The BEC threshold occurs at a phase-space density $n\lambda_d^3B \approx 2.7$. The strong localization threshold occurs at $k\ell \approx 1$ which fixes, at resonance, the density onset for localization at $n^* \approx k^3/6\pi$. For the strontium MOT operating at $\lambda = 461$ *nm*, the obtained phase-space density is $n\lambda_d^3B \approx 5 \times 10^{-12}$, thus far from the BEC onset. The achieved spatial density is $n \approx 4 \times 10^9$ atoms/cm^3 still far from the density onset $n^* \approx 1.5 \times 10^{14}$ atoms/cm^3 at $\lambda = 461$ *nm*.

20.3 Coherent Back Scattering

In this section we present results on CBS experiments. Section 20.3.2 concentrates on light scattering at low saturation parameter *s*. In this case, the excited-state population can be safely ignored and the atomic dipole can be successfully described by a classical damped dipole (elastically-bound

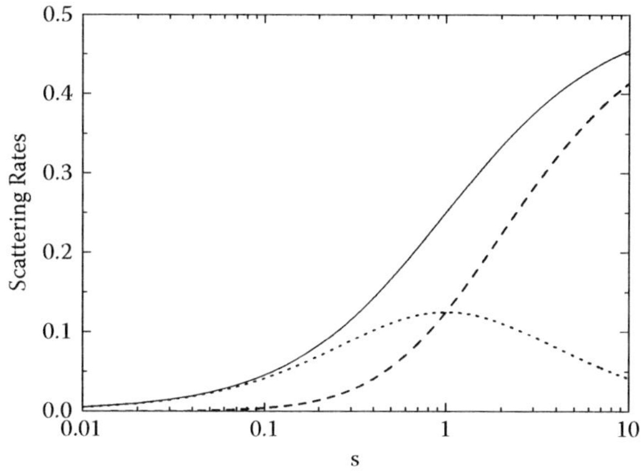

FIGURE 20.7
Light scattering rates of a two-level atom (in units of Γ) as a function of the saturation parameter s. The dotted line represents the elastic scattering rate $\Gamma_{el} = \Gamma_{tot}/(1 + s)$, the dashed line the inelastic one $\Gamma_{inel} = \Gamma_{tot}\, s/(1 + s)$. The plain line is the total scattering rate $\Gamma_{tot} = \Gamma_{el} + \Gamma_{inel} = s/2(1 + s)$.

electron model) [24]. Light scattering by this classical dipole is then purely elastic. This regime is achieved either when $s_0 << 1$ or by suffciently detuning the light frequency to impose $s << 1$.

Section 20.3.3 shows the results obtained when the probe beam saturation parameter s is increased. In this case, the excited-state population is no more negligible and several related effects start to play a significant role. The first one is vacuum-induced fluctuations of the driven atomic dipole. The scattered light spectrum then exhibits a broad inelastic component giving rise to the well-known Mollow triplet at strong fields [25,26]. The total inelastic rate is $\Gamma_{inel} = \Gamma_{tot}\, s\, (1 + s)^{-1}$ whereas the elastic one is $\Gamma_{el} = \Gamma_{tot}\, (1 + s)^{-1}$ where $\Gamma_{tot} = \Gamma/2\, s(1 + s)^{-1}$ is the total scattering rate (see Fig. 20.7). The inelastic component thus dominates over the elastic one as soon as $s > 1$. The correlation time τ_ϕ of the scattered field is then reduced down to the order of the excited-state lifetime $\tau_e = \Gamma^{-1}$. These uncontrolled field phase fluctuations during the multiple scattering events are consequently expected to yield a decoherence mechanism. This decoherence will be effective as soon as τ_ϕ is comparable or shorter than the light transport time τ^*. For resonant scatterers like atoms, $\tau^* \geq \Gamma^{-1}$ [15] and we see that, for inelastic scattering, $\tau^* \gtrsim \tau_\phi$. Theoretical investigations based on a simple toy-model [27,28] have shown that the inelastic spectrum introduces phase-shifts and amplitude imbalance between the CBS interfering paths leading to a CBS enhancement factor reduction. However, even in the limit $s \to \infty$, the CBS enhancement factor α achieves a finite value $\alpha \approx 1.05$ [28]. This is a clear indication that decoherence induced by the inelastic spectrum is not sufficiently strong to fully erase the CBS interference effect. In the experiment, we have indeed observed a CBS reduction (see section 20.3.3). However, our data cannot be directly compared to the prediction of [27,28] who considered just two atoms alone in vacuum.

A second effect is field-induced nonlinearities. Since the excited-state population is no longer negligible, the scattering efficiency, related to the groundstate population, is reduced. This nonlinear effect is embodied in the scattering cross-section which now becomes $\sigma_{NL} = \sigma(1+s)^{-1}$. Equivalently, the atomic susceptibility χ also shows up a dependence on the local saturation parameter s. In turn, light propagation properties are also modified: generation of a nonlinear refractive index for the effective medium (*e.g.* Kerr effect), four-wave mixing, filamentation, *etc* [29]. For classical scatterers, theoretical studies investigating the impact of $\chi^{(2)}$ [30] and $\chi^{(3)}$ [31] nonlinearities do not predict any CBS enhancement factor reduction. This seems to be supported by experimental work on CBS in gain medium [32].

20.3.1 Experimental Procedure and Data Processing

The detailed experimental procedure needed to observe light CBS on a cold atomic cloud has been published elsewhere [11]. For the present experiment, the signal is obtained using a collimated resonant probe laser beam with a waist of 2 mm. The scattered light is collected in the backward direction by placing a CCD camera in the focal plane of an achromatic doublet. The angular resolution of our apparatus is about 0.1 *mrad*, roughly twice the CCD pixel angular resolution. To shield the (weak) CBS signal from the (intense) MOT fluorescence signal, a time-sequenced experiment is developed. The MOT trapping beams and the magnetic field gradient are switched off during the CBS acquisition sequence.

The probe pulse duration is adjusted accordingly (typically from 5 to $70\mu s$) to keep the maximum number of scattered photons per atom below 400. In this way, mechanical effects can be neglected since $400\, kv_{rec} \approx \Gamma/3$, where v_{rec} is the atomic recoil velocity associated with the absorption of a single photon. Once the CBS signal has been recorded, the MOT is switched on again and strontium atoms are thus recaptured. The whole sequence is then repeated as long as necessary (few minutes) to get a good CBS signal-to-noise ratio. The CBS images are finally obtained by subtracting the background image taken without any cold atoms. This background image is recorded in the absence of the magnetic gradient during all the acquisition time. We have thus checked that the fluorescence signal from the residual strontium atoms was indeed negligible.

The CBS parameters (enhancement factor and cone width) are obtained using a two dimensional fitting procedure. In the helicity polarization channels, the CBS cone is isotropic [33]. Thus the signal-to-noise ratio can be significantly improved by first pinpointing the center of the CBS cone and then performing a polar average of the image (see Fig. 20.8). The obtained CBS cone is then fitted by a Monte-Carlo simulation [33] performed in the elastic scattering regime and using the "partial photon" trick [34,35] to extract the scattering contributions at different scattering orders. The amplitude

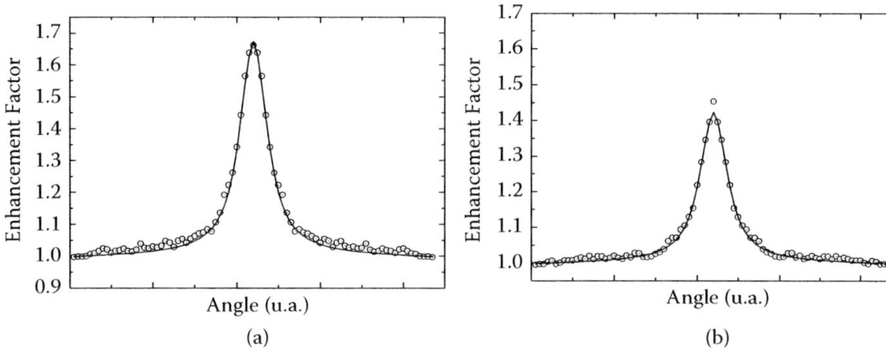

FIGURE 20.8
Plot of the experimental CBS cone as a function of the backscattering angle in the $h||h$ helicity-preserving polarization channel (open circles). Both curves are obtained at $\delta = 0$(a) $s_0 = 0.05$ (elastic regime) and (b) $s_0 = 0.71$ (saturated regime). The solid line corresponds to the Monte-Carlo simulation of the CBS cone (elastic regime) in the $h||h$ polarization channel (see text).

of a multiple scattering path is computed as a function of the initial and final polarizations and of the geometrical positions of the various scatterers which are spatially distributed with a Gaussian of *rms* size L. The spatial variations of the scattering mean free path during the photon propagation are thus faithfully taken into account in our numerical procedure.

At low saturation parameter, the Monte-Carlo calculation is in excellent agreement with our experimental data (see Fig. 20.8a). At higher saturations, the experimental cone shape does not change significantly. Hence we still use the same Monte-Carlo calculation, performed at low saturation parameter, to fit the CBS cone even at larger s (see Fig. 20.8b).

In the fitting procedure, we have also taken into account the finite angular resolution of our detection set-up and of the residual divergence of the CBS probe laser. Thus, prior to the fitting procedure, the Monte-Carlo calculation is convolved by an appropriate Gaussian function. This allows us to remove a systematic error of about 10% on the enhancement factor value.

20.3.2 Elastic Regime

For $J_g = 0$ groundstate atoms, it can be shown that the multiple scattering interference contrast is maximal in the polarization preserving channels ($h||h$ and $lin||lin$ channels). This is the case for strontium. Accordingly, the enhancement factor (peak to background signal ratio) takes its maximal value 2 in the helicity-preserving polarization channel $h||h$ where the single scattering signal is rejected [13]. In all other polarization channels, the enhancement factor is smaller than 2.

Using a resonant probe beam with $s_0 \ll 1$, the experimental enhancement factor is found to be $\alpha = 1.95 \pm 0.03$ at $b \approx 3$ in the helicity preserving channel ($h||h$). With the same probe beam but now detuned at $\Gamma/2$ from the resonance, we found $\alpha = 1.92 \pm 0.04$ at $b \approx 1.5$. These values are very close to the maximal expected value of 2. We think that the small discrepancy is most probably due to contamination of the $h||h$ polarization channel by residual single scattering signal (for more details see [36]). This happens preferentially at low optical thicknesses where single scattering has the largest contribution to the total backscattered signal. For this reason, the enhancement factor is more reduced for $\delta = \Gamma/2$ than for $\delta = 0$. As expected, in the $h \perp h$, $lin||lin$ and $lin \perp lin$ polarization channels, the enhancement factor is much smaller than 2. The results are in very good agreement with the Monte-Carlo calculations.

20.3.3 Saturated Regime

20.3.3.1 Probe Beam Transmission

Beyond the complexity of the situation under consideration (multiple scattering with nonlinear and inelastic scatterers), one has to deal also with nonuniform scattering properties. Indeed, even in an homogeneous slab geometry, the local intensity is not constant, as the incident coherent beam is attenuated when penetrating into the medium. Hence the atoms located deeper inside the medium will not be saturated in the same way as the atoms on the front part of the sample. Thus the saturation, and hence the scattering cross-section, will not be constant along a given multiple scattering path. The importance of the spatial variation of the saturation parameter can be estimated by looking at the attenuation of the coherent beam. In Fig. 20.9, we

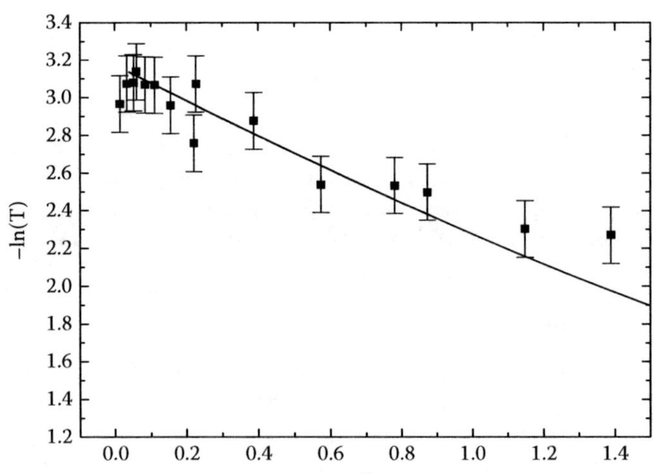

FIGURE 20.9
Resonant ($\delta = 0$) coherent transmission T along a diameter of the cold strontium could as a function of the incident saturation parameter s. The solid line corresponds to the theoretical nonlinear Lambert-Beer prediction (see text). The agreement with the experimental data is good up to $s = 1.2$.

report the measured transmission and we compare it with the Lambert-Beer theoretical prediction taking into account the nonlinear reduction of the cross-section. If one assumes that the local atomic saturation is dominated by the incident field and not by the scattered field, the optical transmission $T(z) = s(I)/s$, with s the incident saturation parameter, is obtained by solving the following equation:

$$(1+sT)\frac{dT}{T} = -\frac{dz}{\ell} \tag{20.6}$$

The factor $(1 + sT)$ features the nonlinear reduction of the scattering efficiency. When $sT >> 1$, $(1 + sT) \approx sT$ and $dT = -dz/s\ell$ leading to a *linear* decrease of the transmission with z

$$T(z) = 1-\frac{z}{z^*}; \quad z^* = s\ell \tag{20.7}$$

When $sT << 1$, $(1 + sT) \approx 1$ and $dT/T = -dz/\ell$ leading to the normal Lambert-Beer law and to its *exponential* attenuation

$$T(z) = \exp(-z/\ell) \tag{20.8}$$

Starting with $s >> 1$, the cross-over between the two regimes $sT >> 1$ and $sT << 1$ occurs around z^*. Noting by $b = L/\ell$ the low-saturation optical thickness, one immediately sees that the medium is fully saturated once $s \geq b$. For $s \leq b$, the medium can be roughly described as composed of a first saturated slice of approximate width z^* followed by a remaining non-saturated slice of width $L - z^*$.

The good agreement between the measured transmission and our simple nonlinear Lambert-Beer model equation 20.6 proves that saturation plays a role in our experimental conditions (since otherwise the transmission would not depend on s) and that the local atomic saturation is indeed dominated by the incident field.

20.3.3.2 Enhancement Factor

Figure 20.10a shows the dependence of the CBS enhancement factor as a function of the incident saturation parameter s when $\delta = 0$. For each value of s, the total number of cold atoms in the cloud is adjusted in order to maintain the coherent transmission T as constant as possible ($T \approx 0.085$ in the experiment). The most striking feature is the rapid quasi-linear decrease of the enhancement factor as s is increased. The slope derived from a *rms*-procedure is $(\delta\alpha/\delta s) \approx -0.6$. The single scattering contribution has been numerically estimated to increase by less than 10% when the saturation parameter is increased up to $s = 0.8$. As the helicity-preserving polarization channel is not perfectly isolated from the single scattering signal, this increase induces a spurious reduction of the enhancement factor. We have estimated it to be of the order of 1%, thus completely negligible compared to the observed reduction.

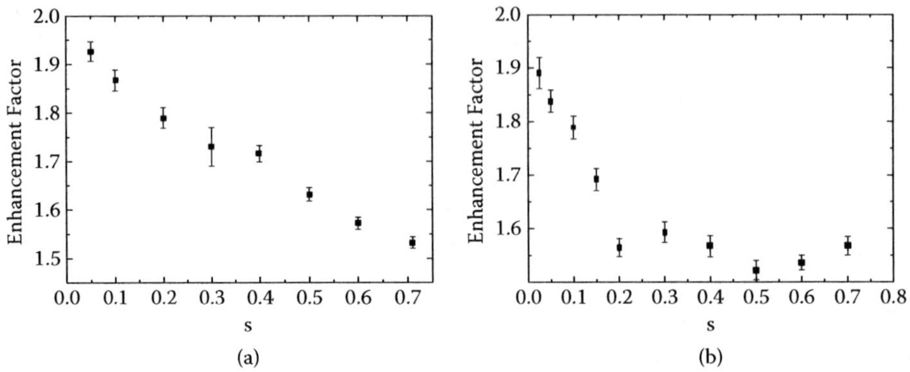

FIGURE 20.10
CBS enhancement factor as a function of the incident saturation parameter s. (a) $\delta = 0$ and(b) $\delta = \Gamma/2$. The coherent transmission value is kept fixed at $T = 0.085$ for (a) and at $T = 0.19$ for(b).

We can then faithfully claim that the observed CBS reduction comes solely from saturation effects in the multiple scattering signal.

In order to see how the atomic resonance modifies the coherence properties probed by CBS, we have performed another experiment at $\delta = \Gamma/2$. The same experimental procedure has been used with a transmission now roughly fixed at $T = 0.19$. As shown in Fig. 20.10b, a different general behavior is observed. First, at low intensity, the linear decreasing is faster since $(\delta\alpha/\delta s) \approx -1.8$. Second, for larger saturation parameters ($0.3 < s < 0.8$) the decrease is then slowed down. The two sets of data in Fig. 20.10a and 20.10b are obtained with a different transmission value, but other studies have shown that the enhancement factor does not sensitively depend on the exact transmission value [37]. Bearing this fact in mind, we are led to the conclusion that s is not the only relevant parameter in our experiment. This can be understood since the exact shape of the in elastic spectrum also depends on the detuning δ. In particular, for the detuned case, part of the in elastic spectrum will overlap the atomic resonance. This resonant in elastic light will thus be scattered again more efficiently then the off-resonant elastic part. This effect is *e.g.*, responsible for an increase of the MOT volume in the multiple scattering regime as we discussed in section 20.2. Finally in our experiment, the ratio between the amount of inelastic and elastic multiply scattered light changes with the detuning. We can then conclude that the CBS reduction is due to the in elastic spectrum.

Conclusion and Perspectives

In this chapter, we have investigated phase coherence properties of a monochromatic light wave propagating in an optically thick disordered sample of laser-cooled strontium atoms. For this purpose, we have used the coherent backscattering effect as a self-aligned zero path-length interferometric tool. The CBS interference contrast is maximal when the phase coherence is fully preserved during transport and is reduced as soon as phase-breaking mechanisms set in.

Concerning light scattering properties, strontium atoms behave as spherically-symmetric resonant point-dipole scatterers. We have seen that, when the incoming light is weakly saturating the atomic internal resonance, coherence is fully preserved and CBS achieves its maximal contrast, 2, in the helicity-preserving polarization channel $h||h$. As soon as the light intensity is increased, the CBS contrast starts to fall down indicating that phase-breaking mechanisms are at work. This is so because vacuum-induced fluctuations of the atomic dipole start to play a role. Nonlinear propagation effects as well as inelastic scattering occur which blur the CBS effect. Understanding, if not circumventing, these spurious effects is important for the quest of strong localization of light in disordered atomic samples.

Indeed, localization is often explained, using hand-waving arguments, as a result of the destructive interference between long scattering paths. As such, maintaining full coherence appears as a strong request. Furthermore, a hypothetical localized optical mode in the atomic bulk may saturate atoms located in its vicinity. This phenomenon may in turn completely modify the Anderson scenario valid for linear waves. As we have seen however, our strontium MOT is far from fulfilling the density requirement to reach the strong localization onset. However subsequent cooling of ^{88}Sr on the $^1S_0 \rightarrow {}^3P_1$ transition at 689 nm allows for a substantial gain in phase-space density and spatial density (see Fig. 20.6). One can even achieve phase-space densities as high as 0.1 [38]. One may then think to use compression techniques to reach the localization onset or even decrease n^* by using infra-red light or a two-photon transition.

Our present studies may also prove valuable in the quest of the random laser regime in cold atoms. Coherent random lasers [39] are probably the most striking systems intrinsically combining both nonlinear effects and disorder. With atoms, gain and nonlinearities are easily induced. In this respect, a key point is thus a proper experimental and theoretical understanding of the mutual effects between multiple interferences and nonlinear scattering.

Acknowledgments

The authors wish to thank D. Delande, C. Müller, T. Wellens, and G. Labeyrie for fruitful discussions. This research was supported by the Centre National de la Recherche Scientifique and by Bureau National de Métrologie Contract 03 3 005.

References

[1] E. Akkermans and G. Montambaux, *Physique mésoscopique des électrons et des photons*, EDP Sciences, France, (2004). English translation is in preparation.
[2] *Mesoscopic Quantum Physics,* edited by E. Akkermans, G. Montambaux, J.-L. Pichard and J. Zinn-Justin (North Holland, Amsterdam, 1995).
[3] A. Akkermans and G. Montambaux, *J. Opt. Soc. Am. B* **21**, 101 (2004).
[4] R. Berkovits and S. Feng, *Phys. Rep.* **238**, 135 (1994).
[5] F. Scheffold and G. Maret, *Phys. Rev. Lett.* **81**, 5800 (1998).
[6] *Diffuse Waves in Complex Media*, NATO Science Series C, Vol. 531, Kluwer, J. P. Fouque, Ed. (1999).
[7] S. John, *Phys. Rev. Lett.* **58**, 2486 (1987).
[8] Wiersma, D. S., Bartolini, P., Lagendijk, A., & Righini, R. *Nature* **390**, 671–673 (1997).
[9] F.Scheffold, R. Lenke, R. Tweer, and G. Maret, *Nature* **398**, 206–207 (1999).
[10] J. Gómez Rivas, R. Sprik, A. Lagendijk, L. D. Noordam, and C. W. *Rella, Phys. Rev.* E. **63**, 046613 (2001).
[11] G. Labeyrie, F. de Tomasi, J. C. Bernard, C. Müller, C. A. Miniatura, and R. Kaiser, *Phys. Rev. Lett.* **83**, 5266 (1999); G. Labeyrie, C. Müller, D.Wiersma, C. Miniatura, and R. Kaiser, *J. Opt. B : Quantum Semiclass. Opt.* **2** , 672 (2000).
[12] D. V. Kupriyanov, I. M. et al. *Phys. Rev. A* **68**, 033816 (2003).
[13] T. Jonckheere, C. A. Müller, R. Kaiser, C. Miniatura, and D. Delande, Phys. Rev. Lett. **85**, 4269 (2000); C. Müller,T. Jonckeere, C. Miniatura and D. Delande, *Phys. Rev. A* **64**, 053804 (2001).
[14] O. Sigwarth *et al., Phys. Rev. Lett.* **93**, 143906 (2004).
[15] G. Labeyrie, E. Vaujour, C. A. Müller, D. Delande, C. Miniatura, D. Wilkowski, and R. Kaiser, *Phys. Rev. Lett.* **91**, 223904 (2003).
[16] D. Wilkowski, Y. Bidel, T. Chanelière, R. Kaiser, B. Klappauf, C. A. Müller, and Christian Miniatura *Phys. B* **328**, 157 (2003).
[17] B. Klappauf, Y. Bidel, D. Wilkowski, T. Chanelière, R. Kaiser, *Appl. Opt.* **43,** 2510 (2004).
[18] D. W. Sesko, T. G. Walker, and C. E. Wieman, *J. Opt. Soc. Am. B* **8**, 946–958 (1991).
[19] T. Chanelière, J.-L. Meunier, R. Kaiser, C. Miniatura, and D. Wilkowski, *J. Opt. Soc. Am. B.* (2005).
[20] C. W. Oates, F. Bondu, and L. Hollberg, Eur. Phys. J. D **7**, 449 (1999).
[21] F. Loo, A. Brusch, S. Sauge, M. Allegrini, E. Arimondo, N. Andersen, and J. Thomsen, J. Opt. B: Quantum Semiclass. Opt. **6**, 81 (2004).
[22] X. Xu, T. Loftus, M. Smith, J. Hall, A. Gallagher, and J. Ye, *Phys. Rev. A* **66,** 011401 (2002).
[23] X. Xu, T. Loftus, J. Halland J., and Ye, *J. Opt. Soc. Am.* **B20,** 968–976 (2003).
[24] We neglect here the recoil frequency shift which is much smaller than the line width $\Gamma/2\pi$.
[25] C. Cohen-Tannoudji, J. Dupont-Roc, G. Grynberg, *Atom-Photon Interactions*, Wiley (1992).
[26] B. Mollow, *Phys. Rev.* **188**, 1969 (1969).
[27] T. Wellens, B. Gremaud, D. Delande, and C. Miniatura, *Phys. Rev. A* **70**, 157 (2004).

[28] V. Shatokhin, C. Müller, and A. Buchleitner, *Phys. Rev. Lett.* **94**, 043603 (2005).
[29] R. W. Boyd, *Nonlinear Optics*, (Academic, San Diego, 1992).
[30] V. Agranovich and V. Kravtsov, *Phys. Rev. B* **43**, 13691 (1991).
[31] A. Heiderich, R. Maynard, and B. van Tiggelen, Opt. Comm. **115**, 392 (1995); S. E. Skipetrov, and R. Maynard *in Wave Scattering in Complex Media: from theory to applications*, NATO Science Series II **107**, eds. B. A. van Tiggelen and S. E. Skipetrov, Kluwer, Dordrecht (2003), p. 75.
[32] D. S. Wiersma, M. P. van Albada, and A. Lagendijk, *Phys. Rev. Lett.* **75**, 1739 (1995).
[33] D. Wilkowski, Y. Bidel, T. Chanelière, D. Delande, T. Jonckheere, B. Klappauf, G. Labeyrie, Christian Miniatura, C. A. Müller, O. Sigmarth, and R. Kaiser, *J. Opt. Soc. Am. B.* **21**,183 (2004).
[34] R. Lenke and G. Maret, in *Scattering in Polymeric and Colloidal Systems*, W. Brown and K. Mortensen, eds. (Gordon and Breach, Reading, 2000), p. 1–72.
[35] D. Delande *et al.*, *Phys. Rev. A* **67**, 033814 (2003).
[36] Y. Bidel, B. Klappauf, J. C. Bernard, D. Delande, G. Labeyrie, C. Miniatura, D. Wilkowski, R. Kaiser, *Phys. Rev. Lett.* **88**, 203902 (2002).
[37] Roughly speaking, inelastic scattering events occur where the saturation is high, *i.e.*, in the first atomic layers. Deeper inside the medium, where saturation is low, elastic scattering events dominate. Thus varying the coherent transmission in the multiple scattering regime should not sensitively modify the inelastic mechanisms at work (and in turn the enhancement factor), at least under our experimental conditions.
[38] T. Ido, Y. Isoya, and H. Katori *Phys. Rev. A* **61**, 061403 (2000).
[39] H. Cao *et al.*, Phys. Rev. Lett. **84**, 5584 (2000); P. Sebbah and C. Vanneste, *Phys. Rev. B* **66**, 144202 (2002).

21

The Nature of Light: Description of Photon Diffraction Based Upon Virtual Particle Exchange

Michael J. Mobley
The Biodesign Institute, Arizona State University, Tempe, AZ 85287, USA

CONTENTS

Abstract

Any discussion of the nature of light must include a reminder that whenever we make the observation of light (photons), we only observe particle-like properties. This chapter provides a reiteration that we don't need wave-like properties to scattered photons to describe phenomena such as diffraction or refraction of light. This chapter updates the original ideas of Duane, later revived by Landé, which provided a description of light diffraction without making reference to a wave nature. These are updated using terminology more common to quantum electrodynamics which describes the interaction of particles in terms of the exchange of virtual photons. Diffraction is described in terms of an ensemble of distinct, probability weighted paths for the scattered photons. The scattering associated with each path results from the quantized momentum exchange with the scattering lattice attributed to the exchange or reflection of virtual photons. The probability for virtual particle exchange/reflection is dependent upon the allowed momentum states of the lattice determined by a Fourier analysis of the lattice geometry. Any scattered photon will exhibit an apparent wavelength inversely proportional to its momentum. Simplified, particle-like descriptions are developed for Young's double slit diffraction, Fraunhofer diffraction and Fresnel diffraction. This description directly accounts for the quantization of momentum transferred to the scattering lattice and the specific eigenvalues of the lattice based upon the constraints to virtual photon exchange set by the Uncertainty Principle, $\Delta p_i = h/\ell_i$.

Key words: diffraction, refraction, double slit experiment, wave-particle duality, virtual photon, Bragg's law, quantum interpretations.

21.1 Introduction

A discussion of the nature of light usually includes mention of diffraction or refraction phenomena and the idea that light exhibits both wave properties and particle properties. It is frequently asserted, particularly in introductory texts,[1,2,3] that diffraction or refraction of light can only be explained invoking wave-like properties. This chapter is intended to remind us that this is simply not true. The traditional picture we get of constructive and destructive interference of waves from optical wave theory is misleading, but it has been difficult to converge on better pictures to describe the nature of light. It is likely that a more integrated picture of light will require a different understanding of the fabric of space, time and matter which does not lean so heavily on simplistic pictures of particles and waves. We've learned, whenever we make the observation of light (photons), we only observe particle-like properties (e.g., momentum or energy transfer).

The wave description of light diffraction holds that the probability of detecting a scattered photon is determined by a wavefunction amplitude at a (potentially very distant) point of detection. There is a fundamental problem with the picture that this probability amplitude is somehow determined by an interference pattern at that detection point. The error in this picture reflects the challenge we often find when interpreting quantum theory. The challenge is to make a distinction between descriptions of phenomena that obey wave-like equations and give us statistical results from those descriptions which attribute wave-like properties to individual free-particles. We can often confuse a statistical probability function determining the potential for observing a particle with an actual wave permeating space. The traditional wave picture for the diffraction of light is inconsistent with our physical laws requiring conservation of momentum and energy. These conservation laws demand that photon scattering is determined at the location of interaction with the scatterer as each scattering pattern (event) must ultimately be related to a distribution of momentum transfers. The angle of scattering is not determined at the point of detection. Any discussion on the nature of light might appropriately reference the influence of traditional wave theory on the historical development of optics and our quantum mechanical formalism, mentioning de Broglie's hypothesis and the Schrödinger equation for how the mathematical formalism has been translated to other quantum particles. However, we must remind ourselves that the wave picture can give us an inherently flawed view of the nature of light. This is not to question the value (or accuracy) of the mathematical formalism of much of wave mechanics and quantum theory. It is primarily to register a caution about the pictures or interpretations we draw for individual, free particles.

Duane[4] in 1923 was the first to demonstrate that the diffraction of X-rays by a crystal lattice could be explained without reference to a wave character based upon a third quantum rule for linear momentum. An extension of Duane's theory by Ehrenfest and Epstein[5] described the diffraction of a particle in terms of the quantal activity of the diffractor. Momentum is changed for the scattered particle (photon) by interaction with the matter distribution of the diffractor which is associated with lattice spacings that can be analyzed according to Fourier's theorem. The momentum increments, $\Delta p_i = h/\ell_i$, form a continuous spectrum which can interact with the scattered particle with a specific probability for a specific lattice/diffractor. In this work, the correspondence principle was assumed which connected momentum eigenvalues for the lattice with sinusoidal terms of the Fourier analysis. A resulting diffraction pattern is generated by the statistical distribution of the individually deflected particles.

These ideas were neglected for many years until their revival by Landé[6,7,8] who used them in his formulation of quantum theory.[9] Landé objected to the perpetuation of the "fact" of a dual nature to elementary particles, although in his formalism he continued to accept a dual nature to light (the photon)—undoubtedly because so much of electromagnetic theory, optical phenomena, and the quantal activity of a diffractor could be explained by reference to waves. Ballentine[10], in presenting his alternate statistical interpretation of quantum mechanics, pointed to Duane's and Landé's work to support abandoning the wave-particle paradigm for sub-luminary particles. Thus, these ideas have played a historical role in new and different approaches to quantum physics. The purpose of this chapter is to both remind us of these ideas and to update them by utilizing nomenclature more common to quantum electrodynamics, QED, which describes the interaction of particles in terms of the exchange of virtual particles.[11] My expectation is that this effort will help us converge on a more accurate picture of the nature of light.

21.2 Photon Diffraction

This description is best understood in connection with more recent formulations of quantum theory such a Feynman's construction of path integrals[12] or the statistical interpretation described by Ballentine.[10] Thus, no new physics is presented, but a picture is drawn for the interaction of light and matter different from the wave picture. Basic to Feynman's path integral formulation is the assertion that a probability is associated with an entire motion of a particle as a function of time. This contrasts with traditional wave mechanical formulations (interpretations) which define probabilities associated with the position or momentum of a particle at a specific time, principally the point of detection. As I have pointed out, in photon diffraction, momentum exchange with a scattered photon must take place at the scattering lattice in order for momentum to be conserved. Scattering probabilities must be determined at the location of scattering. Thus, the path integral formulation or the statistical interpretation of Ballentine would be adaptable to a more consistent phenomenological description of photon diffraction than the traditional formulations derived from optical wave theory.

In this revised description, certain properties for the photon must be assumed such as spatial extension, quantized energy and momentum, and traveling at the speed of light. We should assume the dimensionality of a photon also carries a level of uncertainty in its space and time coordinates. For many phenomena not described in this chapter we would also have to invoke photon properties such as spatial polarization and the ability to exchange angular momentum. This chapter uses a quantum particle model for a free photon to develop updated descriptions of Young's double slit experiment, Fraunhofer diffraction, and Fresnel diffraction, contrasting these with the traditional pictures from optical wave theory to demonstrate such pictures are not necessary.

21.2.1 Young's Experiment

Diffraction has frequently been described through a probabilistic interpretation of the wave amplitude from Kirchhoff's analysis of optical wave theory.[1,13] I will first examine the mathematical description of Young's experiment consistent with this traditional wave theory, then provide the alternate description in terms of momentum exchange with the diffracting slits. Figure 21.1 diagrams the experiment. The source of monochromatic light is assumed to be far from the slits and the incident light beam (photons) traveling along the x-axis is perpendicular to the slits along the y-axis. The figure indicates the light intensity pattern that would be observed on a screen if it were placed at a distance b from the slits, where b is a large compared to the slit separation, d. Lines of intensity maxima can be found in the geometric shadow of the slits at Y_0, and spanning out from this center line at Y_1, $-Y_1$, Y_2, $-Y_2$, … Dark fringes are found between these lines. Optical wave theory explains this pattern in terms of constructive and destructive interference between the light waves emanating from each of the slits.

A first order approximation of the experimental screen intensity, Φ, as a function of the scattering angle, θ, can be expressed by,

$$\Phi = A(1 + \cos 2\pi(d \sin\theta)/\lambda), \tag{21.1}$$

where $d\sin\theta$ is the approximate difference in distance of each slit to the point on the screen, λ is the effective wavelength associated with the light, and

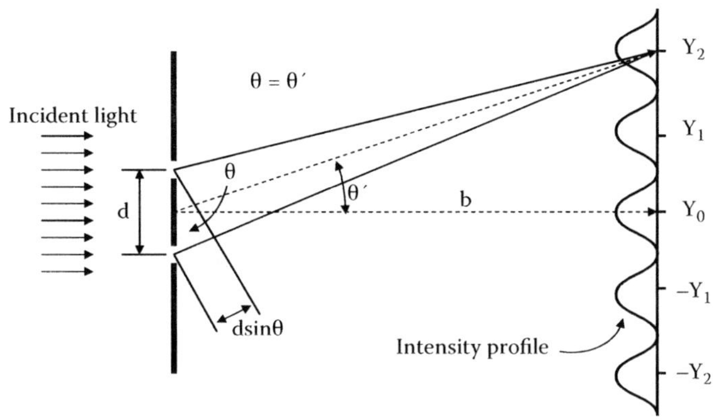

FIGURE 21.1
Young's double slit experiment showing the intensity of light striking a distant screen. The distance between the slit and screen, *b*, is much greater than the slit separation, *d*.

$2\pi(d\sin\theta)/\lambda$ is the phase difference between the assumed waves from each slit. Thus, when $d\sin\theta$ is equal to 0 or an integral multiple of λ there are intensity maxima. However, when $d\sin\theta$ is an odd multiple of $\lambda/2$, we get destructive interference and a dark fringe. This first order approximation describes diffraction within the accuracy of most experiments with appropriate adjustment of the coefficient, *A*, for reduction in wave intensity with increasing angle θ. Wave mechanics interprets the intensity pattern of equation 21.1 as a scattering probability distribution associated with individual photons as the pattern is the same when we have a low intensity of light. The pattern would not be due to the interaction between incident photons.

In a path integral or statistical formalism, the summation of probability weighted particle paths can be used to construct a wavefunction or a probability amplitude able to satisfy the non-relativistic relations of wave mechanics. Adopting such an approach, photon diffraction can be described in terms of elastic scattering of particles with a discrete momentum, *p*. Probability functions are assumed to be constructed from normalized, probability weighted ensembles from different specific paths or states (defined by position and momentum) possible for a scattered photon. In adopting this formalism, the key is determining the probability of each path.

A revised description of diffraction can be derived from an alternate analysis of equation 21.1. Denoting a change in a specific photon's momentum along the *y*-axis parallel to the slits by p_y, we note

$$\sin\theta = p_y/p. \tag{21.2}$$

From Einstein's description of the photoelectric effect, we know a photon's momentum can be described in terms of Planck's constant,

$$p = h/\lambda. \tag{21.3}$$

Here we only need to assume that λ is a parameter inversely proportional to the photon momentum. Using these two relations, we can rewrite equation 21.1 as

$$\Phi = A(1 + \cos 2\pi\, d\, p_y / h), \tag{21.4}$$

which provides a momentum representation of the scattering. Importantly, this transforms this intensity or probability function to one independent of any kinematical properties of the scattered photon. In fact, the form of this probability function would apply whether we were scattering photons or other quantum particles such as electron or neutrons. This probability function is a Fourier transform of a transverse momentum function and it depends only on how much p_y differs from an integral multiple of h/d. We note that scattering and momentum changes are determined at the slits and not at a distant detection point—a result conforming readily with the path integral formalism or the statistical interpretation of quantum mechanics. This observation, though not novel, is rarely pointed out in discussions of the optical wave theory of light.

For a long time scientists have recognized that the scattering amplitude as a function of momentum transfer is the Fourier transform of the scattering potential, just as the angular distribution of light diffracted from an obstacle in classical optics is the Fourier transform of the obstacle.[5,14] Intensity maxima are observed when

$$p_y d/h = n, \tag{21.5}$$

or when

$$p_y = nh/d. \tag{21.6}$$

Equation 21.6 becomes the optimized scattering criterion for these slits. This equation is often used to dimensionalize the Heisenberg Uncertainty Principle.

In QED the interaction between electromagnetic particles is explained in terms of the exchange of "virtual" bosons (e.g., photons) resulting in momentum exchange between the particles. Feynman diagrams describe virtual particles existing in "proper time" (zero time intervals for virtual photons).[11] The coupling between particles can be described without specifying the direction of virtual particle travel. Analogously, we may describe the scattering (diffraction) of a photon by a pair of slits as the exchange of "virtual particles" transferring equivalent energy between the lattice of the slits and the photon. Or equivalently, but perhaps conceptually simpler, we can describe diffraction by the reflection (absorption and emission) of a virtual photon from the lattice by a scattered photon, with additional x-momentum exchange taking place to ensure conservation of energy. QED is normally used to describe the scattering of electrons, but as noted, the scattering of electrons is governed by the same momentum exchange constraints as the scattering of photons.

Thus, QED provides us with a useful model and nomenclature to describe the scattering of photons as well. It is recognized in QED, the quantum mechanical formalism describing the scattering of photons would be different than the scattering of electrons as the cross sections for scattering or their interaction coefficients with the lattice would be different.

In describing diffraction consistent with the conventions of QED, we adopt a momentum representation for the Coulomb potential (electromagnetic field) associated with the lattice. The most probable values for the magnitude of the y-momentum of the virtual photons associated with the scattering potential are integral multiples of $h/2d$. This corresponds to virtual photons with associated energies of $nhc/2d$. We recognize these energies are the eigenvalues of the photon "standing wave" eigenfunctions for the particle-in-a-box problem in quantum mechanics where the length of the box is d. Thus, we might assume that there are probability maxima when the momentum of the exchange particle corresponds to the most probable allowed states for virtual photons within the panel separating the slits. We note that any panel consists of a dense electromagnetic field (of virtual photons) that constrains the individual atoms. The density would be a direct summation over the electromagnetic particles that make up the lattice.

There are minima in the exchange probability when the magnitude of the momentum of the virtual photon would be $(n + 1/2)h/2d$. These momenta might correspond to forbidden eigenfunctions for virtual photons within the panel separating the slits. There is a low probability for a photon to be scattered by virtual photons from the slits with those momentum values. Thus, the scattering probability distribution observed with photon diffraction is derived from a function of the y-momentum exchanged from the scattering by virtual photons of the lattice summed over the probabilities or densities of the virtual photons with the different momentum values.

21.2.2 Bragg Relation

Similar principles to those describing double-slit diffraction can also be applied to photon (X-ray) diffraction by a crystal lattice. Assuming the planar spacings in a lattice are equal to d, the optimum scattering criterion for any particle is

$$q = nh/d, \tag{21.7}$$

where the momentum change, q, is perpendicular to the lattice planes. Again scattering could be described by the exchange of virtual particles with the probability for a change in momentum being dependent on the geometry of the crystal lattice, and $q/2$ is associated with momentum eigenfunctions of the lattice. If we define θ as the complement to the angle of incidence and reflection (the Bragg angle) of a particle scattered from these lattice planes, then $2\sin\theta = q/p$. The angle that fulfills the selection rule is then given by

$$q = p2\sin\theta = nh/d \tag{21.8}$$

or

$$2d\sin\theta = nh/p = n\lambda, \tag{21.9}$$

where $h/p = \lambda$. Thus, we obtain the Bragg relation (Bragg's Law) for lattice diffraction. The parameter $\lambda = h/p$ arises as an apparent wavelength for the scattered particle, but is not derived with reference to any wave property of the particle, only its momentum. Any scattered particle has an apparent wavelength. Thus, de Broglie's hypothesis that any particle will exhibit a wavelength inversely proportional to its momentum may be considered to be a manifestation of the fact that scattering is quantized by the geometry and properties of the scattering lattice.

This derivation has been simplified, but the same conclusions can be drawn from a more rigorous treatment. For example, this description of diffraction can be utilized to predict the scattering intensities from single crystal x-ray diffraction. From a knowledge of the structure factor (dependent on the crystal geometry) and the scattering coefficients of the component atoms (dependent on electron densities), and the x-ray momentum (wavelength), the scattering intensities can be precisely calculated (ignoring the phase problem).[15] In a momentum representation of the lattice, the structure factor and scattering coefficients can be reinterpreted to be directly related to the density of virtual photon momentum states available. These momentum states exchange virtual photons with the scattered x-rays. Thus, the summation of the phased scattering contributions of each of the component atoms in x-ray analysis can be related to the summation over the different momentum eigenstate densities of the lattice.

21.2.3 Fraunhofer Diffraction

Our earlier discussion of double slit diffraction ignored the effect of the width of the slits. This can be accounted for by modifying our optimum scattering criterion, equation 21.6, to

$$p_y = \frac{nh}{d} \pm \frac{2wh}{d_2}, \tag{21.10}$$

where w is an integer and d_2 is the width of the slits. Thus, it is apparent that a modified set of eigenfunctions for the slits determines the probability for scattering when we account for the width of the slits. The second term in equation 21.10 becomes the Fraunhofer diffraction term. If either slit is closed, this would be the only term contributing which would give rise to Fraunhofer diffraction.[16] Similar principles of virtual photon exchange would apply for Fraunhofer diffraction as were applied to describe double slit diffraction. The major difference being that the momentum eigenvalues for the exchange photons are integral multiples of h/d_2. The origin for this difference in the eigenvalues is elucidated in the analysis of Fresnel diffraction.

21.2.4 Fresnel Diffraction

Consider diffraction of light through a single slit of width, d. A long distance from the slit we obtained a Fraunhofer diffraction pattern. By moving the detection point or screen sufficiently close to the slit, a distance γ, we can obtain a Fresnel diffraction pattern with an intensity minimum in the center of the pattern[17] (see Figure 21.2b). Fresnel's criterion for this pattern is

$$\frac{d^2}{4} + \gamma^2 = (1+\lambda)^2. \tag{21.11}$$

Assuming the same scattering criteria applies as for Fraunhofer diffraction, $p_y = 2wh/d$, which only depends on the width of the slit, then we must assume the probability for momentum exchange is not uniform across the slit, but rather must vary as a function of the point within the slit that the photon passes. In order to have a minimum at the center of the slit, photons passing

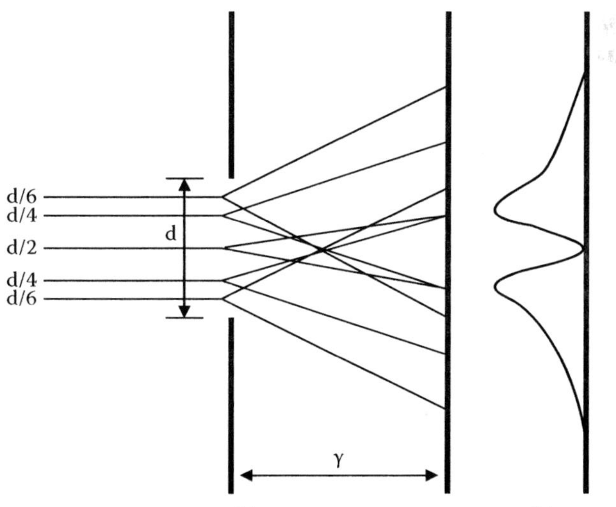

FIGURE 21.2
(a) Predicted most probable scattering paths for photons passing $d/2$, $d/4$, and $d/6$ from the edge of the slit compared to (b) the Fresnel intensity profile at a distance γ from the slit satisfying equation (11).

through (or emanating from) the center of the slit (a point $d/2$ from the edge) must be deflected. There must be a maximum in the probability of scattering for these photons. Photons passing through a point $d/4$ from the edge of the slit, if deflected with a momentum change with magnitude $p_y = 2h/d$, could impinge on the center of the screen where the intensity minimum is. Thus, there must be a low probability for these photons to exchange momentum of $p_y = 2h/d$. Photons passing through points within $d/4$ of the edge must be either be deflected by at least $p_y = 4h/d$ or not deflected.

From these requirements we can deduce that the probability for momentum exchange with the slit is dependent upon the minimum distance between the photon and the edge of the slit; the probability for a total momentum exchange of magnitude $p_y = 2wh/d$ being highest when the distance a photon passes from the edge of a slit is $d/2w$. For exchange to be allowed, the distance between the photon and the edge of the slit, ℓ_y, times the momentum exchanged is equal to h

$$\ell_y p_y = h. \tag{21.12}$$

The momentum exchanged is quantized consistent with the Heisenberg Uncertainty Principle. If we are exchanging virtual photons, then the distance over which a virtual photon is exchanged, ℓ_y, is always equal to half its "effective wavelength" such that its momentum is $h/2\ \ell_y$. We find there are at least two primary factors determining the scattering or diffraction of particles: 1) the eigenvalues possible for the virtual particles exchanged or reflected, and 2) the constraint that the total momentum exchanged times the distance of the exchange equals h. This constraint can be applied to the coupling factor in a summation of scattering probabilities.

This analysis suggests that a generalized description for particle scattering would have the form of a probability function for transfer of momentum along the y-axis, $f(p_y)$ such as:

$$f(p_y) = g(\phi)\rho(p_y)\int \delta\left(\ell_y - \frac{h}{p_y}\right)dy, \tag{21.13}$$

where $\rho(p_y)$ is the probability density of lattice momentum states associated with the momentum differential, p_y, δ is the Dirac delta function and ℓ_y is the difference in y-axis distance between the trajectory of the incident particle and the point of the scattering lattice with which momentum is exchanged. The third factor determining the scattering probability is the coupling factor, $g(\phi)$, which is a function of the angle, ϕ, between the incident particle and the assumed path of the scattering virtual photon of the lattice. This coupling factor reflects the probability that a momentum transfer will occur. This would be dependent upon the momentum that must be imparted to the lattice for scattering to occur.

The probability for transfer is greatest when the angle ϕ is such that we minimize the momentum vector parallel to the particle's incident direction that is absorbed by the scattering lattice. We may note that equation 21.13 resembles one we might generate using the structure factor of a crystal lattice and the coefficients of particle scattering to predict the intensity pattern in crystallography. From our analysis we note that the probability for photon scattering by a single slit would therefore resemble a shark tooth function across the width of the slit with maxima at points $d/2w$ from the edge. The scattering for photons passing through points $d/2w$ from the edge of the slit where $w = 1, 2,$ and 3 are diagrammed in Figure 21.2. This provides a description that can accurately reproduce the Fresnel diffraction pattern if the probabilities are properly parameterized.

From this analysis of Fresnel diffraction, we again see the preeminence of the relation $\ell_i\, p_i = h$ which was identified early as the key relation governing diffraction.[5] It is interesting to note that there are no limits implied in this relation. Any particle separated from other particles by ℓ_i is implied to have imminent potential for momentum exchange of $p_i = h/\ell_i$. Such interconnectedness linking a particle (photon) inextricably with its surroundings and the quantum potential associated with this interaction was the foundation for Bohm's derivation of quantum theory.[18] This relation is foundational to the description of the electromagnetic force in QED. In QED the total momentum exchanged via virtual photons between a positive and negative particle with a separation, ℓ, is $p = h/\ell$. The total force or net potential is determined by the frequency of exchange. The frequency of exchange would realistically be limited by c/ℓ. To limit the magnitude of our force, we must assume that only one momentum (virtual photon) exchange can take place within a period of time. I may note in this respect the exchange particle behaves more like a virtual fermion than a virtual boson. It is not unreasonable to speculate that photons are actually composed of subcomponents and that such subcomponents better describe the virtual particles that exchange momentum. A Feynman diagram depicting the interaction of 2 charged particles via virtual photons is shown in Figure 21.3a. (Unfortunately, the convention used in Feynman diagrams depicts the exchange of virtual photons by a sinusoidal "wave".)

An electromagnetic field has proven to be a useful mathematical convention to describe the collective influence of multiple, charged particles on a theoretical charge (particle) at a point of measurement. The ultimate manifestation of the field is a force (via momentum exchange) on a charged particle. According to QED, this momentum exchange is exerted through virtual photons. Examining one example, the static field across a capacitor is the net result of a difference in the positive and negative charges built up on separated plates. (The number of charged particles can be quite high.) We can assume exchanging virtual photons make up this field. If an oscillating potential is applied to the capacitor, we would generate an oscillating electromagnetic field between our plates which might correctly be depicted by a wavefunction. The propagation of this field is limited by the speed of light as

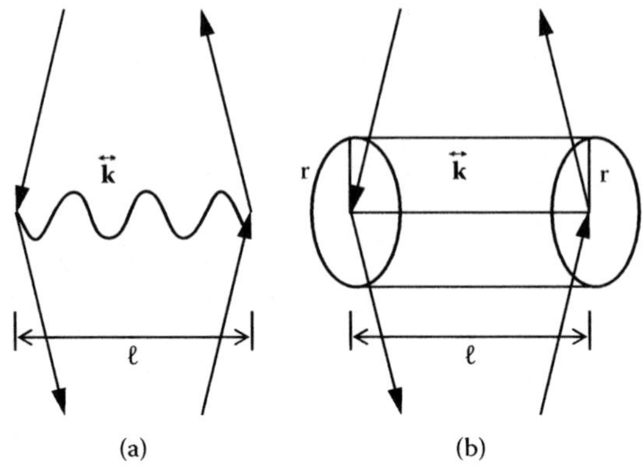

FIGURE 21.3
(a) Feynman diagram of virtual photon, $\ddot{k}$ exchange between two oppositely charged particles with the paths separated by the distance ℓ. (b) Alternate diagram with cylinder of radius r reflecting the region of virtual photon exchange. Spatial polarization can be diagrammed.

this field influence is carried by photons. Thus, while this field will oscillate in space, there is no requirement to assume the photons carrying this field influence must themselves exhibit oscillation. Thus, care must be taken in extrapolating the propagation of electromagnetic waves to a description of individual free photons. If we have 2 oscillating sources of electromagnetic fields, there is the possibility of cooperative phenomenon and the interaction of the photons from the 2 sources. A photon reaching a point from an oscillating electromagnetic field would have a specific polarization which would dictate its interaction with matter at that point. Photons from different sources reaching the same point might have different polarizations. We can accept that each can affect how the other might interact with matter. For example, if the polarizations were opposite, the net polarization at a point might be nullified and we would not get absorption of the photons at that point. Thus, we must recognize the potential for interference and cooperative phenomena. This is critical to observations of coherent light and explaining the behavior of lasers, but again we are not forced to conclude that individual photons behave as waves only that photons are polarized.

This brings us to an objection which might be raised about the picture being presented for photons. The description of diffraction we've set forth calls for specific momentum values for the virtual photons scattered to the lattice. We noted the momentum eigenvalues were those that we might get assuming standing electromagnetic waves within the lattice. Therefore, doesn't this give us a wave picture of (virtual) photons similar to the wave picture being critiqued? The mathematical similarity is clear as periodicity (Fourier transformation) across a spatial dimension is required. I must point out there is a clear difference in a picture of allowed momentum eigenstates in a material lattice, which we often use to describe phenomenon such as Brillouin scattering or Planck's black body radiation, and the propagation of a wave (or wave packet) in free space which somehow carries a detection probability. In the former picture we recognize a structure over which momentum exchange can take place (via the established fields of force involving virtual particle exchange) which is unlike the latter picture. Vigier, *et al.*[19] criticized the strictly particle based derivation by Landé suggesting that it failed to provide an explanation for why the matter distribution of a diffractor should be Fourier analyzed. Why should specific momentum values exist for the photons within a periodic array (without assuming behavior like a wave)?

An explanation or rationale for this might be derived by looking more closely at the quantum rules for momentum exchange which we have been using. From equation 21.12, any array of matter with periodicity in the *i*th direction with spacing ℓ_i, such as a lattice of a crystal, has the potential for virtual photons being exchanged along the array. The most probable momenta of those virtual photons would be $p_i = h/2n\,\ell_i$. As there would be an increasing number of states, the probabilities for specific virtual photons (or phonons) would appear to be reinforced as the array becomes larger (specific Fourier terms become more dominant). As the array reaches a length d, the possible momenta can be $p_i = mh/d$. This provides an explanation for the existence of momentum eigenfunctions for the array without having to imply that the virtual photons of the array must behave as waves. A variation on our traditional Feynman diagram from QED might help with our description. Such a revised diagram is presented in Figure 21.3b. Here, instead of having the exchange of virtual photons depicted by a sinusoidal wave, the gap in the paths between two electromagnetic particles is connected by a cylinder of length ℓ and a radius $r = \ell/\pi$. With this diagram we can also indicate a spatial polarization. The total momentum exchanged in this interaction is $\Delta p = h/\ell$, thus, the exchange virtual photon would have a momentum of

$$p = h/2\ell = \hbar/r. \tag{21.14}$$

We note that only one exchange can take place within this cylinder at a time. From this diagram we might conceptualize cooperative phenomena or an array of electromagnetic particles. An array might be depicted by a line or sequence of concentric cylinders representing virtual exchange photons with the radius, *r*, derived from the various lattice spacings in the array. We do not have to invoke a wave property to the virtual photons to confine them to specific momentum states.

Summary

We are reminded that the nature of light, whenever observed, is always particle-like. Describing light to be like a wave has been an important part of the history of the development of quantum mechanics, but such descriptions have left the erroneous impression that photons and other quantum particles behave as waves. We do not need to invoke wave like properties for a scattered photon to describe phenomena such as diffraction or refraction. I have described diffraction of light in terms of momentum exchange through the exchange of virtual particles between scattered photons and the scattering lattice. The probability for the exchange of a virtual photon with a particular energy and momentum is dependent upon how it matches allowed momentum states of the lattice, which are determined by the lattice geometry. The total momentum exchanged times the distance of exchange is equal to Planck's constant, h.

Probability functions can be constructed as normalized probability weighted ensembles from different specific possible paths for a photon. This description is phenomenologically consistent with the current interpretations of particle interactions from QED. I have provided a revised description to Young's double-slit experiment, Fraunhofer diffraction, Fresnel diffraction and Bragg's relation. In these descriptions the scattered particle does not behave like a wave. We note that the wave properties we historically associate with photons are only manifest if there is an interaction with the photon. If we attempt to define the position or momentum of a photon, we will perturb the photon by an amount defined by the Heisenberg Uncertainty Principle. The significant advantage to these descriptions and the picture we obtain for photon scattering is that they maintain consistency with our law for conservation of momentum as momentum transfer is always defined by the interaction with the scatterer at specific location of scattering. This is not the picture we derive when we rely on traditional wave interpretations.

References

1. H. D. Young, *Fundamentals of Optics and Modern Physics*, McGraw-Hill, New York, 1968.
2. A. Hudson and R. Nelson, *University Physics, 2nd Edition*, Saunders, Philadelphia, 1990.
3. J. D. Wilson and A. J. Buffa, *College Physics, 5th Edition*, Pearson, Upper Saddle River, NJ, 2003.
4. W. Duane, "The Transfer in Quanta of Radiation Momentum to Matter," *Proc. Nat. Acad. Sci.* **9**, 158–164, 1923.
5. P. Ehrenfest and P. Epstein, "The Quantum Theory of the Fraunhofer Diffraction," *Proc. Nat. Acad. Sci.* **10**, 133–139, 1924.
6. A. Landé, "Quantum Fact and Fiction," *Am. J. Phys.* **33**, 123–127, 1965.
7. A. Landé, "Quantum Fact and Fiction. II," *Am. J. Phys.* **34**, 1160–1163, 1966.
8. A. Landé, "Quantum Fact and Fiction. III," *Am. J. Phys.* **37**, 541–548, 1969.
9. A. Landé, *Quantum Mechanics in a New Key*, Exposition Press, New York, 1973.
10. L. E. Ballentine, "The Statistical Interpretation of Quantum Mechanics," *Rev. Mod. Phys.* **42**, 358–381, 1970.
11. R. P. Feynman, *Quantum Electrodynamics*, Benjamin/Cummings, Reading, MA, 1983.

12. R. P. Feynman and A. R. Hibbs, *Quantum Mechanics and Path Integrals*, McGraw-Hill, New York, 1965.
13. E. B. Brown, *Modern Optics*, Reinhold Publishing, New York, 1965.
14. D. H. Perkins, *Introduction to High Energy Physics*, Addison-Wesley, Menlo Park, 1987, p. 12.
15. J. W. Jeffery, *Methods in X-ray Crystallography*, Academic Press, London, 1971.
16. C. Curry, *Wave Optics: Interference and Diffraction*, Edward Arnold, London, 1957.
17. H. Y. Carr and R. T. Weidner, *Physics From the Ground Up*, McGraw-Hill, New York, 1971, p. 437.
18. D. Bohm, *Wholeness and Implicate Order*, Routledge & Kegan Paul, London, 1980.
19. J. P. Vigier, C. Dewdney, P. R. Holland, and A. Kyprianidis, "Causal Particle Trajectories and the Interpretation of Quantum Mechanics," in *Quantum Implications, Essays in Honour of David Bohm*, ed. by B. J. Hiley and F. D. Peat, Routledge & Kegan Paul Ltd., London, 1987, pp. 169–204.

22

What Physics Is Encoded in Maxwell's Equations?

B. P. Kosyakov

Russian Federal Nuclear Center, Sarov, 607190, Russia

CONTENTS

Abstract

We reconstruct Maxwell's equations showing that a major part of the information encoded in them is taken from topological properties of spacetime. The residual information, divorced from geometry, which represents the physical contents of electrodynamics, translates into four assumptions: (i) locality; (ii) linearity; (iii) identity of the charge source and the charge coupling; and (iv) lack of magnetic monopoles. However, a closer inspection of symmetries peculiar to electrodynamics shows that these assumptions may have much to do with geometry. Maxwell's equations tell us that we live in a three-dimensional space with trivial (Euclidean) topology; time is a one-dimensional unidirectional and noncompact continuum; and spacetime is endowed with a light cone structure readable in conformal invariance of electrodynamics. Our geometric feelings relate to the fact that Maxwell's equations are built in our brain. Hence our space and time orientation, our visualization and imagination capabilities are ensured by unceasing instinctive processes of solving Maxwell's equations. People usually agree in their observations of angle relations. For example, a right angle is never confused with an angle slightly different from right. By contrast, we may disagree in metric issues, say, a colour-blind person finds the light wave lengths quite different from those found by a man with normal vision. This lends support to the view that conformal invariance of Maxwell's equations is responsible for producing our notion of space. Assuming that our geometric intuition is guided by our innate realization of electrodynamical laws, some abnormal mental phenomena, such as clairvoyance, may have a rational explanation.

Key words: physical contents of Maxwell's equations, spacetime geometry, perception of space and time.

22.1 Introduction

Since the purpose of this book is to gain new insights into the nature of light, it seems to be of interest to advocate a somewhat odd point of view that the contents of Maxwell's equations is pure geometric. In other words, I will try to argue that *electrodynamics is a mere alternative model of spacetime* expressed in field terms.

Now let us arrange about the meaning of some notions. Our concern here is with the *classical* (non-quantum) description. The reason for this is that space and time are perfectly classical concepts. For electrodynamics to be treated as something tantamount to Minkowski space, it is essential to refer to the classical context.

In order to keep things as simple as possible, we consider a *point particle* whose nature is preserved under time evolution as a primary physical entity. The world lines of such particles are assumed to be timelike smooth infinite curves. Since classical point particles never decay, their world lines cannot bifurcate. Given a particle which moves along a world line oriented from the past to the future, its antiparticle may be thought of as an object identical to it in every respect but moving back in time [1]. That is, the antiparticle world line is oriented from the future to the past, as in Figure 22.1. Accordingly, the annihilation of a pair that occurs at a point $\mathcal{A}$ is depicted as a Λ-shaped world

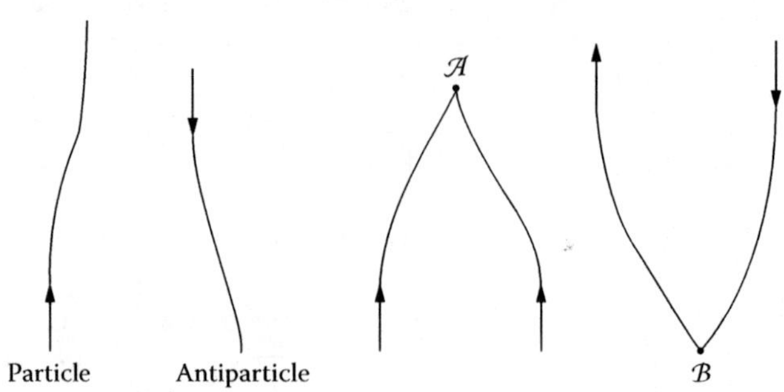

FIGURE 22.1
World lines of particles and antiparticles.

line of a single particle that runs initially from the remote past to the future up to the point $\mathcal{A}$ and then returns to the remote past. Likewise, the birth of a pair at a point $\mathcal{B}$ is given by a V-shaped world line of a single particle that runs initially from the far future to the past up to the point $\mathcal{B}$ and then returns to the far future, as in Figure 22.1.

One should be alert to notions foreign to the classical context. Classical theory leaves room for both particles that experience the proper order of events and antiparticles that follow the reverse order of events, but creations and annihilations of pairs are banned, which precludes the presence of V- and Λ-shaped curves. That is why we select timelike world lines free of cusps. Broken curves are absent from the classical picture since the least action principle does not apply to V- and Λ-shaped world lines. Such curves are automatically excluded if the condition of smoothness is imposed on the allowable world lines.

Let us *define* the electromagnetic field as a physical object that manifests itself through its influence on a particle by the four-force linear in the particle four-velocity. To be more specific, one recognizes the presence of electromagnetic field when particles experience the Lorentz force

$$f^{\mu} = ev_{\nu} F^{\mu\nu}. \tag{22.1}$$

The scalar real parameter e is the electric *charge-coupling*. For a charged particle to remain identical to itself, the coupling of this particle with the electromagnetic field must not vary in time,

$$\dot{e} = 0. \tag{22.2}$$

The state of electromagnetic field at each spacetime point is specified unambiguously by an antisymmetric tensor $F_{\mu\nu}$. In a particular frame, this is equivalent to assigning the electric field intensity $\mathbf{E}$ and the magnetic induction $\mathbf{B}$ to each point.

In Section 22.2, we derive the law governing the behavior of electromagnetic field in the hope to answer the question: to what extent is this law ordered by geometrical features of our world, in particular by the fact that *space has three dimensions*? The complete reconstruction of Maxwell's equations requires the adoption of additional assumptions of non-geometric origin. It would be tempting to think of them as the principles that cover the whole physical content of Maxwell's equations. However, it transpires in Section 22.3, from closer inspection of symmetries peculiar to electrodynamics, that such principles may have much to do with geometry.

The next issue, a plausible mechanism of perception and understanding of space and time, is briefly discussed in Section 22.4. It is shown here that the Kantian apriorism may have a direct relationship to the electromagnetic model of spacetime. Assuming that our geometric intuition is guided by our innate realization electrodynamical laws, some abnormal mental phenomena, such as clairvoyance, may be attributed to conformal invariance.

22.2 Physical Contents of Maxwell's Equations

Let us pretend that we are unaware of Maxwell's equations, and write the general law governing the electromagnetic field in a symbolic form

$$L\,(F) = \Im. \tag{22.3}$$

Here, L is a differential operator that describes local variations of the field state, and $\Im$ is interpreted as the source of these variations. The choice of L as a differential operator relates to the idea of *local action*, by which dynamical variations of fields propagate in space from one point to all nearest with a finite velocity. Partial differential equations of the hyperbolic type are believed to be best suited for the expression of this idea.

We assume that only first derivatives of $F_{\mu\nu}$ enter equation 22.3. This assumption may seem to contradict the situation in mechanics, where Newton's second law is given by a differential equation of the second order with the particle position q_a as the unknown function. But this is only an apparent contradiction because the state of a particle is specified by the pair of variables (q_a, p_a), and the evolution of this system is given by Hamilton equations containing only first derivatives of q_a and p_a. The variables $F_{\mu\nu}$ take into complete account the state of the electromagnetic field, and hence they should be likened to (q_a, p_a), not q_a.

We now choose a particular Lorentz frame and consider the spatial behavior of **E** and **B**. Any smooth vector function **V** can be reconstructed with the knowledge of 9 components of its gradients $\partial_j V_i$. However, to do this requires actually much less information. The tensor $\partial_j V_i$ can be written as the sum of symmetric and antisymmetric terms. In addition, a term proportional to the trace can be separated, rendering the symmetric term traceless,

$$\partial_j V_i = \frac{1}{2}\left(\partial_j V_i + \partial_i V_j - \frac{2}{3}\delta_{ij}\partial_k V_k\right) + \frac{1}{2}(\partial_j V_i - \partial_i V_j) + \frac{1}{3}\delta_{ij}\partial_k V_k, \tag{22.4}$$

where the summation over repeated indices is understood. A remarkable feature of three-dimensional Euclidean space is that the reconstruction of **V** requires only the knowledge of the antisymmetric term $\partial_j V_i - \partial_i V_j$, which is dual to $\nabla \times \mathbf{V}$, namely $\partial_i V_j - \partial_j V_i = \epsilon_{ijk}\,\epsilon_{klm}\partial_l V_m$, and the scalar $\partial_k V_k$, which is $\nabla \cdot \mathbf{V}$, while information on 5 components of the symmetric traceless combination $\partial_j V_i + \partial_i V_j - \frac{2}{3}\delta_{ij}\partial_l V_l$ is unnecessary. This statement is known as the Helmholtz theorem [2]: if a smooth vector function **V** disappears at infinity, it can be reconstructed from its curl, $\mathbf{C} = \nabla \times \mathbf{V}$, and divergence, $D = \nabla \cdot \mathbf{V}$. Indeed, the relation

$$\nabla \times (\nabla \times \mathbf{V}) = \nabla\,(\nabla \cdot \mathbf{V}) - \nabla^2\,\mathbf{V},$$

familiar from any course of the vector analysis, can be rewritten as the Poisson equation

$$\nabla^2 \mathbf{V} = \mathbf{S}$$

with a computable source $\mathbf{S} = \nabla D - \nabla \times \mathbf{C}$. It is easy to show that this equation has a unique solution.

An important implication of this result is that equation 22.3 can be expressed in terms of curls and divergences of **E** and **B**. Therefore, we do not need information on all components of the spacetime derivatives $\partial_\lambda F_{\mu\nu}$; only linear combinations of components containing curls and divergences of **E** and **B** matter. Note that **E** and **B** are related to $F_{\mu\nu}$ as

$$E_i = F_{0i} = F^{i0}, \tag{22.5}$$

$$F_{ij} = F^{ij} = 2\epsilon_{ijk} B_k, \quad B_k = -\frac{1}{2}\epsilon_{klm} F^{lm}, \tag{22.6}$$

where the usual rule of raising and lowering indices holds for tensors in Minkowski space. By equations 22.5 and 22.6,

$$\text{div } \mathbf{E} = \partial_j E_j = \partial_j F^{j0}, \tag{22.7}$$

$$(\text{curl } \mathbf{B})_i = \epsilon_{ijk} \partial_j B_k = \partial_j F^{ji}. \tag{22.8}$$

To express div **B** and curl **E** via linear combinations of $\partial_\lambda F_{\mu\nu}$, we recall that

$$*F^{\mu\nu} = \frac{1}{2}\epsilon^{\mu\nu\alpha\beta} F_{\alpha\beta}, \quad \text{and} \quad \epsilon^{0ijk} = \epsilon_{ijk} \tag{22.9}$$

From equations 22.9, 22.6, and 22.5 we find

$$*F^{i0} = B_i, \qquad *F^{ji} = -\epsilon_{ijk} E_k.$$

Therefore,

$$\text{div } \mathbf{B} = \partial_j \, *F^{j0}, \tag{22.10}$$

and

$$(\text{curl } \mathbf{E})_i = -\partial_j {}^*F^{ji}. \tag{22.11}$$

We see that the desired linear combinations of derivatives are $\partial_\mu F^{\mu\nu}$ and $\partial_\mu^* F^{\mu\nu}$. Indeed, taking into account equations 22.7–22.8, and equations 22.10–22.11, we have

$$\partial_\mu F^{\mu\nu} = (\text{div } \mathbf{E}, -\frac{\partial \mathbf{E}}{\partial t} + \text{curl } \mathbf{B}), \tag{22.12}$$

$$\partial_\mu {}^*F^{\mu\nu} = (\text{div } \mathbf{B}, -\frac{\partial \mathbf{B}}{\partial t} - \text{curl } \mathbf{E}), \tag{22.13}$$

Finally, the symbolic field equation 22.3 become concrete:

$$\partial_\lambda F^{\lambda\mu} = j^\mu, \tag{22.14}$$

$$\partial^*_\lambda F^{\lambda\mu} = l^\mu. \tag{22.15}$$

It remains to clarify what are the sources j^μ and l^μ. To do this requires three additional assumptions which lead directly and unambiguously to Maxwell's equations.

The first assumption is that the field equation 22.3 is *linear.* For this assumption not to seem excessively technical, it can be reformulated as the *superposition principle* (well established experimentally). This principle states: if sources $\Im_1$ and $\Im_2$ generate fields F_1 and F_2, respectively, then source $a\Im_1 + b\Im_2$ generates field $aF_1 + bF_2$. It follows that

$$L(aF_1 + bF_2) = aL(F_1) + bL(F_2)$$

which means that $L(F)$ is a linear operator.

Let us look more closely at the structure of equation 22.14. Linear combinations of the derivatives $\partial_\lambda F_{\mu\nu}$ are already taken into account. Therefore, only terms proportional to $g^\mu F^{\mu\nu}$ where g^μ stands for either the coordinate of Minkowski space x^μ or a fixed vector n^μ or some kinematical variable of some particle, say, the four-velocity at a certain point on the world line $v^\mu(s_*)$, are permitted. However, if it is granted that the system particles plus electromagnetic field is closed, coefficients of all the dynamical equations must be independent of x^μ. The option $g^\mu = n^\mu$ is in conflict with the spacetime isotropy rendering the description not explicitly covariant under rotations or Lorentz boosts. The option $g^\mu = v^\mu(s_*)$ is inadmissible because the instant s_* is selected in contradiction with the time homogeneity.

Thus j^μ is independent of $F^{\mu\nu}$. It may depend only on particle variables. What are those dependences? In order to clarify them, we observe the identity

$$\partial_\mu \partial_\nu F^{\mu\nu} = 0,$$

which is due to the antisymmetry of the tensor $F^{\mu\nu}$. Therefore, to ensure the consistency of equation 22.14, the relation

$$\partial_\mu j^\mu = 0 \tag{22.16}$$

must hold identically. Assuming that j^μ vanishes sufficiently rapidly in spacelike directions as $x^2 \to -\infty$, we have

$$Q = \int_\Sigma d\sigma_\mu j^\mu = \text{const.} \tag{22.17}$$

This equation expresses conservation of the total *charge-source.* The constancy of the charge-source Q would be tempting to relate to the constancy of the charge-coupling e, implied by equation 22.2. How can we do it? Let the hypersurface Σ be intersected by N world lines of charged particles. Our second assumption is that the *total charge-source is the sum of charge-couplings of those particles*,

$$Q = \sum_{I=1}^{N} e_I. \tag{22.18}$$

Imagine for a little that only a single point particle with the coupling e is in the universe, then

$$Q = e. \tag{22.19}$$

The identity of the charge-source and the charge-coupling is a manifestation of the extended *action–reaction* principle. Indeed, the charge-coupling measures the variation of the particle state for a given electromagnetic field state while the charge-source measures the variation of the electromagnetic field state for a given particle state. Both quantities would be reasonable to lump together as the *electric charge* or briefly the *charge*.

A realization of equation 22.18 can be attained, following Dirac [3], by writing $j^\mu(x)$ as

$$j^\mu(x) = \sum_{I=1}^{N} e_I \int_{-\infty}^{\infty} ds_I\, v_I^\mu(s_I)\,\delta^4[x - z_I(s_I)], \tag{22.20}$$

where $v_I^\mu(s_I)$ is the four-velocity of the Ith particle, and $\delta^4(x)$ is the four-dimensional delta-function.

We next turn to equation 22.15. Based on the superposition principle, we reiterate mutatis mutandis the above arguments to conclude that l^μ is independent of the field variables $F^{\mu\nu}$, yet may depend on particle characteristics. The comparison between equations 22.12 and 22.13 shows that the roles of the electric and magnetic fields are interchanged. Therefore, only particles possessing magnetic (pseudoscalar) couplings e_I^* contribute to l^μ. In line with the extended action–reaction principle, the total magnetic charge source

$$Q^\star = \int d\sigma_\mu l^\mu$$

is the sum of magnetic charge couplings:

$$Q^\star = \sum_{I=1}^{N} e_I^\star.$$

Accordingly, we may refer to e_I^* as the *magnetic charge* of the Ith particle.

Our third assumption is the *absence of magnetic charges from nature*, $e_I^* = 0$, and so

$$l^\mu = 0 \tag{22.21}$$

This assumption is based on strong experimental evidence against magnetic monopoles: despite prodigious efforts that went into searching for particles with magnetic charges, no manifestation of them is found. With these observations, the electromagnetic field is governed by the equations

$$\partial_\lambda F^{\lambda\mu} = j^\mu, \tag{22.22}$$

$$\partial_\lambda {}^*F^{\lambda\mu} = 0. \tag{22.23}$$

These equations were first formulated by Maxwell [4], and have been named for him. The interpretation of j^μ as the current of charges is due to Lorentz [5]. Dirac [3] completed the picture by expressing j^μ according to equation 22.20.

To summarize, a major part of the information encoded in equations 22.22 and 22.23 is taken from topological properties of spacetime. The residual information, seemingly divorced from geometry, translates into four assumptions:

(i) Locality;
(ii) Linearity of the dynamical equation, or the superposition principle;
(iii) Identity of the charge-source and the charge-coupling, or the extended action–reaction principle;
(iv) Lack of magnetic monopoles.

22.3 Electromagnetism and Geometry

The general solution to equation 22.23 is

$$F_{\mu\nu} = \partial_\mu A_\nu - \partial_\nu A_\mu. \tag{22.24}$$

Then equation 22.22 becomes

$$\Box A_\mu - \partial_\mu \partial_\lambda A^\lambda = j_\mu. \tag{22.25}$$

Note that $F_{\mu\nu}$ is defined in equation 22.24 only up to a gradient, that is, $F_{\mu\nu}$ remains invariant under the transformation

$$A_\mu \to A'_\mu = A_\mu + \partial_\mu \chi \tag{22.26}$$

where χ is an arbitrary smooth function. Thus, A_μ is the entire equivalence class of vector-valued functions rather than a concrete vector-valued function. Fixing the gauge, we write the general solution to equation 22.25

$$A^\mu(x) = \int d^4y\, G(x-y)\, j^\mu(y). \tag{22.27}$$

Here,

$$G = \bar{G} + G_0, \tag{22.28}$$

with $\bar{G}$ being the inverse of the D'Alembert operator, subject to a particular boundary condition, and G_0 the general Green's function for the homogeneous wave equation. The most commonly used boundary condition is the *retarded* condition which is consistent not only with the causal interrelationship, but also with the fact that time is unidirectional. However, formally, it is possible to invoke the *advanced* condition, and also any linear combination of the retarded and advanced conditions. This freedom in choosing the boundary condition rests on the linearity of the field equation 22.25.

We see that Maxwell's equations 22.22 and 22.23 provide a compact encoding of geometric features of our world. They evidence that we live in three-dimensional space with globally trivial (Euclidean) topology, or, to put it otherwise, in a four-dimensional pseudoeuclidean spacetime $\mathbb{M}_{3,1}$. Indeed, equations 22.22 and 22.23 imply the one-to-one global smooth mapping

$$A^{\mu}: \mathbb{M}_{3,1} \to \mathbb{M}_{3,1} \tag{22.29}$$

defined in equation 22.27. In fact, we have the entire equivalence class of such mappings which results from gauge invariance equation 22.26 and arbitrariness of the boundary condition for $\bar{G}$. The possibility of shuffling A^{μ} corresponds to the fact that both the original and mapped sets $\mathbb{M}_{3,1}$ may be extended to curved (pseudo-Riemannian) manifolds $\mathfrak{R}_{3,1}$, with the understanding that $\mathfrak{R}_{3,1}$ inherits topology from $\mathbb{M}_{3,1}$. As suggested by equation 22.20, "a natural curvilinear coordinate frame for the mapping equation 22.29 is that spanned by a bundle of world lines as time-coordinate lines.

One may wish to abandon assumption (iv). Then this simple topological layout is violated. Indeed, Dirac [6] showed that the field $F_{\mu\nu}$ generated by a magnetic monopole can be expressed in terms of the vector potential A_{μ} through the usual relation equation 22.24, but A_{μ} is singular on a line that issues out of the magnetic monopole, the so-called Dirac string. Cabibbo and Ferrari [7] proposed to express the tensor $F_{\mu\nu}$ in terms of two regular vector potentials A_{μ} and B_{μ} as

$$F_{\mu\nu} = \partial_{\mu} A_{\nu} - \partial_{\nu} A_{\mu} - \epsilon_{\mu\nu\alpha\beta}\, \partial^{\alpha} B^{\beta}. \tag{22.30}$$

Assembling A^{μ} and B^{μ} into a single quantity $P = (A^{\mu}, B^{\mu})$, we arrive at the Cabibbo–Ferrari mapping

$$P: \mathbb{M}_{3,1} \to \mathbb{M}_{3,1} \times \mathbb{M}_{3,1}, \tag{22.31}$$

instead of the singular Dirac mapping. Wu and Yang [8] considered another possibility that $F_{\mu\nu}$ is expressed in terms of a regular vector potential $\mathcal{A}_{\mu}$

through the relation equation 22.24, but $\mathcal{A}_\mu$ is defined on a manifold $\mathcal{M}_{3,1}$ with somewhat involved topology:

$$\mathcal{A}^\mu : \mathcal{M}_{3,1} \to \mathbb{M}_{3,1}. \tag{22.32}$$

Neither of these two regular mappings, equations 22.31 and 22.32, is an isomorphism of $\mathbb{M}_{3,1}$, which implies that the presence of magnetic monopoles has unfitted the model imitating real spacetime topology.

The above reconstruction of equations 22.22 and 22.23 makes it clear that their structure is highly sensitive to the choice of the spacetime dimension $D + 1 = 4$. In order to better appreciate this fact, let us note that Maxwell's electrodynamics is conformally invariant *only for* $D+1=4$. Indeed, the action, from which equations 22.22 and 22.23 can be deduced, is

$$S = 2\int d^{D+1}x\sqrt{-g}\left(\frac{1}{4}g_{\alpha\beta}\,g_{\mu\nu}\,F^{\alpha\mu}\,F^{\beta\nu} + g_{\mu\nu}\,j^\mu\,A^\nu\right), \tag{22.33}$$

where $g = \det g_{\alpha\beta}$. A simple criterion for conformal invariance is that the energy tensor is traceless,

$$T^\mu_\mu = 0, \tag{22.34}$$

can be applied to the theory with the action equation 22.33 for which, in the flat spacetime limit $g_{\mu\nu} \to \eta_{\mu\nu}$, the energy tensor is

$$T_{\mu\nu} = \left\{\left[F_\mu{}^\alpha F_{\alpha\nu} - 2\left(j_\mu A_\nu + j_\nu A_\mu\right)\right] + \eta_{\mu\nu}\left(\frac{1}{4}F^{\alpha\beta}F_{\alpha\beta} + j_\mu A^\mu\right)\right\}. \tag{22.35}$$

Since $\delta^\mu{}_\mu = D + 1$, the condition equation 22.34 is met only for $D + 1 = 4$.

The conformal invariance of equations 22.22 and 22.23 was first discovered by Bateman [9] and Cunningham [10]. It is just the conformal invariance which renders the field equations linear. Indeed, consider the generic nonlinear electrodynamics with

$$S = \int d^4x\sqrt{-g}\,\mathcal{L}(\mathcal{S},\mathcal{P}), \tag{22.36}$$

where $\mathcal{L}$ is an arbitrary analytic function of the invariants of electromagnetic field

$$\mathcal{S} = \frac{1}{2}g^{\mu\alpha}g^{\nu\beta}F_{\mu\nu}F_{\alpha\beta}, \quad \mathcal{P} = \frac{1}{2}\sqrt{-g}\,\epsilon^{\mu\nu\alpha\beta}F_{\mu\nu}F_{\alpha\beta}. \tag{22.37}$$

It is straightforward to show that

$$T^{\mu\nu} = F^\mu{}_\alpha\frac{\partial\mathcal{L}}{\partial F_{\alpha\nu}} - \eta^{\mu\nu}\mathcal{L}, \tag{22.38}$$

and hence

$$T^{\mu}{}_{\mu} = -4(\mathcal{L}_{\mathcal{S}}\mathcal{S} + \mathcal{L}_{\mathcal{P}}\mathcal{P}) + 4\mathcal{L}, \tag{22.39}$$

where $\mathcal{L}_{\mathcal{S}} = \partial\mathcal{L}/\partial\mathcal{S}$ and $\mathcal{L}_{\mathcal{P}} = \partial\mathcal{L}/\partial\mathcal{P}$. The question now arises: what should be $\mathcal{L}$ to make $T^{\mu}{}_{\mu} = 0$? Denoting $l = \log \mathcal{L}$, $s = \log \mathcal{S}$, $p = \log \mathcal{P}$, this gives the partial differential equation for the unknown function l

$$l_s + l_p = 1. \tag{22.40}$$

This equation is satisfied by

$$l = \frac{1}{2}(s+p) + u(s-p), \tag{22.41}$$

where u is an arbitrary function. Turning back to $\mathcal{L}$, $\mathcal{S}$, $\mathcal{P}$, we have

$$\mathcal{L} = \sqrt{\mathcal{S}\mathcal{P}}\; U(\mathcal{S}/\mathcal{P}), \tag{22.42}$$

where U is an arbitrary differentiable function.

Choosing $U(x) = -\frac{1}{2}\sqrt{x}$, we recover Maxwell's electrodynamics $\mathcal{L} = -\frac{1}{2}\mathcal{S}$. Other choices of U are of little importance for models that mimic spacetime topology. The reason for this is that the field propagation in every nonlinear version of electrodynamics may develop shock waves, which makes the mapping equation 22.29 singular. The only exception is the Born–Infeld theory

$$\mathcal{L} = b^2\left(1 - \sqrt{1 + b^{-2}\mathcal{S} - \frac{1}{4}b^{-4}\mathcal{P}^2}\right), \tag{22.43}$$

where b is a constant with dimension of the field strength. Blokhintsev and Orlov [13] showed that the nonlinear system of hyperbolic equations in this theory is unique in that their characteristics do not intersect, and hence no shock wave occurs. However, the Lagrangian equation 22.43 is outside the class of functions covered by equation 22.42. Therefore, this theory is devoid of conformal invariance.

Weyl was the first to establish that only the Maxwellian form of electrodynamics is conformally invariant, and that this symmetry is unique to four dimensions, which offers an argument in support the view that the world of dimension 4 is singled out (see [11], Section 40). It is this topological argument which gives the linear version of electrodynamics its strong intuitive appeal.

At first glance, assumption (ii) is not fundamental and reflects the mere fact that classical electromagnetic fields are so feeble that the linear approximation agrees nicely with the experimental data. However, the linearity is so much a part of this theoretical scheme that one might even sacrifice assumption (i) to it, and come to an alternative theory, the *action-at-a-distance* electrodynamics [14], which has no field degrees of freedom on their own. (This disappearance of electromagnetic degrees of freedom bears some resemblance to the situation in the Maxwell–Lorentz theory where A^{μ} plays the role of the automorphism group for the background $\mathbb{M}_{3,1}$, as equation 22.29 indicates.) It is the linearity which makes it possible to 'derive' the action-at-a-distance electrodynamics, involving retarded and advanced interactions on an equal footing, from the field theory based on the retarded boundary condition.

We see that assumptions (i), (ii), and (iv) have much to do with geometry. Now, where does physics reside? It is clear that the only place where it may be found is the source of the field j^{μ}. Equation 22.20 suggests that physics is determined by the *class of allowable world lines.* If this class is composed of timelike smooth infinite world lines, as in Figure 22.1 left, then the mapping equation 22.29 unravels no other thing than geometry of Minkowski space.

One further assumption is that Λ- and V-shaped curves, shown in Figure 22.1, are also tolerated. Combining fragments of Λ- and V-shaped timelike curves, one can build a zigzag curve corresponding to an effective spacelike worldline. The admissibility of such world lines violates the causal interrelationship, that is, demolishes the light-cone structure. We thus come to an effective four-dimensional *Euclidean* geometry. The retarded Green's function is no longer geometrically justified; instead, the Feynman 'causal boundary condition' proves to be best suited to this geometry because the change between the pseudo-Euclidean and Euclidean metrics is attained by the Wick rotation consistent with the singularity location of the Feynman propagator. Euclidean geometry is in excellent agreement with quantum theory. A photon is a creature of this effective Euclidean world. It was already pointed out in Introduction that the least action principle does not apply to this class of world lines. This, however, is immaterial for quantum theory where the action plays the leading role, while its extremums are of secondary importance. Now, having the Euclidean background, one may pose the question of the experimentally observed electric charge quantization (which is completely ignored in the classical context). Note also that the physics manifests itself as a breakdown of conformal invariance in mechanics of massive particles. However, it is impossible to give here a complete account of these issues for reasons of space. They will be addressed elsewhere.

22.4 Perception of Space and Time

Let us turn to a simple model of perception and understanding of space and time. Let us imagine a machine composed of a radar probing the environment by electromagnetic waves in the optical spectrum (eyes), data link conveying impressions of light (optic nerves), and a computer processing the delivered data (brain). We then assume that Maxwell's equations are built in the brain. These Lord's proprietary software are meant for incessant searching solutions to the Cauchy problem for Maxwell's equations with varying initial data. This is a mere restatement in today's parlance of the famous

Kantian apriorism. Space and time are indeed pure contemplations in that the *electromagnetic* model of spacetime is an integral part of the mind. Our space and time orientation, our visualization and imagination capabilities are ensured by lasting instinctive processes of decoding Maxwell's equations. The appeal of this model of the Maxwell-guided brain may be enhanced if we take an analogue computer, rather than a digital computer, to mean the neural network [15]. Note that even protozoa are equipped with some elements of this machinery.

We may further conceive that these software have capability not only to *solve* Maxwell's equations, but also *transform* the obtained solutions according to electrodynamical symmetries, specifically conformal symmetry.

Recall that the group of conformal transformations [16] consists of Lorentz transformations

$$x^{\mu} \to x'^{\mu} = \Lambda^{\mu}_{\nu} x^{\nu}, \tag{22.44}$$

translations

$$x^{\mu} \to x'^{\mu} = x^{\mu} + a^{\mu}, \tag{22.45}$$

dilatations

$$x^{\mu} \to x'^{\mu} = e^{\rho}\, x^{\mu}, \tag{22.46}$$

and special conformal transformations

$$x^{\mu} \to x'^{\mu} = \frac{x^{\mu} - b^{\mu} x^{2}}{1 - 2b \cdot x + b^{2} x^{2}}. \tag{22.47}$$

The angle between intersecting curves is left invariant under conformal transformations:

$$\cos\varphi = \frac{dx_1 \cdot dx_2}{\sqrt{dx_1^2\, dx_2^2}} = \text{const}, \tag{22.48}$$

hence the name *conformal*, indicating that the shape of any figure is unchanged by such transformations.

Figures can be freely rotated and shifted in our mind. The mental image of any object is readily rescaled. Meanwhile a special conformal transformation is composed of an inversion, translation, and further inversion. Hence its realization in one's head can hardly be conceived. Nevertheless, people are usually agree in their observations of angle relations. For example, a right angle is never confused with an angle slightly different from right. By contrast, we may disagree in metric issues, say, a color-blind person finds the light wave lengths quite different from those found by a man with normal vision. This lends support to the view that conformal invariance of Maxwell's equations is responsible for producing our notion of space.

By equation 22.47

$$x'^2 = \frac{x^2}{1-2b\cdot x+b^2x^2}, \tag{22.49}$$

and

$$dx'^2 = \frac{dx^2}{(1-2b\cdot x+b^2x^2)^2}, \tag{22.50}$$

which shows that the light cone is mapped onto the light cone. However, if x^2 is finite, and $1 - 2b \cdot x + b^2x^2 < 0$, then x'^2 and x^2 are opposite in sign; special conformal transformations can convert a timelike vector into spacelike and vice versa. Does the conformal group violate causality? If we would choose the "active" or "passive" interpretations for these spacetime transformations, we would come to a trouble with causality. Rosen [17] noted that the situation can be improved if we regard the conformal transformations as leaving both spacetime and the coordinate frame unaffected, and map only the world lines and field configuration of a given experimental setting. The same observer then sees different processes and different field configurations in the same flat spacetime background. If every physically valid process is to be transformed into another physically valid process by a given group of transformations, then we have a symmetry of physics [17].

It is clear from equation 22.50 that the sign of the line element is invariant, in particular, timelike has an invariant meaning for tangent vectors. Therefore, special conformal transformations always transform a timelike curve into another timelike curve. However, the transformed curve may have two branches, one being oriented from the past to the future, and the other with opposite orientation. Physically, the case that a single particle is moving along a timelike world line can be converted by an appropriate conformal transformation into the case that the particle is executing quite different motion and is accompanied by an antiparticle (that is, an object with the same characteristics but moving back in time). As a simple illustration, borrowed from [17], we refer to a particle at rest whose world line is shown in Figure 22.2 left. A special conformal transformation characterized by $b^\mu = (0, b, 0, 0)$ converts this straight line into two hyperbolic curves with opposite orientations. Note that the left hyperbolic curve is the image of the domain *AOB* on the original straight line, and the right hyperbolic curve is the image of two disconnected domains *AC* and *BD* of the original straight line.

Thus, a fragment of some movie can be converted into a fragment of a much different movie with other characters in the play and different casting. Were such conversions implemented mentally by some person with highly developed geometric intuition, he would be able to relate some events with

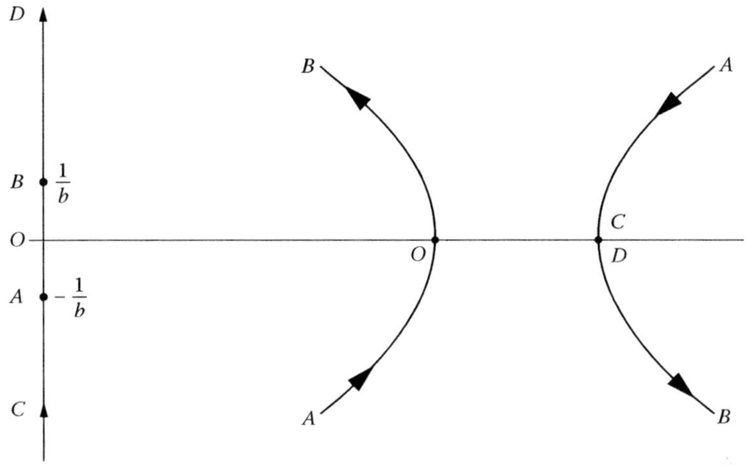

FIGURE 22.2
A straight line is converted to a pair of curves by a conformal transformation.

reverse temporal ordering. It would be appropriate to recognize this person as a "clairvoyant". We see that abnormal mental phenomena, such as clairvoyance, may be attributed to an acute realization of conformal properties of electromagnetic reality.

Acknowledgments

I would like to express my thanks to A. Kracklauer and R. Woodard for discussions.

References

[1] Feynman, R. P. (1966). The development of the space-time view of quantum mechanics. Nobel lecture. *Physics Today*, **19**, 31.

[2] Helmholtz, H. (1858). Ueber Integrale der hydrodynamischen Gleichungen, welche den Wirbelbewegungen entsprechen. *Journal für die reine und angewandte Mathematik*, **55**, 25.

[3] Dirac, P. A. M. (1938). Classical theory of radiating electron. *Proceedings of the Royal Society* (London), **A167**, 148.

[4] Maxwell, J. C. (1873). *A Treatise on Electricity and Magnetism*. Oxford: Clarendon.

[5] Lorentz, H. A. (1909). *The Theory of Electrons and its Applications to the Phenomena of Light and Radiant Heat*. Leipzig: Teubner.

[6] Dirac, P. A. M. (1931). Quantised singularities in the electromagnetic field. *Proceedings of the Royal Society* (London), **133**, 60.

[7] Cabibbo, N. & E. Ferrari (1962). Quantum electrodynamics with Dirac monopoles. *Nuovo Cimento,* **23**, 1147.

[8] Wu, T. P. & C. N. Yang (1975). Concept of nonintegrable phase factor and global formulation of gauge fields. *Physical Review,* **D 12**, 3845.

[9] Bateman, H. (1909). On the conformal transformations of a space of four dimensions and their application to geometric optics. *Proceedings of the London Mathematical Society,* **7**, 70. The transformation of the electrodynamical equations. *ibid.* **8**, 223.

[10] Cunningham, E. (1909). The principle of relativity in electrodynamics and an extension thereof. *Proceedings of the London Mathematical Society,* **8**, 77.

[11] Weyl, H. (1918). *Raum, Zeit, Materie.* 5th edition 1923. Berlin: Springer. *Space, Time, Matter.* Translated from the 4th German edition 1952. New York: Dover.

[12] Born, M. & L. Infeld (1934). Foundations of the new field theory. *Proceedings of the Royal Society* (London), **144**, 425.

[13] Blokhintsev, D. I. and V. V. Orlov (1953). Propagation of signals in the Born–Infeld electrodynamics. *Žhurnal Éksperimental'noi i Teoretitcheskoi Fiziki,* **25**, 513 (in Russian).

[14] Hoyle, F. & J. V. Narlikar (1995). Cosmology and action-at-a-distance electrodynamics. *Reviews of Modern Physics,* **67**, 113.

[15] Kracklauer, A. F. (2005). Private communication.

[16] Fushchich, W. I. and A. G. Nikitin (1987). *Symmetries of Maxwell's Equations.* Dordrecht: Reidel.

[17] Rosen, J. (1968). The conformal group and causality. *Annals of Physics,* **47**, 468.

23

From Quantum to Classical: Watching a Single Photon Become a Wave

Marco Bellini, Alessandro Zavatta, and Silvia Viciani

Istituto Nazionale di Ottica Applicata, Florence, Italy LENS and Department of Physics, University of Florence, Florence, Italy

CONTENTS

Abstract

We present the experimental generation of a new class of non-classical light states and their complete phase-space characterizations by quantum homodyne tomography. These states result from the most elementary amplification process of classical light fields by a single quantum of excitation and can be generated by stimulated emission of a single photon in the mode of a coherent state. Being intermediate between a single-photon Fock state and a coherent one, they offer unique opportunities to closely follow the smooth evolution between the particle-like and the wave-like behaviors of light fields and witness the gradual change from spontaneous to stimulated regimes of light emission.

Key words: single photons, Fock states, coherent states, quantum tomography, Wigner function.

23.1 Introduction

The nature of light has been the subject of intense study and scientific and philosophical debate over several centuries. In the 17th century Newton was convinced of its corpuscular nature. Early in 1800, Young demonstrated that light had to behave like a wave to be compatible with his observations of interference phenomena. Fresnel's diffraction theory and Maxwell's equations later came to strengthen the hypothesis and made scientists believe that the dilemma was finally solved with an apparently complete description of light as a wave of electromagnetic radiation. The situation again became complicated early in the 20th century when Planck and Einstein introduced the quantization of light in elementary particles, the photons, to explain the phenomena of blackbody radiation and photoelectric effects. Although the quantum theory of matter with the discretization of atomic energy levels introduced in the 1920s has explained such phenomena even without the need of light particles, the advent of new sources and more efficient detectors over the past 30 years led to the flourishing of experimental proofs of the strictly corpuscular nature of light generated under particular conditions.

The wave-particle duality is now a firm point of modern quantum physics and solves the question affirming that, not only light but also matter in its various forms, exhibit a wave-like or particle-like aspect depending on how we generate and observe it. In particular, light is seen to assume a typical particle-like behavior when it is generated in the so-called Fock states, the eigenstates of the number operator $\hat{n} = \hat{a}^{\dagger}\hat{a}$, where $\hat{a}^{\dagger}$ and $\hat{a}$ are the photon creation and destruction operators for a single field mode considered here for simplicity. The perfectly defined number of excitation quanta (or intensity) of the field in such states implies a complete lack of determination of the value of the phase. The corresponding Wigner function, a quasi-probability distribution which fully describes the state of the quantum system in the field quadrature space, consequently exhibits a perfect cylindrical symmetry around the origin of the quadrature axes. Single-photon Fock states in a well-defined spatio-temporal mode have been recently generated experimentally, and a quantum tomographic analysis via time-domain balanced homodyne detection has been used to recover the density matrix elements of the states and to reconstruct their Wigner functions.[1,2] Such reconstructions have clearly confirmed the strict quantum-mechanical nature of the Fock states as indicated by classically impossible negative values of the phase-invariant Wigner function around the origin.

On the other hand, the wave-like light regime is best represented by the so-called coherent states $|\alpha\rangle$, the eigenstates of the photon destruction operator $\hat{a}$, such that $\hat{a}\,|\alpha\rangle = \alpha\,|\alpha\rangle$. Such states are the closest analogues to a classical oscillating light field with amplitude and phase determined within the bounds of Heisenberg's uncertainty principle. The number of quanta in such states is subject to fluctuations. If its average is $N = |\alpha|^2$, the phase

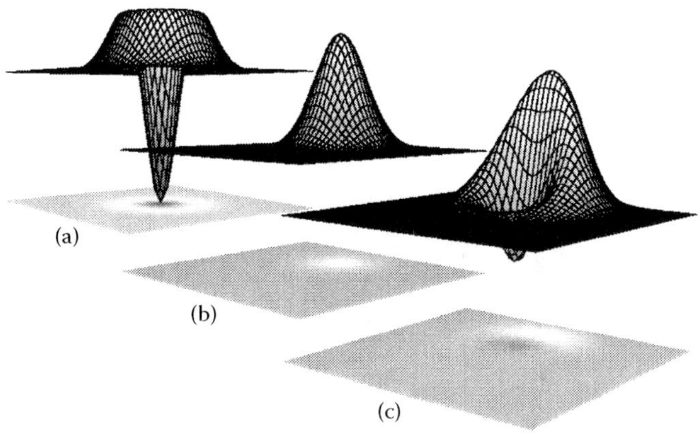

FIGURE 23.1
Calculated Wigner functions of a) the single-photon Fock state, b) the coherent state with $|\alpha| = 1$, and c) the corresponding single-photon-added coherent state (SPACS) obtained by stimulated emission of a single photon in the same mode of $|\alpha\rangle$.

uncertainty is of the order of $1/\sqrt{N}$. The corresponding Wigner function is simply seen to correspond to a displaced (by α) version of the Gaussian function of the vacuum state. It is always positive and exhibits equal variances along any given quadrature.

Intermediate conditions however exist between the two extreme situations described above. In particular, when a classical coherent state is excited by exactly a single photon, the result is an hybrid state exhibiting a mix of the characteristics of two such different parents. If the amplitude of the initial coherent state is gradually increased starting from the vacuum, the character of the final state can be continuously tuned between that of a purely quantum-mechanical form of light, the single-photon Fock state, toward that of an almost classical coherent one exhibiting a wave-like behavior with well-defined amplitude and phase. See Fig. 23.1.

In the following we will briefly illustrate some of the most important properties of these single-photon-added coherent states (SPACSs),[3] present their experimental generation by means of conditional preparation methods, and then show the results of a complete characterization performed with a high-frequency, time-domain, quantum tomographic technique recently developed by our group.[4]

23.2 Properties of Single Photon-Added Coherent States (SPACSs)

Single photon-added coherent states, first described in a general form in 1991 by Agarwal and Tara,[5] result from a single application of the photon creation operator $\hat{a}^\dagger$ on a classical coherent state $|\alpha\rangle$ and, in their normalized form, read as:

$$|\alpha,1\rangle = \frac{\hat{a}^{\dagger}|\alpha\rangle}{\sqrt{1+|\alpha|^2}}. \qquad (23.1)$$

From the expansion of SPACSs in terms of Fock states

$$|\alpha,1\rangle = \frac{e^{-\frac{|\alpha|^2}{2}}}{\sqrt{1+|\alpha|^2}} \sum_{n=0}^{\infty} \frac{\alpha^n}{\sqrt{n!}} \sqrt{n+1}\,|n+1\rangle. \qquad (23.2)$$

it is evident that they lack the vacuum term contribution and thus differ quite heavily from an ordinary coherent state, especially for low amplitudes, where the missing contribution has a stronger impact. Indeed, while the application of the photon destruction operator does not change a coherent state, its single-photon excitation transforms it into a very non-classical object. The non-classical character of SPACSs can be readily illustrated by the evaluation of their Wigner function which, for arbitrary amplitude α, can be expressed as:

$$W(z) = \frac{-2(1-|2z-\alpha|^2)}{\pi(1+|\alpha|^2)} e^{-2|z-\alpha|^2}. \qquad (23.3)$$

The distribution can become negative, a proof of its quantum character, whenever the condition

$$|2z-\alpha|^2 < 1 \qquad (23.4)$$

is satisfied. Interestingly, differently from Fock states, SPACSs possess another key feature normally associated to quantum states: the reduced fluctuations (or squeezing) in one of their quadratures. Given a field quadrature $\hat{x}_\theta = \frac{1}{2}(\hat{a}e^{-i\theta} + \hat{a}^{\dagger}e^{i\theta})$, it is easy to find that its fluctuations amount to:

$$[\Delta x_\theta]^2_\alpha = \langle x^2_\theta \rangle_\alpha - \langle x_\theta \rangle^2_\alpha = \frac{1}{4} + \frac{1-|\alpha|^2\cos(2\theta)}{2(1+|\alpha|^2)^2} \qquad (23.5)$$

and that the quadrature obtained by choosing $\theta = 0$ exhibits reduced fluctuations, and is thus squeezed with respect to the coherent state, whenever $|\alpha| > 1$.

23.3 How to Produce SPACSs

SPACSs are produced by injecting a coherent state $|\alpha\rangle$ as a seed into the signal mode of an optical parametric amplifier and exploiting the stimulated emission of a single down-converted photon into the same mode.[3]

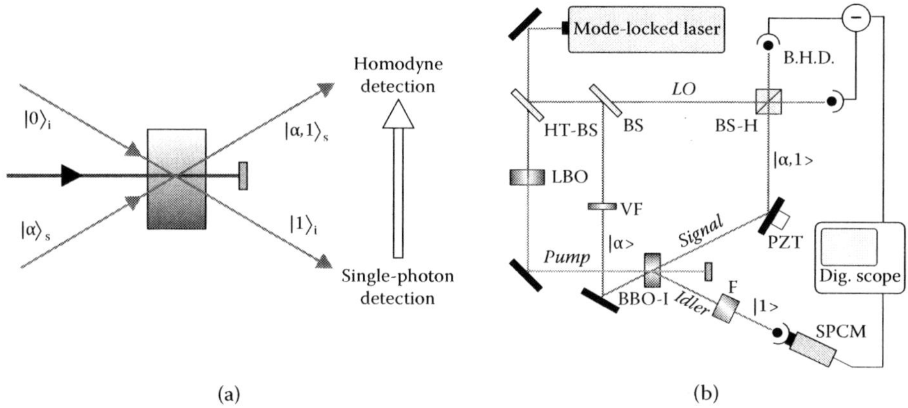

FIGURE 23.2
(a) Schematic view of the processes involved in the conditional preparation of single-photon-added coherent states. b) Experimental apparatus: HT-BS high transmission beam-splitter, LBO lithium triborate crystal, BS and BS-H 50% beam-splitters, VF variable attenuation filter, BBO-I type-I β-barium borate down-converter crystal, PZT piezoelectric transducer, B.H.D. balanced homodyne detector, F spectral and spatial filters, SPCM single photon counting module.

Single-photon emission in the signal channel involves the generation of the desired target state and takes place every time that a single photon is detected in the correlated idler mode (see a schematic view of the process in Fig. 23.2a). With the low parametric gain of our experimental situation, the final output state can be approximated as

$$
\begin{aligned}
|\psi\rangle &\approx [1+g(\hat{a}_s^\dagger \hat{a}_i^\dagger - \hat{a}_s \hat{a}_i)]\,|\alpha\rangle_s\,|0\rangle_i = \\
&= |\alpha\rangle_s\,|0\rangle_i + g\hat{a}_s^\dagger\,|\alpha\rangle_s\,|1\rangle_i
\end{aligned}
\tag{23.6}
$$

where g is a gain constant with $|g| \ll 1$ and the coherent field $|\alpha\rangle_s$ enters the parametric crystal in the signal mode, while vacuum ($|0\rangle_i$) enters the idler channel. The output signal mode will thus mostly contain the original coherent state, except for the few cases when the state $|1\rangle_i$ is detected in the idler output mode. These relatively rare detection events, which take place with a probability proportional to $|g|^2(1 + |\alpha|^2)$, project the signal state onto the desired SPACS $|\alpha, 1\rangle_s$, corresponding to the stimulated emission of one photon in the same mode of $|\alpha\rangle$.

Note that when the input state is of the form $|0\rangle_s\,|0\rangle_i$, i.e., no seed coherent field is injected into the crystal, spontaneous parametric down-conversion takes place starting from the input vacuum fields, and pairs of entangled signal and idler photons with random (but mutually correlated) phases are produced in the crystal in the state $|1\rangle_s\,|1\rangle_i$ with a low probability proportional to $|g|^2$. In this case, the detection of a single photon in the idler mode projects the signal state onto a single-photon Fock state, hence, by following the evolution of the final quantum state while the amplitude α increases from zero, one can thus witness the gradual transition from the spontaneous to the stimulated regimes of light emission with the smooth transformation of a single photon (particle-like) state towards a coherent (wave-like) one.

Interestingly, one can obtain an absolute calibration of the amplitude of the seed coherent field $|\alpha\rangle_s$ injected in the SPDC signal mode by measuring the rate of counts in the idler channel and comparing it to the un-seeded case. As stated above, the ratio of such rates equals $(1 + |\alpha|^2)$ and this is clearly due to the enhancement of emission probability characteristic of stimulated emission in bosonic fields.

The experimental apparatus used to generate the SPACS is schematically drawn in Fig. 23.2b. A mode-locked Ti:sapphire laser, emitting 1–2 ps long pulses at 786 nm and at a repetition rate of 82 MHz is used as the primary source. The laser pulses are frequency doubled to 393 nm in a 13-mm long LBO crystal which thus produces the pump pulses for parametric down-conversion in a 3-mm thick, type-I BBO crystal slightly tilted from the collinear configuration in order to obtain an exit cone beam with an angle of ~ 3° from which symmetric signal and idler modes are roughly selected by means of irises placed at about 70 cm from the crystal.

In order to non-locally select a pure state on the signal channel,[6–9] idler photons undergo narrow spatial (single-mode fiber) and frequency (a pair of etalon interference filters) selection before detection by a single photon counting module (Perkin-Elmer SPCM AQR-14). The weak coherent state $|\alpha\rangle$ is obtained by controlled attenuation of a small portion of the laser emission which is fed into the signal mode of the parametric crystal.

23.4 Time-Domain Quantum Tomography and State Reconstruction

Balanced homodyne detection provides the measurements of field quadratures $\hat{x}_\theta = \frac{1}{2}(\hat{a}e^{-i\theta} + \hat{a}^\dagger e^{i\theta})$[10–12] allowing the characterization of a quantum field mode by the reconstruction of its density matrix elements and Wigner function.[13] Here it is performed by mixing the target field state with an intense classical local oscillator (LO, again obtained from a portion of the original laser pulses) onto a 50% beam-splitter (BS-H in figure) whose outputs are then detected by proportional photodetectors (Hamamatsu S3883). The difference in the photocurrents produced by the two detectors is amplified and sent to a fast digital oscilloscope whose acquisition is triggered by the detection events in the idler channel.

Such a signal is proportional to the SPACS quadrature selected by varying the relative phase θ between the LO and the signal field by means of a mirror mounted on a piezoelectric transducer (PZT).[14,15] Note that, in this time-domain version of the homodyne detection technique, the difference in the photocurrents for each laser pulse has to be singularly analyzed in order to extract the quadrature measurements. A very high frequency bandwidth is thus necessary for the whole detection system in order to cope with the 82 MHz repetition rate of the mode-locked laser oscillator. Our experimental apparatus is, to our knowledge, the only existing system capable of such a high frequency.[4]

About 5000 pulse area acquisitions can be stored in the scope at a maximum rate of 160,000 frames per second before being transferred to a personal computer where the areas of the pulses are measured and their statistic distributions are analyzed in real time. Typical rates of state preparation for vacuum input are about 300 s^{-1}, with less than 1% contribution from accidental counts. A typical sequence of about 5000 acquisition frames can thus be captured and analyzed in about 20–30 s. It is interesting to remind that the probability of detecting an idler photon is proportional to $|\hat{a}^{\dagger}|\alpha\rangle|^2$, hence, as soon as α is increased and stimulated emission starts taking place, the rate of trigger events grows proportional to $(1 + |\alpha|^2)$, thus making the acquisition rate much higher.

Figure 23.3 presents the raw acquired homodyne data as a function of the PZT position for different values of the seed amplitude $|\alpha|$. The first plot (Fig. 23.3(a)), obtained with a blocked signal input, corresponds to the single-photon Fock state: it is clearly phase-independent and shows the typical "hole" in the center of the distribution which is responsible for the negativity of the corresponding Wigner function at the origin.[1,2] When the coherent seed is initially switched on at very low intensity, the phase-invariance is broken and data show the appearance of higher density regions due to the gradual appearance of a defined phase in the field (Fig. 23.3(b and c)). Finally, for increasing seed amplitudes (Fig. 23.3(d)), the signal distribution becomes more and more similar to that of a classical coherent field, with a clear oscillating behavior and well defined amplitude and phase.

These series of homodyne measurements yield the quadrature probability distributions $p(x, \theta)$ corresponding to the marginals of the Wigner quasi-probability distribution $W(x, y)$[13]:

$$p(x,\theta) = \int_{-\infty}^{+\infty} W(x\cos\theta - y\sin\theta, x\sin\theta + y\cos\theta)dy. \qquad (23.7)$$

Given a sufficient number of quadrature distributions at different values of the phase $\theta \in [0, \pi]$, one is therefore able to reconstruct the quantum state of the field under study.[15,16] We reconstructed the elements of the density matrix $\hat{\rho}$ of the state in the number-state representation by averaging the so called "pattern functions" $f_{nm}(x, \theta)$ over the outcomes of the quadrature operator and over the phase θ as

$$\langle n|\hat{\rho}|m\rangle = \frac{1}{\pi}\int_0^{\pi} d\theta \int_{-\infty}^{+\infty} dx\, p(x,\theta) f_{nm}(x,\theta). \qquad (23.8)$$

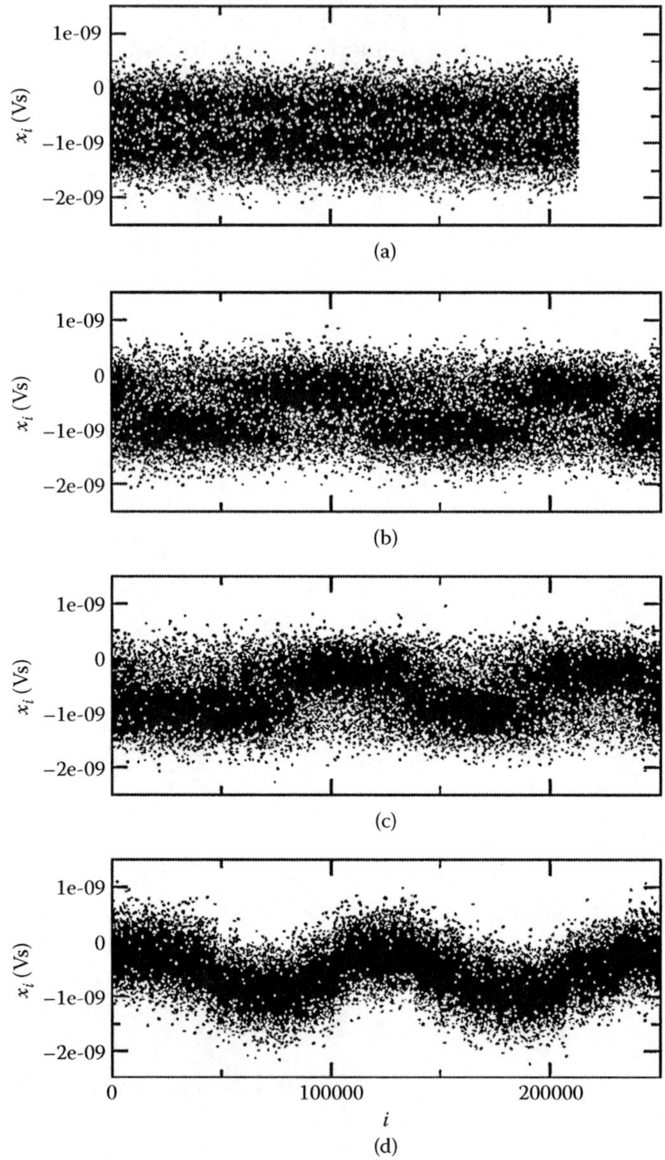

FIGURE 23.3
Raw homodyne data for the single-photon-added coherent states at increasing seed amplitudes: a) $|\alpha| = 0$, i.e., the single photon Fock state is generated; b) $|\alpha| = 0.387$; c) $|\alpha| = 0.723$; d) $|\alpha| = 3.74$.

The Wigner function can then be obtained by means of the following transformation:

$$W(x,y)=\sum_{n,m}^{M}\rho_{n,m}W_{n,m}(x,y) \tag{23.9}$$

where $W_{n,m}(x, y)$ is the Wigner function of the operator $|n\rangle\langle m|$. Note that, using this procedure, the Wigner function of the state is reconstructed from a truncated density matrix of dimension $M \times M$. This implies a finite resolution in the reconstructed function which, however, can be adapted to the particular physical situation of interest in order to avoid loss of information on the state.

Figure 23.4 shows Wigner functions obtained from such truncated density matrices for increasing seed amplitudes. The first one (a) again corresponds to the single-photon Fock state obtained by conditional preparation from the two-photon wavefunction of SPDC,[1,2] and clearly exhibits classically

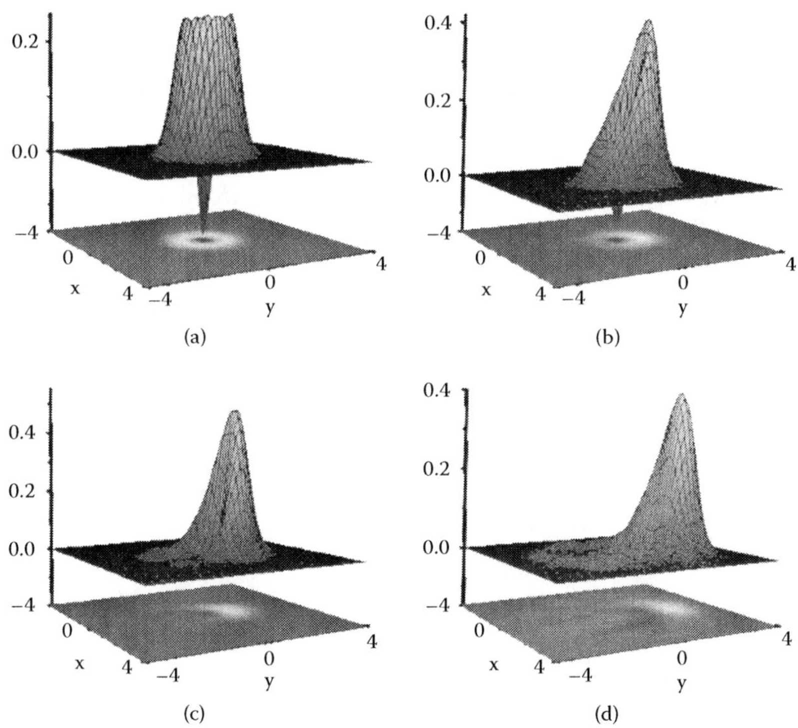

FIGURE 23.4
Wigner functions of the SPACSs as derived from the experimental data: a) $|\alpha| = 0$ i.e., single photon Fock state, calculated from a 6×6 reconstructed density matrix; b) $|\alpha| = 0.387$, with a 7×7 matrix; c) $|\alpha| = 0.955$, 8×8; d) $|\alpha| = 2.61$, 14×14.

impossible negative values around the center of the cylindrically symmetric (due to the undefined value of the phase) distribution. For a weak coherent seed of very low intensity ($|\alpha| \approx 0.4$, i.e., an average of one photon every 7 pulses), the Wigner function is seen to lose its cylindrical symmetry while moving away from the origin due to the gradual appearance of a defined phase, but it still exhibits a clear non-classical nature as indicated by its partial negativity (b). Then the negativity gradually gets less evident (c) and the ring-like wings in the distribution start to disappear making it more and more similar to the Gaussian typical of a classical coherent field (d). Interestingly, even at relatively high input amplitude $|\alpha|$, the Wigner distribution for the SPACS $|\alpha,1\rangle$ keeps showing the effect of the one-photon excitation when compared to the corresponding, slightly displaced, un-excited $|\alpha\rangle$ state.[3]

When comparing the reconstructed Wigner functions and density matrix elements to the theoretical ones for the corresponding quantum states, one has to take into account the limited efficiency of the homodyne detection apparatus which does not allow one to generate and analyze pure states but always involves some mixing with the vacuum. The limited efficiency enters both in the preparation of the quantum state, where the non-ideal conditioning performed on the idler channel does not allow one to generate a completely pure state in the signal channel, both in the homodyne detection process itself, due to the limited efficiency of the photodiodes and to the imperfect mode-matching of the signal field with the LO.[2]

From a fit of the experimental marginal distributions for the Fock state to the corresponding theoretical curves, we obtain an overall detection efficiency of $\eta = 0.602 \pm 0.002$. The expression for the Wigner function of SPACSs in the presence of limited efficiency η is the following:

$$W(z) = \frac{-2}{\pi(1+|\alpha|^2)}[2\eta - 1 - |2\sqrt{\eta}z - \alpha(2\eta - 1)|^2]e^{-2|z-\sqrt{\eta}\alpha|^2} \tag{23.10}$$

and it can be simply seen that the variance in its quadratures as a function of θ becomes:

$$[\Delta x_\theta]^2_{\alpha,\eta} = \frac{1}{4} + \frac{\eta[1 - |\alpha|^2 \cos(2\theta)]}{2(1 + |\alpha|^2)^2} \tag{23.11}$$

clearly showing that, while the original coherent state has equal fluctuations in the different quadratures independently from its amplitude, the one-photon-excited state exhibits a squeezing in one of the quadratures and larger fluctuations in the orthogonal one as soon as $|\alpha| > 1$. An intuitive interpretation of this behavior can be connected with the reduction in the intensity noise of the coherent state when excited by a perfectly defined number of quanta with the corresponding increase in the phase noise due to the intrinsic lack of phase information of the Fock state. This effect starts to become evident in the reconstructed Wigner function of Fig. 23.4(c) (which is however still at the border of the un-squeezed region), where a somewhat reduced width appears along the radial direction, while the increase in the phase noise is indicated by the appearance of the ring-like wings in the tangential direction of the Wigner distribution.

Even in the presence of an imperfect preparation and detection ($\eta < 1$), the generated states can thus clearly exhibit two typical features of a quantum character, i.e., the negativity of the Wigner function combined with a quadrature squeezing which, to our knowledge, have never been detected simultaneously in the same light state. Figure 23.5 presents the measured value of the Wigner function in its minimum and the variance in the squeezed quadrature (corresponding to the case with $\theta = 0$ in Eq. 23.11) for a range of amplitudes of the coherent seed pulse. The reconstructed Wigner function clearly exhibits negative values in a range of seed amplitudes limited by the noise of the reconstructed data and by the non-unit efficiency of the system. Correspondingly, the experimental variances for the $x_{(\theta=0)}$ quadrature get smaller than those of the corresponding coherent state (also shown in the graph and independent of the seed intensity) as soon as the amplitude exceeds unity, and a maximum squeezing of about 15% is obtained for $|\alpha| = 1.85$.

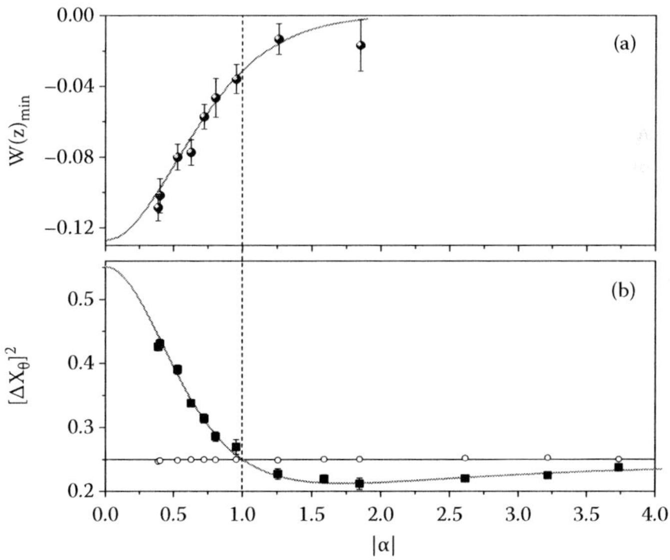

FIGURE 23.5
a) Minimum value of the reconstructed Wigner function and b) variance of the squeezed quadrature (filled squares) of the SPACS for different coherent state amplitudes. Solid lines are obtained from Eq. 23.11 and Eq. 23.10 with $\theta = 0$ and with a global efficiency set to $\eta = 0.6$. Also shown are the experimental data (empty circles) and the theoretical curve (horizontal line at 1/4) for the variance of the coherent state. The vertical line at $|\alpha| = 1$ sets the threshold for the squeezing appearance.

Conclusions

We have generated a new class of light states whose degree of non-classicality can be continuously tuned between the extreme situations of pure quantum states and almost classical ones. Such single-photon-added coherent states are particularly interesting from a fundamental point of view as they represent the result of the most elementary excitation of a classical light field and clearly show the passage from the spontaneous to the stimulated regimes of light emission.

The demonstrated possibility to follow their evolution so closely will certainly push the experimental research towards the investigation of other interesting and equally fundamental quantum processes.

Acknowledgments

This work has been performed as part of the "Spettroscopia laser e ottica quantistica" project of the Department of Physics of the University of Florence and was partially supported by the Italian Ministry of University and Scientific Research (MIUR) under FIRB Contract RBNE01KZ94.

References

1. A. I. Lvovsky, H. Hansen, T. Aichele, O. Benson, J. Mlynek, and S. Schiller, "Quantum state reconstruction of the single-photon Fock state," *Phys. Rev. Lett.* **87**, p. 050402, 2001.
2. A. Zavatta, S. Viciani, and M. Bellini, "Tomographic reconstruction of the single-photon Fock state by high-frequency homodyne detection," *Phys. Rev. A* **70**, p. 053821, 2004.
3. A. Zavatta, S. Viciani, and M. Bellini, "Quantum-to-classical transition with single-photon-added coherent states of light," *Science* **306**, p. 660, 2004.
4. A. Zavatta, M. Bellini, P. L. Ramazza, F. Marin, and F. T. Arecchi, "Time-domain analysis of quantum states of light: noise characterization and homodyne tomography," *J. Opt. Soc. Am. B* **19**, p. 1189, 2002.
5. G. S. Agarwal and K. Tara, "Nonclassical properties of states generated by the excitations on a coherent state," *Phys. Rev. A* **43**, p. 492, 1991.
6. Z. Y. Ou, "Parametric down-conversion with coherent pulse pumping and quantum interference between independent fields," *Quantum Semiclass. Opt.* **9**, p. 599, 1997.
7. T. Aichele, A. I. Lvovsky, and S. Schiller, "Optical mode characterization of single photons prepared by means of conditional measurements on a biphoton state," *Eur. Phys. J. D* **18**, p. 237, 2002.
8. M. Bellini, F. Marin, S. Viciani, A. Zavatta, and F. T. Arecchi, "Nonlocal pulse shaping with entangled photon pairs," *Phys. Rev. Lett.* **90**, p. 043602, 2003.

9. S. Viciani, A. Zavatta, and M. Bellini, "Nonlocal modulations on the temporal and spectral profiles of an entangled photon pair," *Phys. Rev. A* **69**, p. 053801, 2004.
10. H. P. Yuen and J. H. Shapiro in *Coherence and Quantum Optics IV*, L. Mandel and E. Wolf, eds., Plenum, (New York), 1978.
11. H. P. Yuen and V. W. S. Chan, "Noise in homodyne and heterodyne detection," *Opt. Lett.* **8**, p. 177, 1983.
12. G. L. Abbas, V. W. S. Chan, and T. K. Yee, "Local-oscillator excess-noise suppression for homodyne and heterodyne detection," *Opt. Lett.* **8**, p. 419, 1983.
13. K. Vogel and H. Risken, "Determination of quasiprobability distributions in terms of probability distributions for the rotated quadrature phase," *Phys. Rev. A* **40**, p. 2847, 1989.
14. S. Reynaud, A. Heidmann, E. Giacobino, and C. Fabre in *Progress in Optics*, E. Wolf, ed., **30**, p. 1, Elsevier, (Amsterdam), 1992.
15. U. Leonhardt, *Measuring the Quantum State of Light*, Cambridge University Press (Cambridge, UK), 1997.
16. G. M. D'Ariano in *Quantum Optics and Spectroscopy of Solids*, T. Hakioğlu and A. S. Shumovsky, eds., pp. 175–202, Kluwer (Dordrecht), 1997.

24

If Superposed Light Beams Do not Re-Distribute Their Energy in the Absence of Detectors (Material Dipoles), Can a Single Indivisible Photon Interfere?

Chandrasekhar Roychoudhuri

Photonics Laboratory, Physics Department, University of Connecticut, Storrs, Connecticut 06269, USA

CONTENTS

Abstract

The intention of this chapter is to underscore that to understand fundamentally new properties of light beams, we must first find the limits of semi classical model to explain optical interference phenomena. We claim that we have not yet reached that limit. Careful analysis of the processes behind detecting fringes indicate that the effect of superposition of multiple optical beams can become manifest only through the mediation of the detecting dipoles. Since the detectors are quantum mechanical, (i) the observed effects are different for different detectors for the same superposed light beams, and further, (ii) they are only capable of registering discrete number of "clicks", whose rate will vary with the incident intensity. A reduced rate of "clicks" at very low intensity does not prove that light consists of indivisible packets of energy. We have also experimentally demonstrated that (i) neither Fourier synthesis, nor, (ii) Fourier decomposition actually model the behavior of EM fields under all possible circumstances. Superposed light beams of different frequencies do not *synthesize* a new average optical frequency. A pure amplitude modulated pulse does not contain any of the mathematical, Fourier analyzed frequencies. The QED definition of photon being a Fourier mode in the vacuum, it necessarily becomes non-local. Since we have demonstrated that the Fourier theorem has various limitations in classical physics, its indiscriminate use in quantum mechanics should also be critically reviewed.

Key words: single photon interference; semi classical approach to interference; limitations of Fourier theorem; non-interference of light beams.

24.1 Introduction

It is not possible to provide a conclusive answer to a rather controversial question raised by the title of this chapter. Our approach would be to highlight the conceptual continuity and differences between the use of the principle of superposition (PS) in classical and the quantum physics while carefully looking at the detection processes behind recording of optical interference phenomenon. The apparent conceptual break down occurs when one attempts to visualize "single photon interference" while reducing the intensity from classically comfortable values to arbitrarily low value. We will make an attempt to make the enquiring minds aware that in spite of staggering successes of both the classical and the quantum optics, the true nature of light still may not yet be completely revealed to us [1; see also chapters 1–5], because we "see" light only through the "eyes" of the detectors.

Our starting assumption is that the universe is one continuum and the nature is undergoing incessant, creative and causal evolution from micron-size single living cells to the inanimate galaxies spanning over many light-years based on the same set of laws of forces. Such a universe cannot be artificially divided into classical and quantum worlds. The apparent division is a reflection of our current limitation in our ability to create a unified mathematical formulation supported by visualizable model (paradigm) for the actual, causal and local processes (all forces exert influence over a finite range). The principle of superposition (PS) is the strongest operational principle that is common for both classical and quantum physics. Interestingly, PS was formulated, developed and validated in classical physics before the birth of quantum physics. Yet, unlike classical physics, PS is not only an essential driving force behind quantum physics, but it also has a very different interpretation, almost to the level of mysticism, as can be appreciated from the prevailing interpretations like non-locality, non-causality, delayed choice, many worlds, teleportation, etc., to interpret interference and diffraction fringes at very low light levels and particle flux levels.

Our long-term goal is to revisit the detailed detection processes behind the basic but the simplest classical and quantum measurement experiments. All phenomena evolve through superposition of two or more, similar or different real entities of nature (and their force fields) followed by energy exchange and some transformation. The measured (observed) transformations are the reports given to us by one or more of these entities which are "colored" by their own uniquely different characteristics of interactions. And, none of these entities are known to us completely. We are always challenged to continuously develop and extend conceptual continuities between the various classical and quantum phenomena.

PS provides the commonality between all interactions, although not all transformational energy exchanges are quantized, or requires initiation through the intrinsic amplitude of undulation of the entities concerned. In this chapter, we will remain focused on the measurements of the effects of superposition of light beams since this provides the most important bridge between the classical and quantum PS. The classical world assumes light consists of spreading wave packets emitted by atoms and molecules that can shape and re-shape themselves as they propagate and evolve through diffraction and interference. While the quantum world assumes the photons to be discrete, independent, indivisible packets of energy those propagate as modes of the vacuum (cosmic medium). Thus we have a "clash of cultures" when the total intensity (flow of EM energy per unit time per unit area) in the interfering or diffracting beams is reduced equivalent to a single "click" in the detector at any particular moment.

We can safely assume that both the cultures accept that when a single atom or a molecule undergoes a single de-excitation (downward transition), it emits a photon, a packet of EM energy given by $\Delta E = h\nu$. In classical physics, it is a space and time finite wave packet that evolves and propagates following Huygens-Fresnel principle validated by classical theory of diffraction, including van Cittert-Zernike theorem that correctly models the enhancement of spatial coherence by diffraction (propagation) of light from non-laser sources [2]. The wave packet has a precise carrier frequency ν, which was heuristically prescribed by Planck's radiation law and later more systematically by quantum mechanics. However, in general, QED claims this wave packet to be simultaneously an indivisible packet of energy and a unique Fourier frequency mode of oscillation of the vacuum medium [1,3–5] and hence it can behave both as a local and a non-local entity depending upon the design of the experiments [6].

However, there is some form of tacit commonality between the classical and quantum worlds' assumptions as to how the energy is redistributed in the plane of recording of the interference or diffraction fringes. In classical physics, the tacit assumption is that the local field energy is redistributed due to the superposition of the fields themselves. In quantum mechanics, the explicit assumption is that the probability of the rate of arrival of the indivisible photons on the detector locations is dictated by the superposition equation determined by the entire instrument, inherently accepting interpretations like non-locality, delayed choice, etc. Both approaches have remained focused on interpreting the final mathematically predicted results, validated by the measurements, but ignoring the need to explore the actual processes (the real physics) behind detecting the EM energy. Our objective is to establish the fact that all measurements indicate that light beams, containing conveniently measurable energy, simply do not interfere by themselves to create fringes in the absence of detectors. Thus the various claims of single photon interference are fundamentally in doubt. However, the absorption of energy in the presence of EM fields in steps of discrete packets, ΔE (= $h\nu$), by any and all detectors, which are necessarily quantum mechanical, is not in question at all.

Publications in the mainstream literature [6] clearly imply that the issue of single photon interference is resolved, just as the definition of a photon is resolved. Let us first acknowledge that we never "see" light. What we observe or measure is what some transformation is experienced by a detector in the presence of light, which always constitute some quantum mechanical (QM) dipole (single or aggregate) in some form or another. In photo emissive devices, electrons are bound quantum mechanically and the released electrons are quantized particles and can never be fractional. In photo conducting devices, again discrete electrons are stimulated from the valence to the conduction band, which generate photo current under imposed potential difference. In photographic plates, the silver halide molecules in microscopic crystals, which are again quantum mechanical devices, are broken up and follow on chemical processing establishes silver atoms as discrete black spots. Thus, in the final analyses, any and all photo detection, whether at very low or at very high intensity, will always appear as summation of many discrete events, only the rate of accumulation will be different.

Such discreteness only validates that our model of atoms and molecules as quantum mechanical devices is correct. This does not un-ambiguously validate the existence of EM field packets emitted by atoms as indivisible particle-like. The quantum condition of energy absorption $\Delta E = h\nu$ only dictates that ΔE amount of energy can be absorbed from any and all locally available E-fields undulating (and stimulating the detecting dipole) at the desired frequency ν. Further, different quantum detectors have different quantum properties with very narrow or very broad frequency band passes of different central frequencies as in (i) fixed energy gaps defined by sharp energy levels for atoms in gaseous states, (ii) fixed but broad energy bands for photo conducting solid state detectors, or (iii) a fixed binding energy with allowed continuum as in photo induced ionizations, or molecular dissociations. This is why our retinal molecules or a silicon detector will report being in "dark" even when illuminated by γ-ray, x-ray or UV-photons, while suffering some damages. We claim that whether light exists only as indivisible and non-local states is not conclusively resolved by discrete "clicks" or "spots" that we observe at low light levels. We should be careful in separating the inherent properties of light from those of the detectors. We will also discuss the necessity of employing critical review in using the ever present Fourier theorem in optics underscoring pitfalls, as well as successes.

24.2 Does Light Really Interfere with Light as Implied by Fourier's Theorem and Maxwell's Wave Equation?

Maxwell's free space wave equation is given by:

$$\nabla^2 \bar{E} = (1/c^2)\partial^2 \bar{E}/\partial t^2 \tag{24.1}$$

A simple CW solution, neglecting the arbitrary phase factor, is $\exp[-i2\pi\nu t]$. Mathematically, any linear combination of this solution $\Sigma_n b_n \exp[-i2\pi\nu_n t]$ will also satisfy Maxwell's wave equation. Well before Maxwell, Fourier established a very useful theorem for handling a time finite signal by its transform in the frequency space using the well-known integral,

$$a(t) = \int \tilde{a}(f)\exp[-i2\pi ft]df \tag{24.2}$$

Notice the similarity between the summation (integral) between the Fourier theorem and the acceptability by the Maxwell's wave equation of the linear combination of its simple solutions. This congruency, as if, strengthens and validates the reality of the superposition of EM waves. Unfortunately, in the absence of any material medium and specifically, in the absence of detectors, well formed light beams pass through each other completely unperturbed. A well formed light beam can be defined as when the local diffraction effect is negligible. This is true when the spatial variation of the amplitude and phase on its wave front is much slower than the characteristic dimension of it wavelength.

Such slowly diffracting light beams do not operate on each other to redistribute each others' energy and/or frequencies, even when they physically cross through each other. But insertion of proper detector within the physical domain of superposition will record fringes as we do for holography and other interferometry. The bright and dark fringes represent the locations where the resultant electric vectors are in phase or out of phase. A dark fringe indicates that the detecting dipole cannot be stimulated to absorb energy from the fields as it is locally zero; it is not due to non-arrival of photons. If well formed light beams were to perturb each others energy distributions then, with light pouring in from trillions of stars from every directions, (i) the visual universe, instead of appearing steady, would have always been full of glittering speckles in space and time; (ii) the instrumental spectroscopy could not have discerned the Doppler shifts of individual star light crossed by trillions of other star light and predict the "expanding universe".

Terrestrially speaking, (iii) the wavelength domain multiplexed (WDM) communication, the back bone of our internet revolution, would not have worked; all the useful data would have evolved into random temporal, light beating pulses, and (iv) the Fourier transform spectroscopy would have never worked if light of different frequencies really interfered with each other on slow detector (we always drop the interference cross-terms between different frequencies). The effects of linear superposition of multiple light beams, supported by Fourier theorem and Maxwell's wave equation, becomes manifest only in the presence of interacting materials (dipoles).

Here we should underscore the difference between the two phenomena of diffraction and "interference" of light beams. In classical optics, light always propagates through diffraction process, given by Huygens-Fresnel (H-F) diffraction integral [2]. It has been successfully predicting all possible propagation of light from the evolution of spatial coherence from distant star light, to the formation of simple or complex cavity modes in lasers, and to the evolution of wave fronts in most recent and complex nano photonics wave guides. The H-F principle is mathematically congruent with Maxwell's wave equation since it accepts superposition of H-F secondary wavelets!

Whether emitted by thermal sources or by laser cavities, the atomic and molecular emissions evolve by diffraction toward an angularly sustainable beam with increasing spatial coherence. The near field diffraction clearly indicates spatial re-grouping potential of EM field energies belonging to the same E-vector frequency, which becomes evident as the diffraction pattern evolves into the angularly stable far field pattern.

The confusing issue of diffraction vs, superposition of independent light beams can be further appreciated from the classic double slit "interference pattern", which is routinely used to underscore the "strange wave-particle duality" of "single photon interference". This "interference pattern" has always been studied in the far field where the two superposed single-slit far-field patterns are of the form given by (sinx/x) function. People tend to focus on the periodic cosine fringe pattern produced on a detector due to the superposition of the two "sinc" beams, ignoring the two a-periodic but well formed sinc diffraction patterns, which again evolved from very complex and rapidly changing near field patterns.

24.3 Do EM Fields Synthesize New Composite Fields Under Simple Superposition?

We review [7] here a simple experiment that we have carried out by superposing two CW light beams carrying two distinctly different carrier frequencies separated by 2 GHz, symmetrically centered on one of the Rb-resonance lines. When the superposed beams are sent through an Rb-vapor tube, it did not show any resonance fluorescence, even though by simple trigonometry (according to two terms Fourier synthesis), we were supposed to get the matching resonance frequency (mean of the sum of the two superposed frequencies) [see also Fig. 24.1]:

$$\vec{a}_{total}(t) = \vec{a}_1 \cos 2\pi\nu_1 t + \vec{a}_1 \cos 2\pi\nu_2 t = 2\vec{a}_1 \cos 2\pi\frac{\nu_1 - \nu_2}{2}t . \cos 2\pi\frac{\nu_1 + \nu_2}{2}t \quad (24.3)$$

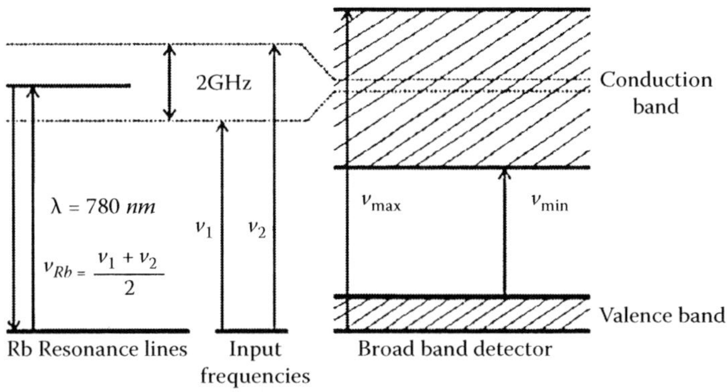

FIGURE 24.1
Comparison of energy diagrams of one pair of the Rb-resonance lines, one pair of input frequencies and one pair of valance-conduction band diagrams of a photo conductor. When the input frequencies of the superposed light beams are symmetrically above and below the Rb-excitation line, Rb-dipoles do not experience their presence in the linear domain and fails to respond to the superposed light beams. In contrast, the assembly of the dipole molecules of the photo conductors is quantum mechanically allowed to respond to both the frequencies. As they do so, their amplitude of excitation undulates at the difference frequency (not the mean of the sum), creating an undulatory rate of transfer of discrete number of electrons from the valence to the conduction band [7].

This revalidates that light beams do not operate on each other by themselves. However, when we sent this same superposed beam on to a high-speed photo conductor, we found the traditional AC current undulating at the difference (beat) frequency. The valence and the conduction bands of the photo detector are broad. This allows the detecting dipoles to simultaneously respond to all the allowed frequencies (here two), and the resultant current becomes:

$$I(t) = \left|\vec{d}e^{-i2\pi\nu_2 t} + \vec{d}e^{-i2\pi\nu_1 t}\right|^2 = 2d^2[1 + \cos 2\pi(\nu_2 - \nu_1)t] \tag{24.4}$$

Here $\vec{d}$ is the dipole undulation vector induced on the detecting dipoles by the $\bar{E}$-vector ($\vec{a}$). The detailed detecting process ("picture") in our view is that the undulating electric vector of the EM field induces the material dipoles to undulate with it. If the frequency matches with the quantum mechanically allowed transition frequency, then only there is absorption of energy. For the superposition effects to be manifest, the detecting dipoles must be collectively allowed to respond to all the light beams simultaneously. When the superposed light beams have multiple frequencies, the detecting dipoles must have broad quantum mechanical bands to be able to register the superposition effects [see Fig. 24.1].

If two superposed light beams are of orthogonal polarizations, the detectors cannot register the superposition effects. The dot product of orthogonal vectors is zero, whose "visual image" translation is that the same dipole (or, a collective set) cannot simultaneously carry out two independent and orthogonal undulations at the same instant in the linear regime of stimulation. Basically, the detecting dipoles respond to all the local E-vectors. If the E-vectors are orthogonal, then the dipoles respond to one or the other E-vector if they are embedded in isotropic medium. If the dipoles are embedded in a crystalline solid state, then the crystal axes dictate the allowed direction of dipole undulation.

There are important physical processes hidden behind the Eq. 24.4. The final energy transfer during a photo detecting process is correctly given by the square modulus of the *linear superposition* of all possible (quantum mechanically allowed) dipole undulations in *complex representation* as in Eq. 24.4. If we have a simple EM field represented by a real function, $\vec{a}\cos 2\pi\nu t$ the induced dipole undulation can be represented by $\vec{d}\cos 2\pi\nu t$. However, the measured detector current, for optical fields, is proportional to d^2 and not $d^2\cos^2 2\pi\nu t$. The complex representation hides a short time averaging process that we normally tend to ignore.

We are hypothesizing that this hidden time averaging process is physically real. The detecting dipole is actually undulating under the influence of all the E-vectors while the mutual quantum compatibility for energy exchange is being ascertained (the availability of necessary amount of energy ΔE, and the right stimulating frequency ν). This point can be further supported from the following arguments. If the two superposed field amplitudes for the above experiment (Fig. 24.1) are represented by real fields as in Eq. 24.3, one gets two unphysical frequencies, mean of the sum and the mean of difference frequencies. We have carried out systematic measurements with a very high resolution Fabry-Perot spectrometer in conjunction with very high speed detectors, scopes and electronics spectrum analyzers. We were not able to detect any of these two frequencies. They are not physically observable quantities. The two superposed light beam amplitudes did not interfere to synthesize new light field amplitude represented by the Eq. 24.3. They remained as two non-interacting, independent fields ($\vec{a}_1\cos 2\pi\nu_1 t;\vec{a}_1\cos 2\pi\nu_2 t$), albeit being collinearly superposed.

The summation sign in Eq. 24.3 does not represent a valid physical operation as these beams do not operate on (interact with) each other. However, in the presence of appropriate detector with broad excitation bands [Fig.24.1], the dipoles collectively attempt to respond to both the fields. When quantum mechanically allowed, they carry out the quantum compatibility sensing undulations simultaneously with both the fields and effectively sums their superposed effects while exchanging energy from both the fields. The result of Eq. 24.4 can be recovered using the Eq. 24.3 by time averaging the square of the superposed real dipole undulations induced by the two real fields:

$$I(t)=\frac{1}{T}\int_{-T/2}^{T/2} d_{total}^2(t)dt=\frac{d^2}{T}\int_{-T/2}^{T/2}[\cos 2\pi\nu_1 t+\cos 2\pi\nu_2]^2 dt \approx d^2[1+\cos 2\pi(\nu_1-\nu_2)t] \tag{24.5}$$

Thus, (i) the superposition effect to become manifest (measurable), multiple light beams must be present simultaneously both in space and in time, on the microscopic detecting dipoles. Further, (ii) the quantum rules (broad bands) of detecting dipoles must allow them to simultaneously respond to all the superposed frequencies; and (iii) there is embedded time averaging in the detection step. Explicit recognition of all these processes behind detecting photons does not support mysterious interpretations like delayed choice (superposition), teleportation, etc. The detection process requires and preserves the strict causality.

24.4 Do Amplitude-Modulated EM Fields Contain Fourier Analyzed Frequencies?

In the last section we demonstrated that the energy of light beams corresponding to different frequencies did not regroup as pulses on their own with a new average frequency. Fourier synthesis did not take place by simple physical superposition of light beams. In this section, we test the inverse process, the Fourier analysis—whether amplitude modulated light beams physically contains Fourier decomposed frequencies.

We tried a variety of high resolution spectrometric experiments, but the beat spectroscopy turned out to be the conceptually simplest [8–11]. We used two 1550 nm communication lasers. One laser had a fixed frequency, a DFB-type with about 20 MHz line width. The second laser was a tunable external cavity type with line width less than 100 KHz. The DFB laser was used both as a CW source and as an amplitude modulated source (by using an external, 10 GHz Mach-Zehnder modulator). The two laser beams were combined on to a very high speed, broad band (30 GHz) detector, connected parallel to a high speed scope and an electronic spectrum analyzer (ESA). The function of ESA is to present the oscillating currents it receives in terms of harmonics.

Among a wide variety of experiments on the basic theme, we are presenting two sets of data in Fig. 24.2a, b. For both the cases the optical frequencies of the two lasers were detuned from each other by about 15 GHz. For Fig. 24.2a, both the lasers are running CW, and for Fig. 24.2b, one of the lasers, the DFB, is undergoing AM at about 2.5 GHz [pseudo random super Gaussian (almost square) data pulses of width 0.4 ns]. When the two lasers are running CW, the beat spectrum is a narrow line located at 15 GHz as shown in Fig. 24.2a since the detector current is literally a sinusoid at this 15 GHz difference frequency [see Eq. 24.4]. When the DFB laser is amplitude modulated, the corresponding ESA display of the beat signal (Fig. 24.2b) is again very much like that for the CW case. No new E-vector frequencies have been generated by the external AM. But, since the ESA now receives the 15 GHz sinusoid with random duration of 0.4 ns square pulses, it represents these random square pulses of current by its Fourier transformed spectral intensity distribution, which is a sinc^2 function with its first zero at 2.5 GHz.

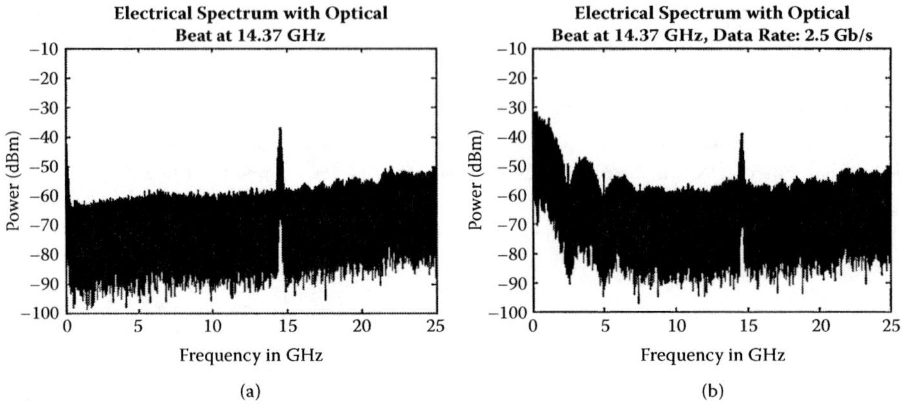

FIGURE 24.2
Output from an electronic spectrum analyzer (ESA) fed by the photo current from a high speed detector illuminated by the superposed light beams of two different frequencies. The left photo corresponds to two CW light beams separated by about 15 GHz, the beat frequency. The right photo corresponds to the external amplitude modulation of one of the lasers by 0.4 ns super Gaussian (square-like) pulses (2.5 GHz pseudo random data). The carrier frequency (beat) signal remains essentially unchanged, while the presence of AM is separately displayed as the Fourier transform of the square-like pulses, sinc^2-like harmonic distribution with the first zero close to 2.5 GHz location [10].

If the modulation truly generated new Fourier frequencies, the half-width of the beat frequency line would have become 2.5 GHz; instead it has remained almost the same (probably 20 MHz, not discernable in the data presented). Notice that the vertical scale is logarithmic and the half-width point (3 dB below the peak) for the beat signal line does not show any measurable change, especially compared to the first zero of the sinc^2 curve at 2.5 GHz. We must conclude that simple amplitude modulation does not generate new optical frequencies.

The Fourier frequencies for a square pulse are not present at the optical beat signal location. Thus, the traditionally accepted "time frequency bandwidth product", $\delta v \delta t \geq 1$, is not a fundamental limit of nature. We have validated that analytically [12, 13] and experimentally [10, 11]. One can recover the actual width of the carrier frequency content of a light pulse with ultra precision, limited only by the stability and intrinsic width of the CW reference signal. The width of $\delta\nu$ of the beat line in Fig. 24.2b is orders of magnitude narrower than demanded by the Fourier analyzed width, 2.5 GHz. The mathematical representation of the detector current is very similar to Eq. 24.4, but partially complicated by the fact that one of the superposed signals gets turned on and off intermittently; we are considering a single pulse for mathematical simplicity:

$$I(t) = \left| \vec{d}_{cw} e^{-i2\pi\nu_{cw}t} + \vec{d}_p e^{-i2\pi\nu_p t} \right|^2 = d_{cw}^2 + d_p^2 + 2\vec{d}_{cw}.\vec{d}_p \cos 2\pi(\nu_{cw} - \nu_p)^t \quad (24.6)$$

Here, $\vec{d}_{cw}, \vec{d}_p$ are the dipole undulations induced by the CW reference signal (v_{cw}) and the pulsed signal (v_p) respectively. When the superposed light beams are of parallel polarizations, the magnitude of the dipole undulations induced by the super Gaussian (square-like) light pulses can be expressed by Eq. 24.7, where m is an integer greater than 2 and τ is the pulse half width:

$$d_p(t) = \exp[-(t/2\tau)^{2m}] \quad (24.7)$$

The electrical signal of Eq. 24.6 is analyzed by an HP-ESA (#8593E). It is able to discern the harmonic undulation, $\cos 2\pi(v_{cw} - v_p)t$ as a sharp line whether it is CW or cut off randomly by $d_p(t)$. The ESA is designed with memory and software to store the pulsating currents and analyze them in terms of sinusoids. Note that due to continuous and pseudo random (data) presence of $d_p(t)$, its ESA representation is a continuous sinc^2-like function. If it were perfectly periodic, the ESA would have produced a periodic array of spikes under the sinc^2 envelope. We have recorded similar results when the input pulses were periodic.

The key significance of this experiment is that the Fourier decomposed frequencies of a pulse do not represent actual optical frequencies. We have directly demonstrated that a short optical pulse can carry its unique carrier frequency and is not burdened by the Fourier analyzed frequencies. Thus, when an excited atomic dipole spontaneously releases semi-classical "photon" as a discrete packet of energy ΔE in the vacuum (cosmic medium), the classical model of the evolution of the photon as a time finite EM wave packet out of it with a uniquely defined carrier frequency ν, is congruent with the QM postulate $\Delta E = h\nu$.

It is not necessary to define the photon as an indivisible, non-causal, non-local, Fourier frequency mode of the vacuum. However, we must rush to underscore that when the atoms and EM fields are confined inside a micro cavity by enforced boundary conditions, the situations are different from free space evolution of photons [14].

24.5 Discussion

The purpose of the chapter has been to raise rational doubt on the current paradigm that light propagates as indivisible particle-like entities while preserving its wave behavior, requiring explicit acknowledgement that interference effects have to be explained as a non-local phenomenon. We have argued, through the exploration of the detection processes behind detecting superposition ("interference") phenomenon that light beams really do not interfere with each other. Phenomenologically, indivisible single photons cannot give rise to interference effects, unless one assumes that the single photon interference (at extreme low light level) is a distinctly different phenomenon compared to when one has abundant light energy. It is the paradigm of indivisible-photon that is forcing us to introduce a host of non-causal hypotheses.

The problems have been further complicated by the assumption that the Fourier theorem, although an elegant and very successful mathematical tool in its own right, represents actual physical processes experienced by light fields (interference). However, we have experimentally demonstrated that neither Fourier synthesis, nor Fourier decomposition represent physical realities for light. The Fourier theorem is extensively used in modeling natural processes both in classical and quantum physics. Because of its extended limits of integration, it has the potential to bring in non-causality into the analytical processes that people have been aware of [15] since its inception by Fourier. In fact, the definition of "what is a physical spectrum?" has been an evolving debate for over a century, although, the prevailing view is that if the light is pulsed, the Fourier spectrum is the right representation [16–19]. But, this is probably the first time that we are claiming that superposing EM radiations of infinite extent, in the name of Fourier theorem, neglecting even causality violation, does not represent any physical reality. This is simply because light does not interfere with light. Thus, if the applicability of a mathematical theorem can be seriously questioned in one application, it should be critically reviewed for all other applications in physics that includes QED definition of a photon.

The uncertainty principle should be revisited [20] since its essential platform is the product of the half-widths of a pair of functions related by Fourier transform. These widths may not necessarily represent any physical reality. Diffraction fringe patterns are analytically given when the aperture function is known, and the de-convolution provides spatial super resolution [21]. For the classical time-frequency domain, we have shown analytically [22,23] that the corresponding Fourier band width product, is not a fundamental limit in classical spectrometry in determining the carrier frequency content in a pulse. The experiment of Fig. 24.2 above directly validates this assertion. Ref. 23 shows that the extra width of the final time integrated "spectral" fringe is due to "time diffraction" and spatial spread of the energy corresponding to the same carrier frequency. This extra, time-integrated fringe width is mathematically shown to be derivable as the convolution of the CW intensity impulse response with the Fourier (transformed) spectral intensity function of the time pulse. This coincidence may have lulled us to accept the Fourier spectrum of an amplitude pulse as real "spectrum" without a critical review. The mathematical equivalency comes by using Parseval's energy conservation theorem.

It is at the same time important to underscore at least two causally self consistent applications of the Fourier theorem in optics. The first one is in diffraction. When the light duration is sufficiently long (effectively CW), the far-filed diffraction pattern is correctly given by the spatial Fourier transform of the diffracting aperture [2]. However, this is based on the identification of the structural similarities between the Fourier transform integral and the Huygens-Fresnel space-space, diffraction integral (a recognized principle of physics) as it drops the quadratic curvatures of the Huygens' secondary wavelets in the far-field in favor of plane waves.

The Fourier transform conjugate variables are between two physical space coordinates (two spatial planes). Unlike for the time-frequency Fourier transform, no causality is violated in this space-space Fourier transform if the signal duration is much longer than the maximum relative phase delay between the center and the edge of the diffraction pattern. The second one is the Fourier transform spectroscopy. Again, this is based on the identification of the fringe intensity pattern as Fourier inverse transformable sinusoidal undulations (after removal of the "dc" bias from the recorded intensity) based on the correct physics hypothesis that on slow detector there are no superposition effects between different optical frequencies (no cross terms). The two conjugate variables are the actual carrier frequency and the interferometer delay time (not the real running time) constituting the recorded sinusoidal fringe function [24].

The strength of our strictly causal and local model behind recording fringes due to superposition of multiple light beams is that it is congruent with the semi-classical model [25–27]. So, the possibility of extending this model to explain the superposition of truly indivisible quantum mechanical particles should be encouraging. Accordingly, the author is developing conceptual continuity in interpreting such superposition effects to be published elsewhere.

Some readers may find the observations presented in this chapter not sufficiently convincing and insist on preserving the paradigms (i) that the EM energy packets emitted by atoms and molecules are simultaneously non-local and indivisible and (ii) that the indivisible single "photons" do interfere. For such readers, we would like to refer to the following references [28–31] where the authors argue against the single photon interference. The famous Bell's inequality does not strengthen the case for non-locality either [32,33]. Ref. 28 has experimentally demonstrated that both the photographic plate and the photo detectors become sub-linear in their detecting efficiencies at very low light levels, clearly raising serious doubt as to the validity of the claim behind "single photon interference" and that only a "single photon" at a time was present in the entire interferometer system. In fact, it is well known that a minimum of 3 to 4 photons equivalent energy exposure is needed before a photographic grain can be successfully developed as a black grain.

We hope that this chapter will inspire new developments in mathematical modeling of photons. Atoms and molecules being space and time finite, any form of energy released by them have also to be finite in space, time and energy value, if we simply accept conservation of energy, even if one is ignorant of the existence of atom quantization. It is no wonder that Newton insisted on "corpuscular" nature of light in its emission. The question is how does this space and time finite energy packet evolve and propagate in a causal fashion without the need to introduce any non-causal behavior?

Acknowledgments

This research has been supported in part by the Nippon Sheet Glass Corporation. Qing Peng drew the Figure 24.1 sketch.

References

1. C. Roychoudhuri and R. Roy, Optics and Photonics News, October 2003: *The Nature of Light: What is a Photon?* http://www.osa-opn.org/abstract.cfm?URI=OPN-14-10-49.
2. M. Born and E. Wolf, *Principles of Optics*, Ch. 8–10 (Cambridge University Press, 1999). L. Mandel and E. Wolf, *Optical Coherence and Quantum Optics*, Ch..4 and 7 (Cambridge University Press, 1995).
3. M. O. Scully and M. S. Zubairy, *Quantum Optics*, Cambridge University Press (1997).
4. W. P. Schleich, *Quantum Optics in Phase Space*, Wiley-VCH (2001).
5. R. Loudon, *The Quantum Theory of Light*, Oxford University Press (2000).
6. A. Zeilinger, *et al.*, *Nature* **433**, 230–238, 2005, "Happy centenary, Photon".
7. Dongik Lee and C. Roychoudhuri, *Optics Express* **11**(8), 944–51, (2003); http://www.opticsexpress.org/abstract.cfm?URI=OPEX-11-8-944.
8. C. Roychoudhuri, *Boletin Inst. Tonantzintla*, **2**(2), 101(1976).
9. C. Roychoudhuri, J. Siqueiros & E. Landgrave, in *Optics in Four Dimensions;* M. A. Machado Gama and L. M. Narducci, Eds. American Institute of Physics, 1981, p. 87.
10. C. Roychoudhuri and M. Tayahi; *Intern. J. of Microwave and Optics Tech.*, July 2006, "Spectral super-resolution by understanding superposition principle & detection processess", ID# IJMOT 2006-5-A6: http://www.ijmot.com/papers/papermain.asp.
11. C. Roychoudhuri, "Consequences of EM fields not operating on each other", 35th Annual Conference on Physics of Quantum Electronics, Snowbird, UT 2005.
12. C. Roychoudhuri; "Overcoming the resolution limit for pulsed light by deconvolving the instrumental pulse response function"; Photonics-India Conference, December, 2004; Paper P2.97.
13. C. Roychoudhuri, *SPIE Proc.* Vol. **5531**, pp. 450–461 (2004).
14. H. Walther, "Cavity QED" in *Encyclopedia of Modern Optics*, R. D. Guenther, D. G. Steel, and L. Bayvel, Eds., Elsevier, Oxford, 2004, pp. 218–223.
15. H. P. Hsu, *Fourier Analysis*, Simon & Schuster (1970).
16. N. Wiener, *Acta Math.* **55**, 117 (1930).
17. J. H. Eberly & K. Wodkiewicz, *JOSA* **67**(9), 1252–1261 (1977).
18. L. Mandel, *Am. J. Phys.* **42**(11), 840–846 (1974).
19. M. S. Gupta, *Am. J. Phys.*, **43** (12), 1087–1088 (1975).
20. C. Roychoudhuri, *Foundations of Physics* **8** (11/12), 845 (1978).
21. P. A. Jansson, Ed., *Deconvolution of Images and Spectra*, 2nd ed., Academic Press (1984).
22. C. Roychoudhuri et al. *Proc. SPIE* **5246**, 333–344, (2003).

23. C. Roychoudhuri, *SPIE Proc.* Vol. **5531**, 450–461 (2004).
24. A. A. Michelson, *Studies in Optics*, University of Chicago Press (1968).
25. E. T. Jaynes, "Is QED Necessary?" in *Proceedings of Second Rochester Conference on Coherence and Quantum Optics,* L. Mandel and E. Wolf, Eds., Plenum, New York, 1966, p. 21; Jaynes, E. T., and F. W. Cummings, *Proc. IEEE.* **51**, 89 (1063). http://bayes.wustl.edu/etj/node1.html#quantum.beats.
26. W. E. Lamb, *Appl. Phys.* **B60**, 77–84 (1995).
27. Willis E. Lamb, Jr. and Marlan O. Scully, in *Polarization, Matter and Radiation;* Presses Universitaires de France (1969), pp. 363–369.
28. E. Panarella, in *Quantum Uncertainties: Recent and Future Experiments and Interpretations,* W. M. Honig, D. W. Kraft, and E. Panarella, Eds., Plenum Press (1987).
29. S. Sulcs, *Foundation of Science* **8**, 365–391, 2003.
30. T. W. Marshall, "The zero-point field: no longer a ghost", (2002) http://arxiv.org/PS_cache/quant-ph/pdf/9712/9712050.pdf.
31. B. C. Gilbert and S. Sulcs, *Found. Phys.* **26**, 1401 (1996).
32. A. F. Kracklauer and N. A. Kracklauer, *Phys. Essays* **15** (2), 162 (2002); A. F. Kracklauer, J. Opt. B Quantum *Semiclas. Opt.* **6, S544,** (2004); A. F. Kracklauer, J. Opt. B Quantum *Semiclas. Opt.* **6, S544** (2004).
33. N. David Mermin, *Found. Phys.* **29**, 571 (1999).

25

What Processes Are Behind Energy Re-Direction and Re-Distribution in Interference and Diffraction?

Chandrasekhar Roychoudhuri
Physics Department, University of Connecticut, Storrs, Connecticut 06269, USA

CONTENTS

Abstract

The interpretation of the detection of very slow rate of photo counts in interference and diffraction experiments have given rise to the prevailing interpretation that photons interfere by themselves and they are indivisible, albeit non-local. The purpose of this chapter is to inspire the development of alternate models for the photons by underscoring that, in reality, light does not interfere with light. The effects of superposition, registered as interference fringes, can become manifest only when a suitable detector can respond simultaneously to all the superposed light beams separately arriving from all the paths (or, slits). It should be a strictly causal process. In fact, different detectors with different quantum properties, report different results while exposed to the same superposed fields. Interference and diffraction effects are always observed as fringes through the processes of re-distribution and/or re-direction of the measured energy of the superimposed fields.

Accordingly, we present a number of experiments, actual and conceptual, which highlight the contradictions built into the notion of non-locality in interference. A closer examination of these experiments can guide us to develop a conceptually congruent and causal model for both the evolution of photons and the interference (diffraction) effects by adapting to the classical diffraction theory. This theory has been correctly predicting the characteristics of light whether it is star light propagating through the inter galactic space, or nano tip generated light propagating through complex nano photonic waveguides.

Key words: non-interference of light beams; locality and causality of interference; single photon interference; semi classical approach to interference.

25.1 Introduction

Background. The predominant view [1, 2] of the nature of light is that it constitutes indivisible packets of electromagnetic energy $\Delta E = h\nu$, where ν is the Fourier monochromatic mode of oscillation of the vacuum field (cosmic medium that sustains everything). But this paradigm is forced to accept self contradictory interpretation that a photon is simultaneously indivisible and non-local (represented the by infinite extent Fourier monochromatic oscillation). This, of course, has nurtured a wide variety of non-causal interpretations for the "quantum world", not observed in the "classical world", like "delayed choice", "many worlds", "teleportation", etc. [1, 2]. The indivisibility interpretation comes from the combined "necessary and sufficient" assumption that discrete "clicks" registered by our quantum mechanical detectors constitutes the ultimate proof of indivisible photons. Even though semiclassical treatments have successfully demonstrated the analytical explanation of photoelectric effects based on classical electromagnetic fields and quantum detectors [3, 4, 5], including very low counts influenced by background fluctuations [6–8], the dominant opinion remains in favor of indivisible but non-local photons because of disagreements on interpreting micro cavity QED effects [9, 10] and coincidence counting originating from entangled "photon" producing sources [11].

Reality Ontology. The epistemological assumption behind this chapter is that we cannot have an unbridgeable "causal classical world" built out of the "non-causal quantum world". The macro universe, from inanimate sand particles, and animate single cells on the Earth to the stars and galaxies in space, all are evolving with a high degree of causality and yet they are sustained through incessant interactions between the molecules, atoms, elementary particles and "photons" of the micro universe. Our position is that we should be able to find some conceptual continuity (congruency) between the micro and the macro universes as they are one and the same.

We have been interpreting experimental observations, especially, the interference and diffraction fringes, without explicit attention to comprehending the actual, physical, processes behind our recording the discrete "clicks" and their accumulation as observable fringes. The thesis of this chapter is that a critical exploration of the processes behind the fringe formation as *local* redistribution and/or re-direction of the collective field energy (in interference and diffraction experiments) as a result of the detector response, could lead us to find the conceptual congruency between the classical and the quantum worlds.

Detector hypothesis. Our detectors that register the observable fringes are "classical" in size but quantum mechanical in action as they constitute many quantum mechanical devices (array of atoms or assembly of atoms). Each of these component QM detectors is highly localized within the macro detector and also within their own quantum mechanically defined average physical, nanometric size, while carrying out quantum mechanical undulations and other agitations due to ever present thermal and other variety of background fluctuations like zero point energy, dark energy, dark matter, etc. (that we do not yet fully comprehend). The spatially modulated field energies, constituting the superposition of actual multiple fields, must simultaneously stimulate these highly local and microscopic detector elements for them to undergo observable transformations. Then we can raise the following two questions.

First, (i) does the original incident field have the mysterious capacity to sense the distribution and orientation of all the parts of an interferometric or a diffractive apparatus and accordingly re-direct and/or re-distribute its energy spatially on the detector array? All natural entities, undulating fields or particles alike, must contain finite amount of energy and accordingly must have a finite space and time duration and finite velocity. It is the assumption that the indivisible and independent photons arrive only at the bright fringes, sensing the entire apparatus non-locally, gives rise to the non-causal possibilities like "delayed choice", "teleportation", etc.

Second, (ii) does the original field divide itself, as per classical wave model, into multiple field entities and after causal propagation and superposition, and collectively re-distribute their field energy to be recorded as orderly fringes by the detector? While this apparently causal model is centuries old, it has not succeeded in resolving the non-causal interpretations simply because we have been ignoring the blatant fact that light beams do not interfere with each other in the absence of materials (dipoles); actually they propagate through each other without influencing each other. So, we propose that the exploration and understanding the actual physical processes behind detection could restore the causality. *In the causal and real world, the principle of superposition can become manifest to us only through the material dipoles while they experience and respond to the simultaneous presence of multiple field entities on them.* This is why the debate on in-determinability of "which way through the interferometer the photon has traveled" has remained as a blind alley.

What is a Photon? Is it possible to find a single self consistent description of the processes behind fringe formation (i) whether the superposed fields contain energy equivalent to one or very many units of $\Delta E = h\nu$, and (ii) whether the units behave collectively or as independent and indivisible entities? We do assume that space and time finite atoms and molecules emit discrete packets of EM energy, the photons, as has been correctly formulated by QM. However, we are going to follow the success pattern of Huygens-Fresnel (HF) principle (with its mathematically self consistent modern improvements [12]). We are assuming that all photons start with their own quantum of energy $\Delta E = h\nu$ as a space and time finite wave packet, which is a mode of oscillation of the vacuum with the unique carrier frequency ν. The wave packets evolve and propagate following the H-F diffraction integral, allowing association with other wave packets of the same carrier frequency. Atoms and elementary particles with non-zero rest mass are localized entities and accordingly require a different model for interference and diffraction, which will be dealt with elsewhere.

The range of success of HF integral in conjunction with Maxwell's wave equation is staggering. It accurately predicts the transformation of diffraction patterns from very complex and rapidly changing near filed patterns to angularly stable and sustainable far filed patterns in free space when simple or most complex apertures rupture spatially coherent wave fronts. However, the evolution of diffraction patterns (fringes) are more complex and enigmatic compared to interference fringes due to superposed beams accompanied by negligible diffraction. Spatial near filed patterns are rather complex and evolve rapidly, as if the various ruptured wave fronts produced by a grating from a single coherent wave front propagate without modifying (or, operating) on each other. But, toward the far field, the evolution of the pattern becomes slow and eventually it assumes an angularly stable and sustainable pattern as if the diffracted wave fronts have collectively remolded themselves into an angularly stable and sustainable new wave packet (or, multiple wave packets as in grating orders) to minimize the energy loss as it propagates further.

HF principle correctly predicts the emergence of spatial coherence out of incoherent complex sources like discharge tubes in labs. or distant stars (van Cittert-Zernike theorem) [12]. It correctly derives the spatial eigen mode structures of most complex laser cavities where the wave front emerges through the collective diffraction of randomly emitted spontaneous photons (wave packets) that gets selectively amplified through stimulated emissions [13]. It is now correctly predicting the propagation modes of near and far field patterns due to nano photonic waveguides or nano photonics tips [14]. The study of diffraction phenomenon indicates that EM wave is a collective and cooperative phenomenon in the vacuum as the classical wave equation implies. Whether emitted spontaneously or stimulated from many individual atoms or disrupted by diffracting apertures, the multitudes of wave packets collectively and cooperatively evolve into an angularly sustainable but well defined wave form. Understanding this complex process of evolution of new filed pattern from a ruptured coherent field may lead us to better understand the evolution of complex photon wave fronts starting from multitudes of statistically random photons.

We believe that the diverse and complex variations in "photon counting statistics" reported in the literature [1], can be derived by semi classical theory if one allows the statistically finite number of wave-packet photons to diffract from the source through mutual superposition and derives the effective field on the detector at a finite distance from the source. In fact, we predict that since the very near field and the far field diffraction patterns are dramatically different, the corresponding temporal "photon counting" statistics will also vary for a typical "thermal" source due to collective evolution (propagation) of photons. However, the situations in micro cavity QED experiments are very different where the photons do not have the space and time to evolve as free space EM waves [9, 10]. It may be very instructive to find out all the situations where this model of classical wave packet for the photons clearly breaks down.

Contents. We present a series of actual and contrived experiments, both in interference and in diffraction, to underscore that the interference and diffraction fringes require signals to divide and travel through all the available paths and be present on the detector simultaneously. The detectors require actual superposition of multiple waves carrying multiple phase information on them to be able to report any "superposition effect". In fact, the observed effects of superposition for the same set of fields differ with different detectors [15] based on their differing quantum response properties, like energy gaps and energy levels and their widths.

25.2 Local Energy Re-Distribution Belonging to Different Laser Modes at High Resolution by Multiple Beam Superposition

This is a conceptually simple experiment that we have carried out [16] to demonstrate that it takes real physical superposition of a number beams with a periodic delay by replicating the original beam to be analyzed for its frequency content. In general, the energy separation (re-distribution) becomes apparent only when detected. This is to underscore the point that the principle

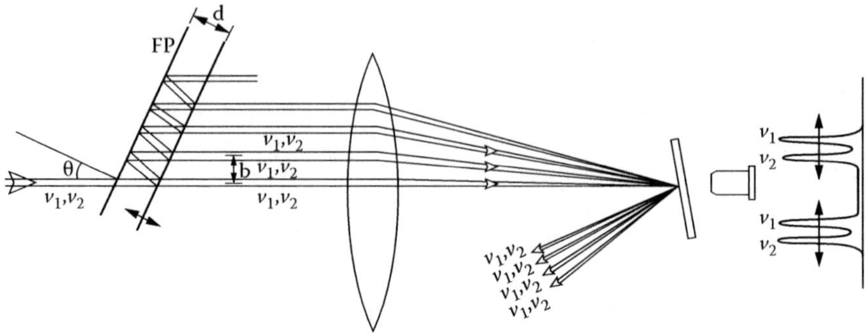

FIGURE 25.1
Experimental demonstration of non-interference of light beams in spite of crossing each other at the focal plane, while at the same time, delivering the classical spectrometric information when a detector is placed in the plane of superposition [16].

of superposition becomes manifest through the active participation of a detector. Figure 25.1 shows the schematic diagram of the experiment. A two-mode (two frequencies) He-Ne laser beam was directed at an angle toward a high resolution Fabry-Perot interferometer (FP) with plane parallel mirrors.

The beam was replicated into a set of spatially displaced beams, as if they were coming out of a grating. The beams were then physically superposed by a focusing lens on a tilted glass plate. When the transmitted beam was used to sharply re-image and enlarge the focal plane by a microscope objective, one could see the repeated fringes due to the two laser frequencies when the FP was set properly [16]. However, the reflected portion of the focused beam diverged out as spatially separated and independent beams, mirroring their origin. When we separately analyzed any one of these fanned out beams by another FP, they showed to contain both the laser mode frequencies. Conceptually there are no surprises if one things along the line of classical geometrical or physical optics. However, if the energy re-distribution were determined non-locally by the entire apparatus based on the paradigm of arrival and non-arrival of indivisible photons, then the re-emergence of all the focused beams as unperturbed, independent beams would not have been possible.

Only detectors can experience the apparent energy separation corresponding to the two different frequencies; the focused light beams did not redistribute their energy in the focal plane. The photons directed to travel through an FP at an angle experience it only as a pair of beam splitters, but not as a frequency sensitive resonator. Note also that if the incident light beam is a pulse shorter than the round trip delay between the mirrors, the train of pulses will never exist simultaneously at the focal plane and correspondingly there will be no interference (spectral) fringes [17], even though the single incident wave packet will be split into N-delayed packets, will travel through the N-distinct paths and cross the focal plane.

25.3 Locking Independent Laser Array by Near-Field Talbot Diffraction

More than 100 years ago Talbot discovered that an amplitude grating reproduces itself as a perfect image at a distance ($2D^2/\lambda$), where D is the grating periodicity [18]. We have exploited this near field diffraction phenomenon to phase lock (enforced collaborative, laser oscillation) on a periodic array of independent diode lasers [19]. The relevance of this experiment in the context of this chapter is again the causality and locality of the interference and diffraction phenomena. Figure 25.2 presents the summary of the effects and some results of mode control. A flat mirror at the half-Talbot distance can enforce spatial mode locking because the feedback into the independent laser element becomes maximum when their individual image falls back on themselves. This becomes possible only when their statistically random

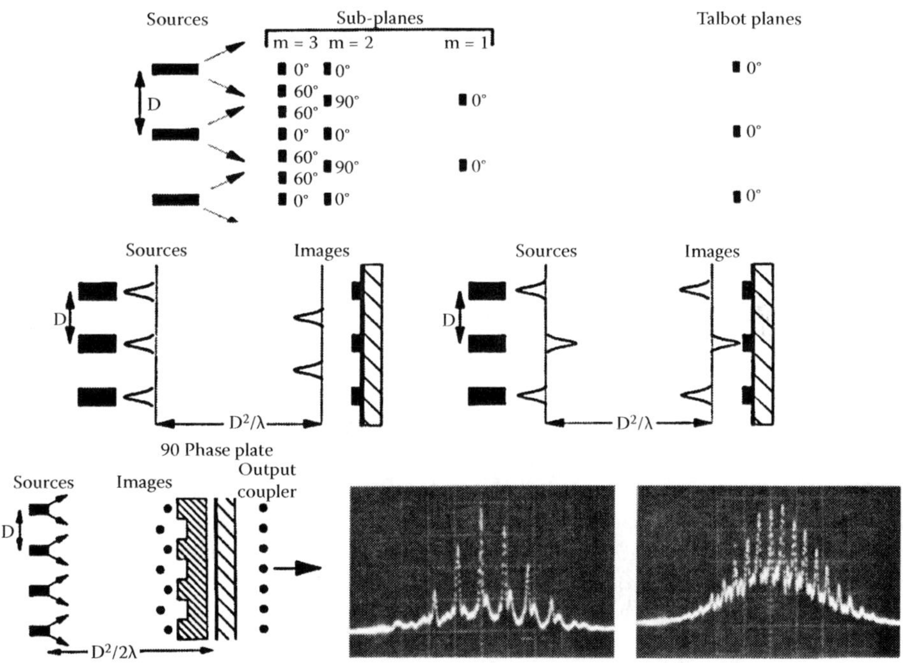

FIGURE 25.2
Exploitation of the complex periodicity in the near filed diffraction to phase lock an array of independent lasers. Top: Various Talbot images due to diffraction of a coherently illuminated grating or a coherent laser array. Middle left: Laser array oscillating in the fundamental mode and the corresponding Talbot image. Middle right: Laser array oscillating in the highest order spatial mode and the corresponding Talbot image. Bottom left: A phase filter in the sub-Talbot cavity to impose oscillation in the fundamental mode. Bottom middle: The far field of a 30-element diode array oscillating in the fundamental mode with spatial filter. Bottom right: Far field for the same array oscillating in the higher order mode without the spatial filter [19].

spontaneous emissions start accidentally to match up in their phase and their local superposed effects strengthen the stimulated emission.

The excited laser molecules act as the material detectors to make the superposition effects become manifest. If the mirror is displaced from the (D^2/λ) position, the "superposed" diffraction pattern does not match the phase condition on the laser array and they do not get phase locked. Further, if the Talbot mirror is removed, the "diffraction" pattern evolve as incoherent superposition of the N individual laser beams. The Talbot images in the near field are actually quite complex along with phase shifts and there are actually multiple Talbot planes, shown in Fig. 25.2 (top) [18]. We have exploited the second sub-image plane to discriminate against higher order spatial modes by inserting appropriate phase aperture. The model of photon as an indivisible but non-local vacuum oscillation (Fourier monochromatic mode) brings conceptual confusion as to how it can undergo such rapid spatial variations across such a large angular and spatial domain in the near field without invoking classical diffraction theory.

Lande's quantized scattering model [20] will require arbitrary changes in the quantization of the angles to different sets of values depending upon where one places the detector plane (various Talbot images or the far field). This implies precognition capability by the photons as to where the experimenter places the detector. In contrast, the model of photon as a classical, time-finite wave packet with a unique carrier frequency, propagating out as per classical diffraction and superposition theory while freely associating with other wave packets, gives us complex results along with a causally congruent picture.

25.4 Simple Two-Beam Holography Experiment

We know that when two light beams cross each other, they propagate out unperturbed by each other. Light does not interfere with light. But, when we place a holographic plate to record the fringes, we perturb the two wave fronts due to spatially differential absorption of energy during the time of exposure [21]. So, the two beams should suffer amplitude modulations, giving rise to some diffraction effects. To our "first order" accuracy we could not detect any diffraction during the live detection process of the fringes with a hologram at the beam intersection (Fig. 25.3 "Bottom-right"). We repeated the experiments from $1/30^{th}$ of a second to 180 seconds of exposure by reducing the beam intensity by a factor of 5.4×10^3 to keep the hologram density (after similar development conditions) the same for reconstruction purposes.

Unlike photo refractive and photo chromic materials, photographic plates do not experience any appreciable index change with low light exposure in the absence of development. From this stand point, the absence of diffraction by any of the crossing beam during live exposure may be acceptable. However, we are asking a more subtle question. How do the light beams propagate unperturbed even during the process when the beams are depositing spatially varying energy? *Is it because the finite time that it takes for the detecting dipoles to absorb energy provides the light beams the time to readjusts their original wave front integrity?*

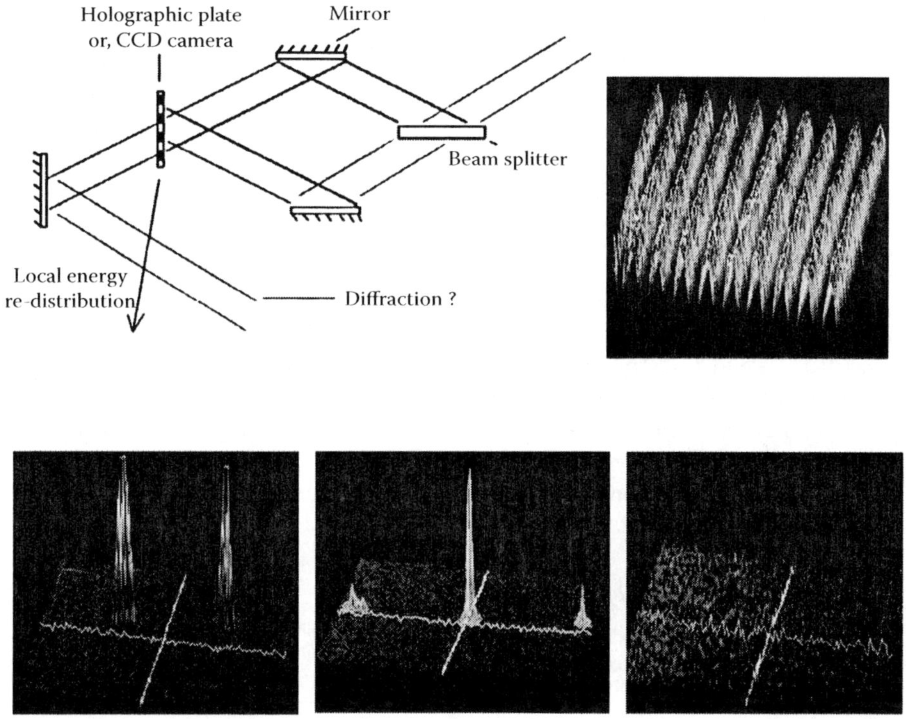

FIGURE 25.3
How is the energy re-distribution managed by intersecting light beams? Light does not interfere with light but detector array records local energy re-distribution when placed in the intersection. The transmitted beams appear to remain unperturbed even when active detection remains operational in the intersection! Top left: Schematic diagram for a two beam holographic set up. The angle of intersection of the two beams was only a few degrees; it is greatly exaggerated in the sketch. Top right: CCD camera record of the two beam fringes when the camera screen is symmetrically placed where the two beams intersect. It shows the local energy re-distribution Bottom left: The two intersecting beams are focused on the CCD camera as two separate spots beyond their point of crossing. They are un-influenced by each other even though they crossed each other earlier. Bottom middle: A hologram of the fringes (like top-right) was recorded and then replaced at the original intersection plane and illuminated by the beam-1 (assumed reference beam); the beam-2 was blocked. The CCD picture shows the focused spot for the directly transmitted reference beam-1 and two of the very weak, multiple diffracted orders from the hologram. [Notice apparent narrowing of the central light spot at low light level compared to the directly focused strong beams in bottom-left picture [see Ref. 6]. Bottom right: Failed attempt to record diffracted orders when a long term, live exposure for a hologram was going on at the intersection of the two beams. The direct beams were care fully blocked off the camera screen. A very small amount of scattered light from the blocked direct beam can be seen on the left, but no diffracted orders.

We know that the beams suffer diffraction when the developed hologram is placed back in the original position as it imposes stationary amplitude or phase perturbations on the beams. It appears to us that this set of experiments may be very important and that it is worth repeating it with a lot more care and quantitative measurements at every step that we did not carry out. The spatial intensity distribution is given by the square modulus of the two amplitudes. The unbalanced amplitudes in real experiments are indicated by a_1 and a_2. The variable phase delay along the spatial axis is given by τ.

$$I(\tau) = \left|a_1 e^{i2\pi\nu t} + a_2 e^{i2\pi\nu(t+\nu)}\right|^2 = A[1 + B\cos 2\pi\nu\tau] \tag{25.1}$$

The fringe visibility is degraded by the factor $B = 2a_1a_2/(a_1^2 + a_2^2)$; *where*, $A = (a_1^2 + a_2^2)$. The traditional complex representation, while very convenient to derive a quantity proportional to the absorption of light energy, hides a very important detection process, a short time that is required by the detecting dipole to respond to the field and carry out the absorption process. This can be appreciated by re-writing the field amplitudes in real terms as $\{a_1 \cos 2\pi v t;\ a_2 \cos 2\pi v\ (t + \tau)\}$. Then the recovery of the RHS of Eq. 25.1 will require accounting for a finite exposure time over a few cycles. The time integration is also physically justifiable because the EM field energy is always moving with the finite velocity, c.

$$I(\tau) \propto \int_0^T \{a_1 \cos 2\pi\nu t + a_2 \cos 2\pi v(t+\tau)\}^2\, dt \approx A[1 + B\cos\, 2\pi\nu\tau] \tag{25.2}$$

25.5 Double-Slit Fringes by Holographically Recording One Slit at a Time

From the view point of classical physics, the conceptual model behind this experiment is quite standard; this is classical holographic interferometry! However, the paradigm of indivisible photon will encounter some conceptual challenge here because the photons are now required to have a pre-cognition of the existence of an obstruction behind one of the two slits before propagating through. Accordingly, the indivisible photons must statistically distribute themselves on the Fraunhofer (spatial Fourier transform) plane in (sinc^2)-form rather than in the form of a product, ($\cos^2$)(sinc^2). This has to be true because we do not let the obstruction touch the double-slit screen, which allows the photon to cross through the slit to determine that it cannot travel all the way to the Fraunhofer plane!

The experimental results [22], shown in Fig. 25.4, were recorded in two different ways. (i) By double exposure holography, which records both the single slit patterns separately and then reconstructs the fringes holographically by keeping both the slits blocked. (ii) By real time holography, which first records only one of the two slits, say slit-2, and reconstructs the double-slit pattern by real physical superposition of the signal arriving directly from slit-1 with the holographically reconstructed signal for slit-2 (while the actual slit-2 remains blocked). In our experiment we have used a 10mW He-Ne laser (~3.10^{16} photons/second). If indivisible single photon beams really existed

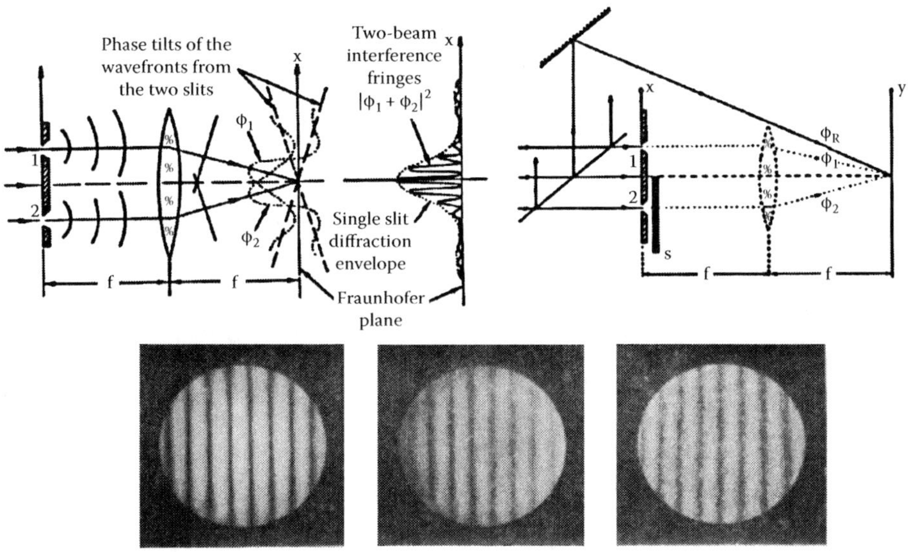

FIGURE 25.4
Signals from each one of the double slits can be recorded holographically one at a time and then the standard double-slit pattern can be reconstructed. Top left: Geometric drawing of the classical interpretation as to how the signals from each slit arrives on the far-filed as a sinc-enevelope (spatial Fourier transform, FT, of each slit) with a finite tilt to generate the standard cosine fringes. Top right: Holographic set up consistent with the sketch shown in top-left. Bottom left: Direct record of the traditional double-slit pattern recorded at the FT (far field) plane. Bottom middle: Holographic reconstruction of the double-slit pattern from a hologram that separately recorded the two single-slit patterns separately. The process is also known as double exposure holography. Bottom right: Re-generating the double-slit fringes by real-time holographic interferometry – the signal from the slit-1 arrives directly on the hologram and the signal from the slit-2 is reconstructed from the holographic record (actual slit-2 remains closed during this observation) [22].

(3.1×10^{-19} W), can one really record such holograms? Based on Panarella [6], a minimum of 3 to 4 photons equivalent of energy must be simultaneously present to trigger a single photographic grain to become "exposed" (chemically developable). Would these 3 or 4 photons go through a single slit as a single "clump" [6] and arrive at the right spot, or we need multiple photons arrive at the same spot but traveling through the two slits?

25.6 Slowly Moving Double-Slit Fringes with Small Doppler Shift on One Slit

This is a conceptual experiment [23] designed to challenge the assertion that any attempt to determine which slit the light passes through will always destroy the formation of the interference fringes. The apparatus of Fig. 25.5

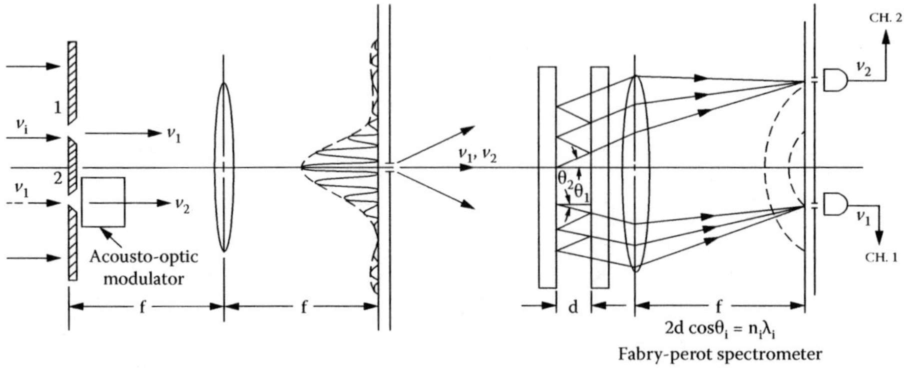

FIGURE 25.5
It is possible to determine that the double slit pattern is actually due to the superposition of two signals traveling separately through each slit and arriving at the detector plane with different relative phase delays. In the above experiment the identifier is a Doppler frequency shifter, ν_1 to ν_2. This makes the double-slit fringes at the Fraunhofer plane spatially move through a point at a rate of the beat frequency, $\delta\nu = (\nu_1 - \nu_2)$. A high resolution spectrometer behind the Fraunhofer plane can separately count the photons corresponding to each frequency and the counting will show precise coincidence. A spatial segment on the Fraunhofer plane can be intercepted by a fast Streak Camera to record the fringes, albeit moving spatially [23].

consists of several separate smaller experiments that we routinely carry out in the laboratory. We have a pinhole at the center of the plane that can record the standard double-slit Fraunhofer pattern to allow the collection of light for high resolution spectrometric analysis by a Fabry-Perot interferometer (FP) operating in the fringe mode. When the double-slit is illuminated by a coherent beam carrying a frequency ν_1, one can observe the stationary cosine fringes on the Fraunhofer plane and the detector, named Ch. 1, will register some count since the location has been chosen where the FP forms the fringe for frequency ν_1, with a constructive interference condition, $2d\cos\theta_1 = m\lambda_1$. If one switches the carrier frequency of the incident beam to be ν_2 [condition, $2d\cos\theta_2 = m\lambda_2$], then only the detector, Ch. 2, will register counts. Let us now illuminate the double-slit with a light beam of frequency ν_1, but insert an acousto optic modulator behind the slit-1 that generates a frequency ν_2. The cosine fringes on the Fraunhofer plane will now be given by:

$$I(\tau) = |e^{i2\pi\nu_1 t} + e^{i2\pi\nu_2(t+\tau)}|^2 = 2[1 + \cos 2\pi\{(\nu_1 - \nu_2)t - \nu_2\tau\}] \qquad (25.3)$$

These spatial fringes, as usual, defined by the spatial delay τ along the spatial axis [see Eq. 25.1], are temporally modulated by the difference frequency, $(\nu_1 - \nu_2)$, which is the traditional beat frequency. A pico second streak camera, covering a segment of the Fraunhofer plane can easily record these moving fringes as long as the beat frequency is in the domain of GHz or less. Now, if we pay attention to the detectors, Ch. 1 and Ch. 2, behind the FP spectrometer, we should be able to identify the ν_2-photons as those coming through the slit-2 after undergoing Doppler shift by the AOM and the ν_1-photons coming through the slit-1.

This is not a "Gedanken" experiment. This is an experiment that does not challenge the current technology at all. Does it resolve the paradigm of "single-photon interference" unambiguously? No, but the purpose of this chapter is to underscore that interference is always the result of real physical superposition of more than one signal on a quantum detector carrying more than one phase information (traveling through more than one path). "Which way" can be determined without destroying the fringes, if we use a fast enough detector.

25.7 Spatial Localization of Mach-Zehnder Fringes Using Polarization

This experiment, actually carried out in our laboratory, exploits the quantum properties of the detectors that the same dipole cannot execute two orthogonal undulations at the same moment in the linear domain. Since light does not interfere with light, the absence of fringes (local re-distribution in detected energy) due to the superposition of orthogonally polarized light beams, has to be atributed to the intrinsic properties of the detectors, not that of light. Fig. 25.6 gives the schematic diagram of the Mach-Zehnder interferometer (MZ), the recorded fringes and the schematic representation of the presence and absence of spatial fringes over the screen.

Good visibility fringes are recorded when the state of linear polarization is deliberately set to be parallel. Then, turning the two parallel polarizers in the two arms of the MZ by 45° in the opposite directions, the fringes are completely destroyed. But, insertion of a linear polarizer, exactly bisecting the 90° restores the interference fringes. To underscore the locality of interference (detectors carry out the superposition process), we deliberately made the fringe restoring polarizer physically smaller than the total beam size.

Only behind the polarizer the two transmitted beams are now polarized parallel and the detecting dipoles now can oscillate either strongly (bright fringes) wherever the superposed two E-vectors are in phase, or they do not oscillate (dark fringes), wherever the superposed two E-vectors are out of phase. Outside the polarizer on the detector screen, the dipoles can respond to either one of the E-vectors, not to both, irrespective of their phases; accordingly, the energy absorption is uniform without modulation. Mathematically, this is traditionally taken care of by the vector product of the dipole undulations:

$$I(\tau) = |\vec{d}e^{i2\pi\nu t} + \vec{d}e^{i2\pi\nu(t+\tau)}|^2 = 2[d^2 + \vec{d}\cdot\vec{d}\cos 2\pi\nu\tau] \tag{25.4}$$

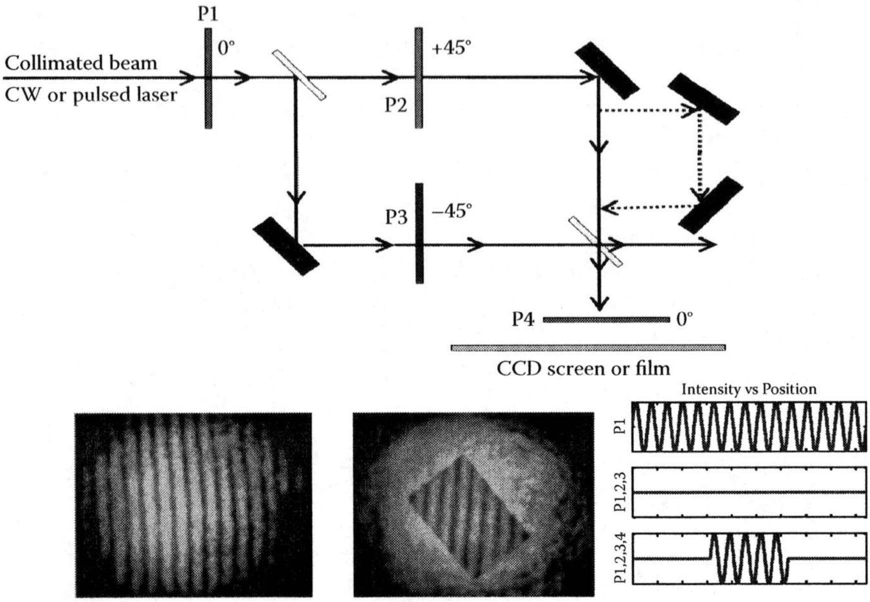

FIGURE 25.6
Mach-Zehnder (MZ) fringes are underscored using a small piece of Polaroid in front of the detector screen when the two superposed MZ beams are deliberately made orthogonally polarized. Top: MZ interferometer with four polarizers to assure proper manipulation of the state of polarization while keeping the amplitudes of the two beams very closely equal. Bottom left: The two states of polarizations are parallel in the two MZ arms. Bottom middle: The two states of polarizations are orthogonal to each other in the two MZ arms indicating complete loss of fringe effect, except in the middle where a linear polarizer is placed right on the detector plane bisecting the two orthogonal directions. Bottom right: Three different depictions of the intensity record on the detector plane. The top curve describes the situation shown at bottom left. The straight middle curve depicts the situation for the bottom middle figure outside the Polaroid. Its bottom curve indicates the re-appearance of the fringes just behind the small Polaroid (bottom middle).

Here $\vec{d}$ is the electric field induced dipole vector. The interference cross term vanishes when the two orthogonal fields try to stimulate the same detecting dipole at the same instant.

25.8 Spatial and Temporal Localization of Mach-Zehnder Fringes by Superposing Train of Translated Pulses with Separate Beam Diameters

The purpose of this experiment is to raise further doubts on the concept of non-locality of photons when one can easily confine the energy of electromagnetic fields, both in space and in time, simply by using optical components and modulators. This is another experiment not yet carried out but quite feasible with the standard off-the-shelf technologies. Consider the MZ of Fig. 25.6 ("Top") illuminated by a train of square pulses, derived from a stabilized CW laser by a high speed amplitude modulator. The pulses can be combined at the output of the MZ with variable temporal delay (i) either to exactly match the simultaneous temporal superposition of the pulses from the two arms on the detector, (ii) or, to completely mismatch their time of arrival on the detector.

When the pulses are time synchronous on the detector, one can record perfect fringes with simple slow detectors like photographic plate that integrates the signal over the entire period of exposure. Remember that due to delay, this interference is due to simultaneous presence of different pulses, and hence due to superposition of different time delayed photons. When the pulses are exactly asynchronous (never simultaneously present together on the detector), there will be uniform intensity record but no fringes. The cross term between the two amplitudes is absent because the detector dipoles could not experience the simultaneous stimulation by the two amplitudes at the same time. This point also underscores again that the effect of superposition becomes manifest only through the participation of the detector dipoles.

If one now drastically reduces the photo count by reducing the input beam energy, the appearance of the fringes will require long time integration. Does this classic "click-by-click" integration to build up the fringe pattern imply the indivisible, non-local photons could anticipate the arrangement of the entire apparatus to arrive at the right location of the potential fringe? This cannot be right because now the photons in the time domain have been confined within the pulse width and one can validate that the "clicks" can be registered only within this allowed periodic time intervals. Further, one can choose an interferometer many nano seconds long while the photons can be kept confined within the pulse width of a few pico seconds. The implication is that the paradigm of "non-local photon" is self-contradictory.

If the beam size in one of the two MZ arms is telescoped down to a smaller size than the other one, the fringes will be visible only over the smaller beam size; the out side will register energy without fringes. This spatial confinement is some what similar to the experiment of Fig. 25.6 where the fringes were restored just behind a small polarization parallelizing element.

If the MZ beams are collimated and are of exactly the same amplitude and physical shape, and further, if they are superposed on the final beam splitter surface at an angle such that they create perfect co-linearity between the transmitted beam from one beam with the reflected counter part of the other beam, then the total energy contained in both the beams will be re-directed only in one of the two allowed directions, based on the relative phase conditions. Again, this energy re-direction can take place only through the mediation of the dipoles on the surface of the beam splitters and the derivation was done more than a century ago using Lorentzian dipole model and Maxwell's equations, without the advantage of quantum mechanics.

Summary

We do not see light without the mediation of other materials. Light beams do not interfere with each other without the mediation of interacting materials (detectors). It is logically inconsistent that we should be able to produce a new phenomenon of interference between light beams without the mediation of detecting materials simply by reducing the intensities to arbitrary low values. So, the paradigm of non-local and indivisible single photon producing interference effect should be carefully revisited [3–8]. Even the use of Bell's inequality to justify non-locality has been logically questioned [24; see also Chapter 6].

We presented a number of actual and potential experiments to underscore that the effects of superposition of light beams can become observable only when some appropriate detector is capable of simultaneously responding to all the superposed fields arriving through all the allowed paths. All the fields must also be physically present simultaneously on the detector (both in space and in time) so the detector has the causal opportunity to act on all of them (or be simultaneously influenced by all of them) and register the effect of superposition. Photons definitely contain a sharply defined quantum of energy $\Delta E = h\nu$ at their birth. But, how do they evolve as they propagate? It is worth modeling their evolution (propagation) as classical wave packets following the classical diffraction theory that allows them to evolve collaboratively (superposition principle) into new wave packets by sharing energies in space and time such that their energy loss by diffraction is minimized in their long journey! Without first finding validated failure of classical, causal diffraction theory, it is premature to accept a non-causal model for photons that is simultaneously indivisible and non-local.

Acknowledgments

This research was supported in part by the Nippon Sheet Glass Corporation. Kenneth Bernier carried out the experiment depicted in Figure 25.3. Aristotle Parasko carried out the experiment depicted in Figure 25.6 and Qing Peng drew the sketch.

References

1. A. Zeilinger, *et al.*, *Nature* **433**, 230–238, 2005.
2. C. Roychoudhuri and R. Roy, Optics and Photonics News, October 2003; *The Nature of Light: What is a Photon?* http://www.osa-opn.org/abstract.cfm?URI=OPN-14-10-49.

3. W. E. Lamb, *Appl. Phys.* **B60**, p. 77–84 (1995).
4. W. E. Lamb, Jr. and M. O. Scully, in *Polarization, matter and radiation*; Presses Universitaires de France (1969), pp. 363–369.
5. E. T. Jaynes, "Is QED Necessary?" in *Proceedings of the Second Rochester Conference on Coherence and Quantum Optics*, L. Mandel and E. Wolf, Eds., Plenum, New York, 1966, p. 21; E. T. Jaynes, "Clearing up mysteries: the original goals", in *Maximum Entropy and Bayesian Methods*, J. Skilling, Ed., Kluwer, Dordrecht (1989), pp. 53–71; E. T. Jaynes and F. W. Cummings, Proc. IEEE. **51**, 89 (1063). http://bayes.wustl.edu/etj/node1.html#quantum.beats.
6. E. Panarella, in *Quantum Uncertainties: Recent and Future Experiments and Interpretations*, W. M. Honig, D. W. Kraft, and E. Panarella, Eds., Plenum Press (1987), p. 105.
7. S. Sulcs, *Foundation of Science* **8**, 365–391, 2003.
8. T. W. Marshall and E. Santos, **18**, 185 (1988); *Recent Res. Devel, Opt.* **2**, 683 (2002).
9. H. J. Kimble, *Phil. Trans. R. Soc. Lond.* **A355**, 2327–2342 (1977).
10. H. Walther, in *Encyclopedia of Modern Optics*; R. D. Guenther, D. G. Steel, and L. Bayvel, Eds., Elsevier, Oxford (2004) pp. 218–223.
11. B. C. Gilbert and S. Sulcs, *J. Opt. B: Quantum Semiclass. Opt.* **3**, 268–274(2001).
12. M. Born and E. Wolf, *Principles of Optics*, Cambridge University Press (1999), Ch. 8–10; L. Mandel and E. Wolf, *Optical Coherence and Quantum Optics*, Ch. 4 and 7, Cambridge University Press (1995).
13. A. E. Siegman, *Lasers*, University Science Books (1986).
14. P. N. Prasad, *Nanophotonics*, Wiley Interscience (2004).
15. D. I. Lee and C. Roychoudhuri, *Optics Express* **11**(8), 944–51, (2003). http://www.opticsexpress.org/abstract.cfm?URI=OPEX-11-8-944.
16. C. Roychoudhuri; *Am. J. Phys.* 43 (12), 1054 (1975); Bol. Inst. Tonant. 2 (2), 101 (1976).
17. C. Roychoudhuri; *J. Opt. Soc. Am.;* **65** (12), 1418 (1976); *SPIE Proc. Vol.* **5531**, 450–461 (2004).
18. M. V. Berry and S. Klein, *J. of Mod. Opt.*, **43**, 2139–2164 (1996).
19. X. D'Amato, E. T. Siebart, and C. Roychoudhuri; *Appl. Phys. Lets.*, **55** (9), 816-818 (1989); *SPIE* **1043**, (1989).
20. A. Landé, *Am. J. Phys.* **33**, 123–127, 1965; **34**, 1160–1163, 1966; **37**, 541–548, 1969.
21. C. Roychoudhuri, ETOP Conference, 2003, http://spie.org/etop/ETOP2003_EMIIO.pdf.
22. C. Roychoudhuri, R. Machorro, and M. Cervantes, *Bol. Inst. Tonant.* 2 (1), 55 (1976).
23. C. Roychoudhuri, invited talk, Einstein Centenary Celebration, Mexican Physical Society, 1979.
24. A. F. Kracklauer and N. A. Kracklauer, *Physics Essays* **15** (2), 162 (2002); A. F. Kracklauer, SPIE Proceeding **5866**, Paper 14 (2005).

26

Do We Count Indivisible Photons or Discrete Quantum Events Experienced by Detectors?

Chandrasekhar Roychoudhuri
University of Connecticut, Department of Physics, Storrs, CT 06269, and Femto Macro Continuum, 7 Fieldstone Dr., Storrs, CT, 06268
chandra@phys.uconn.edu; croychoudhri@earthlink.net

Negussie Tirfessa
Manchester Community College, Manchester, CT 06045
NTirfessa@mcc.commnet.edu

CONTENTS

Abstract

As low light detection technologies are advancing, novel experiments like single molecule spectroscopy, quantum computation, quantum encryption are proliferating. Quantum mechanical detectors can produce only discrete "clicks" at different rates based on the propagating field energy flux through them, irrespective of whether the photons are divisible or indivisible packets of energy. This is because electrons are quantized elementary particles and they are always bound in quantized energy levels in different quantum systems. Highly successful quantum formalism is not capable of providing the microscopic picture of the processes undergoing during QM interactions; that is left to human imaginations allowing for sustained controversies and misinterpretations. This chapter underscores the paradoxes that arise with the assumption that photons are indivisible elementary particles based on the obvious but generally ignored fact that EM fields do not operate on (interfere with) each other. Then we propose that atomic or molecular emissions emerge and propagate out as space and time finite classical wave packets. We also suggest experiments to validate that the amplitude of a photon wave packet can be split and combined by classical optical components using the specific example of an N-slit grating.

26.1 Introduction

The current scientific culture accepts that light energy constitutes discrete indivisible packets of energy, we call photons. The concept is supported by underscoring that in all photoelectric emission experiments only an integral number of electrons are emitted. But electrons being quantized themselves and always bound to quantized energy levels, discrete photoelectron emission does not establish beyond doubt that the EM field energy constitutes only indivisible packets of energy. Let us briefly review the origin of the quantized photon concept. A little over a century ago in 1903 Planck introduced the concept that light energy is emitted and absorbed by atoms and molecules with discrete quantized amount of energy $h\nu$ and a unique carrier frequency ν. His idea was to correctly map the measured energy distribution of frequency-continuous blackbody radiation. His proposal also easily accommodated the measured discrete frequency spectrum of many gas-discharge emissions, both terrestrial and cosmic, given by already known Rydberg formula. But Planck never accepted that the photons themselves, containing quantized energy at emission, were indivisible packets as they propagate out. Einstein proposed in 1905 that the photons might behave like indivisible packets of energy to explain the contemporary photoelectric emission experiments. However, he was strongly doubtful in the later part of his life whether he understood what a photon is [see Chapter 1]. Because of such prevailing doubts, we took the effort to publish the reference-1 that brings together the views of five global experts in quantum optics. Recently Goulielmakis et al. [2, or Chapter 27] has published a paper describing the successful direct measurement of the sinusoidal undulation of the electric field strength of a carefully generated laser pulse with Gaussian-like envelope containing barely five cycles of light. If this pulse consisted of indivisible photons, then the electric vectors of the photons in the pulse were marching in remarkable unison to each other mimicking Maxwell's classical description of an EM pulse. Since laser pulses are manipulatable by various established techniques, one can conclude that the photons can have flexible temporal amplitude envelopes. Then we face the contradiction that a photon with a uniquely defined frequency ν at the moment of emission can have different temporal envelopes as it propagates through different optical systems that manipulates the pulse shapes. This would conflict with the time-frequency Fourier theorem that customarily dictates what the spectrum of a time-finite signal should be. Lamb, whose work gave credence to the quantum electrodynamics, also has shown consistent critical views against associating a discrete photon with the emission of a discrete photo electron [3, 4]. Further, Panarella [5] has experimentally demonstrated that a minimum of four photon equivalent energy is required to detect discernable diffraction pattern at very low light levels. This clearly raises doubt regarding one-to-one correspondence for photoelectron emission. Comprehensive classical and quantum treatments of photo detection processes are given by Mandel and Wolf [6].

This chapter underscores the reasons for holding healthy doubts against the concept of photon as an indivisible elementary particle. *We propose that photons are space and time finite classical wave packets that propagate out from light emitting atoms and molecules following Huygens-Fresnel principle.* Our key logical platform derives form the commonsense fact, neglected in the books and literature that electromagnetic fields do not interfere with or operate on each other. Well formed light beams cross through each other without redistributing their spatial or temporal energy distributions. *The effects of superposition of EM fields become manifest when the right detector molecule, allowed by QM rules, is able to respond to all the fields superposed on it, there by summing all the filed induced effects and absorbing proportionate amount of energy.* QM formalism does not restrict simultaneous energy absorption from multiple sources. In fact, that is what the prescription given by the Superposition Principle. We have spent a considerable amount of time looking at the various aspects of optical phenomena where two or more optical beams are simultaneously superposed, but the superposed EM fields do not interfere [7–17].

We discuss first Einstein's photoelectric equation to emphasize the role played by detectors (atoms and molecules). We present the semi classical description of the photo detection process. After that we give some examples of paradoxes if we use the notion that light beams interfere with each other by themselves. Next we present results and implications of an important experimental observation made by Panarella [5] using low level light. In the next section we discuss our photon wave packet model (rapidly rising exponential pulse envelope amplitude) and compare with a pure exponential model. The finite time and finite energy associated with photo induced transitions is then discussed. Finally, we discuss the implications of our divisible photon model.

26.2 Einstein's Photoelectric Equation

Our position is that Einstein's photoelectric equation does not establish photons as indivisible packets of energy beyond any doubt. Since electrons are quantized elementary particles, they can be detected only as indivisible particles. Also electron transition (binding) energy is always quantized to a characteristic value $\Delta E = h\nu$ in all quantum systems. A particular quantum system must first undulate like a dipole at a frequency ν while holding the electron before it can absorb energy ΔE and release the electron. Einstein's 1905 paper on photoelectric effect reflects the experimental observations of Hertz (1887) and others after him. In all these early experiments electrons were released free from metal plates and measured as a current through a collection plate whose voltage was manipulated to measure the kinetic energy of the free electrons. Einstein correctly formulated the observed results as if a photon carries a packet of energy $h\nu$ which is expended to provide the binding energy of the electron in the metal (work function) and the rest is used by the electron as its kinetic energy (KE) as a free particle. This is a bound-free transition:

$$h\nu = \text{Work function} + \text{Electron KE} \tag{26.1}$$

In contrast, electrons in modern photo detectors undergo bound-bound transition (Fig. 26.1). These detectors, including "single photon" counters, are essentially semiconductor p-n junction devices where electrons experience quantum mechanical (QM) level transition from valance to the conduction band after absorbing energy from an incident EM field. The conduction band electrons are then measured as a photoelectric current by applying external voltage across the p-n junction. In this bound-bound QM transition kinetic

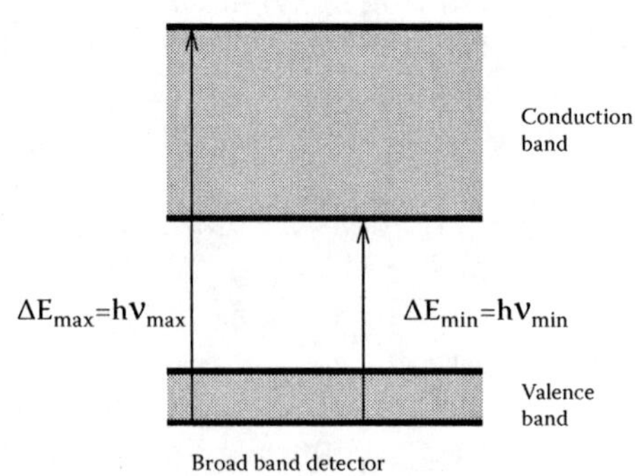

FIGURE 26.1
A photon with a higher energy than $h\nu_{max}$ will not transfer an electron to the conduction band. Unlike Einstein's photoelectric equation, higher frequency (energy) "photon" does not get counted.

energy does not play any explicit role. The transition can take place as long as the incident EM field frequency is such that the equivalent photon energy $h\nu$ is bounded by:

$$(\Delta E_{max}=)h\nu_{min} \leq h\nu \leq h\nu_{max}(= \Delta E_{max}) \tag{26.2}$$

A photon with higher frequency than $h\nu_{max}$ will not help transfer an electron to the conduction band. Such EM radiation will not be detected by the photo detector. A silicon detector can be damaged by intense x-rays, but as a device it will keep on reporting that it is in "dark".

The physical process behind Einstein's photoelectric emission in free space is very different from photo induced photoconduction inside semiconductors (p-n, p-i-n, APD, etc.). In the first case, electrons are stimulated to acquire kinetic energy from the field and then use a portion of that energy to overcome the binding energy of the metal; the rest of the kinetic energy remains measurable externally. In the second case, electrons undergo pure band-to-band QM transition without acquiring any freely available kinetic energy. In fact, avalanche photo diodes (APD) have been constructed where one applies voltage gradient across the detector to provide extra kinetic energy to the conduction electron such that it can generate more charges via collision to provide photoconductive gain within the same structure [18]. Let us carefully recapitulate: (i) Electrons are quantized, (ii) their binding energies within the material are quantized and (iii) their release or QM level change is always stimulated by dipole-like stimulations requiring unique frequency ν of the EM fields (relations 1 and 2). Thus, photoelectric emission or photoconduction current will always consist of discrete number of electrons requiring trigger by unique frequency of the EM field. Accordingly, we cannot unambiguously claim that propagating EM field energy definitely consists of discrete, indivisible packets. Quantized energy exchange behavior $h\nu$ and their dipolar behavior with characteristic frequency ν may be sufficient to explain relations 1 and 2 without quantizing the EM field itself [3,4,19].

26.3 Semiclassical Model Adequately Explains Photo Induced Transitions

It is well recognized that for most of the normal photoelectric detection, the semiclassical model (without quantization of the EM field) is adequate (3, 4, 19). Here we will underscore the key processes behind photo induced transition that are obvious in the semiclassical model. Any EM field incident on a material body will attempt to induce dipolar undulation in the constituent atoms and molecules. The total polarization $\vec{P}(t)$ is the sum of linear polarizability χ_1 and all the non-linear polarizability χ_n (n > 1), which are intrinsic properties of the medium dictated by the quantum properties of the constituents.

$$\vec{P}(t) = \chi_1 \vec{E}(t) + \chi_2 \vec{E}^2(t) + \chi_3 \vec{E}^3(t) + \cdots \tag{26.3}$$

where $\vec{E}(t) = \vec{a}(t)\exp[2\pi \nu t]$.

Normally $\chi_n << 1$ for quantum mechanically un-allowed frequencies. When the field frequency ν matches with the required energy exchange relation,

$$\Delta E = h\nu \tag{26.4}$$

the polarizability χ_1 is strong and the atom undergoes through the quantum transitions by absorbing the required amount of energy $h\nu$ if it is available from the field within its vicinity. The detector current is then given by the standard square modulus of the field:

$$D(t) = |\chi_1 a(t)\exp[2\pi \nu t]|^2 = \chi_1{}^2 a^2(t) \tag{26.5}$$

26.4 Paradox of Non-Interference of Light

It is quite common to explain that no photons arrive at the location of dark fringes in a two beam interferometer (Mach-Zehnder, Michelson, Young's double slit, etc.). The implication is that it does not matter whether the light beam contains one or multitude of indivisible photons, the outcome will always be the same. If photons are really indivisible packets of energy and "photon interferes only with itself", then why do we need phase and frequency coherence properties between different parts of a light beam? Our viewpoint is that the belief in "single photon interference" is a highly flawed simply because light beams do not interfere with each other, whether they contain one photon or trillions of photons. Both classical and QM mathematical formulations tacitly assume that EM fields do not interact with (operate on) each other. Then how can crossing light beams redistribute the field energy by themselves? Our model of expanding universe is based upon the measurement of Doppler frequency shifts of light from distant stars. Light from specific stars and galaxies from many light years distance away are always crossed by trillions of the light beam from other stars. Yet the Doppler shift remains unchanged characteristic signature of each individual star. In our daily life, we have no problem recognizing a face from a distance even though the image carrying beam had to cross multitudes of other the light beams going in different directions. Well formed light beams do not interfere with each other. They pass through each other unperturbed in the absence of interacting molecules (detectors). Light does not interfere with light. This is why the WDM communication system works. We combine a large number of communication channels by wavelength domain multiplexing (WDM) using light beam with a distinct set of frequencies and send them through a common path of hair-thin fiber of tens of kilometer and we separate each channel by demultiplexing without loosing any data. If light beams of different frequencies interacted on each other by themselves, the output signal would have become chaotic pulses.

But we do record and measure the absence of any EM field energy at the dark fringes due to superposition of coherent beams on a detector array or a photographic plate. For two superposed coherent beams of equal amplitude with a delay τ, the detector response produces sinusoidal fringes:

$$D = |\chi_1 a e^{i2\pi\nu t} + \chi_1 a e^{i2\pi\nu(t+\tau)}|^2 = 2\chi_1{}^2 a^2[1+\cos 2\pi\nu\tau] \tag{26.6}$$

At a location where the two equal amplitudes fields are undulating with opposite phases, the detector dipoles cannot execute opposing dipolar undulations at the same time. So they are not stimulated and hence they cannot absorb energy from superposed fields. EM field energy passes through them since they cannot redistribute their field energy by themselves [11].

26.5 Panarella's Low Light Level Experiment

In view of the persisting claims of "single photon interference" for almost a century, we want to draw attention of the readers to a publication by Panarella [5]. He carried out the measurements of the diffraction patterns due to a pin hole illuminated by a CW He-Ne laser beam whose intensity was systematically reduced by carefully calibrated steps. He found out that when the beam power drops below four-photon equivalent energy, the side lobes of diffraction rings cannot be recorded even with prolonged integration time. This result conforms to our semiclassical view. The detectors first stimulated as dipoles by the superposed fields can undergo QM transition provided there was enough field energy within their vicinity to absorb $h\nu$ amount of energy. However, Panarella's experiment brings up another important question. Why does his experiment require the simultaneous presence of more than 4-photons to register a "click"? We believe that it is because photons, after being emitted by atoms and molecules, propagate as expanding (diffractive) wave packets with reduced energy densities.

26.6 Photons Are Divisible Classical Wave Packets

The field of optics has been successfully modeling the propagation of light beams using the mathematically advanced version of the Huygens-Fresnel (HF) principle [20]. The HF integral correctly predicts (i) the emergence of spatial coherence out of completely incoherent thermal light (Van Cittert-Zernicke theorem), (ii) near field and far field diffraction patterns due to any simple and complex diffracting aperture, (iii) generation inside a laser and propagation outside a laser of Gaussian transverse mode pattern, (iv) evolution of spatial modes and the propagation characteristics in exquisite details inside simple single mode waveguides and the most complex nano-photonic waveguides. Quantum Mechanics has not produced any better substitute for HF integral. HF integral does not require quantization of EM fields. It is worth noting that the quantization of atoms has revolutionized our understanding of the material world by providing us with a staggering amount of new knowledge about the material world. In contrast, the quantization of the EM field has actually suppressed the exploration of the real physical process taking place during the detection process of superposed light beams and gave birth to non-casual and non-local interpretation of superposition phenomenon. Embedded in HF integral are two profoundly important but dialectical characteristics of all wave phenomena. A wave is a collective phenomenon that will always have a finite space and time extension. The waves propagate as a group even though they constantly expand as if they have a built in propensity to diverge but evolve into a space-finite sustainable far-field pattern whose divergence angle remains constant [20]. Yet, if such a self-sustainable wave front is disrupted, the broken wave fronts always regroup themselves into a new pattern whose near field pattern and angular divergence evolve again into a new sustainable space-finite far field pattern. Thus, the field pattern or amplitudes distribution of a wave front is constantly evolving, which is equivalent to an evolution of available energy re-distribution of the field. Describing a light beam as consisting of multitudes indivisible photons and make them conform to these changing angular redistribution from near field into far field, are beyond casual description. Accordingly, we are forced to impose non-casual, non-local behavior on the indivisible photons.

We define photons as classical wave packets that evolve after atoms and molecules release their quantum of energy $\Delta E = h\nu$ into the cosmic medium as a time finite pulse with a carrier frequency exactly equal to ν:

$$\vec{E}(t) = \vec{a}(t)\cos 2\pi\nu t = \mathrm{Re}[\vec{a}(t)e^{i2\pi\nu t}] \tag{26.7}$$

In the far field from the atoms and molecules, the wave packet would have the physical shape of a Gaussian spatial wave front and a semi-exponential temporal envelope (Fig. 26.2 top curve). We are choosing Gaussian spatial cross-section in analogy with the spatially stable mode that always evolves in laser cavities and in long single mode wave guides [18, see Chapters. 7,

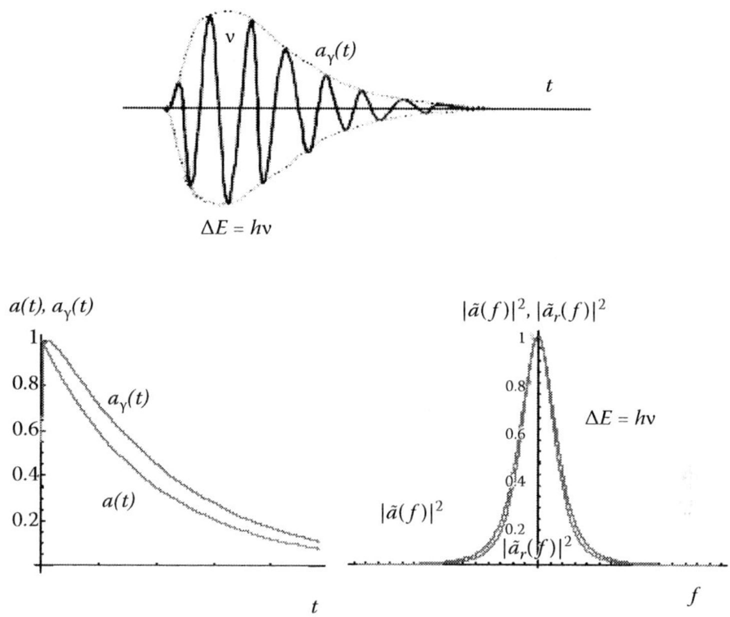

FIGURE 26.2
Top: A model for a rapidly rising and exponentially dying photon wave packet envelope with carrier frequency ν. Bottom Left: Pure exponential (lower curve) and rapidly rising but exponentially dying (upper curve) photon wave packet amplitudes. The pure exponential $a(t)$ used here is given by $a(t) = e^{-t/2t}$, where $\tau \simeq 1ns$. The rapidly rising and exponential dying amplitude $a_r(t)$ model is given by $a_r(t) = t^r e^{-t/2t}$, where r = 0.05. Bottom Right: Fourier transform of the pure exponential has a slightly larger FWHM (upper curve) than the rapidly rising but exponentially dying amplitude model (lower curve). We have used the frequency of red cadmium light as the resonance frequency.

8, 9]. The choice of semi-exponential temporal wave envelope derives from the well established and measured spectral envelope of the so-called natural line width of spontaneous emission. Exponential and Lorentzian curves form a Fourier transform pair (Fig. 26.2 lower set of curves). It is important to recognize that the experimental time integrated spectral fringe shape due to a pulse can also be mathematically shown to be the Fourier transform of the pulse envelope while the carrier frequency of the pulse determines the central location of the spectral fringe [8, 14, 17]. But why choose a semi-exponential pulse envelope? We believe that nothing in the universe can happen instantaneously or continue over an infinite duration. So it is physically impossible to start the rise of a pulse envelope at the peak exponential value instantaneously. It must start from zero value and very rapidly rise to the required exponential peak value and die down exponentially. We are also assuming that this rise time to exponential peak value is extremely short so that the Fourier transform of this semi-exponential envelope is still a small deviation from the true Lorentzian, the shape of the natural linewidth that a traditional spectrometer measures. Our final assumption in constructing this semi-exponential pulse is that the electromagnetic energy carried under this envelope is exactly $\Delta E = h\nu$.

26.7 Finite Time and Energy for A Single Photo Induced Transition

Both the proponents and opponents of photons (spontaneous emission from individual atoms or molecules) as indivisible packets of energy concur with the experimental observations that the transition time required for a photo induced transition is extremely short. For visible range ($\nu \sim 10^{15}$ Hz) it is in the domain of 10^{-15} seconds or around one femto second. They also concur that even at very low intensity, if there is any photo induced transition, it always happens within the fs time constant; only the rates of clicks are very low. In this context we find the observation of Panarella [5] very interesting. At extremely low intensity he was unable to detect the secondary diffraction rings even after very long time integration when the low count rate for the central disc was still measurable. While Panarella has proposed a "photon clump" theory to explain his observation, we are proposing that it is due to photons being *divisible, diffractively spreading classical wave packets,* they present much weaker field energy densities at larger diffraction angles.

For photo induced transition to take place, the quantum device must be bathed in sea of EM field energy with $\Delta E = h\nu$ amount of energy within its immediate vicinity whose E-vector undulation frequency ν matches with that for the quantum transition. This will allow the field to induce dipole undulation on the detecting device and trigger the required amount of energy absorption provided it is available in its immediate vicinity. It will take the EM field at least one cycle, if not more, of time to find its compatibility with the QM required dipole frequency ν to trigger the quantum transition and energy absorption. While this time is finite, it is very short, a few fs, in the domain of visible light. So, Panarella's experiment implies that when the field energy density (due to diffraction or wave front spreading) falls below some density, the detecting dipoles fail to absorb any energy. So one of the conclusions is that dipoles cannot keep on integrating energy from the flowing weak field over a very long period to accumulate ΔE amount of energy. This is in congruence with the photo detecting community. Since we can never produce any abruptly rising sharp pulse, we may be ignoring the possibility that low energy tails of weak pulses prepare the detectors to undergo rapid transition when a sufficient amount of energy becomes available around its vicinity.

To test this possibility, we suggest the following experiment using a planar grating that produces multiple higher order diffraction spots with diminishing intensity. Each measurement should be carried out by illuminating the grating with a single short pulse whose input intensity is gradually diminished in a series of experiments to see which diffraction orders stop producing photoelectrons. There is an advantage in using a single pulse and many diffraction orders with an array of identical detectors. Once a laser-optical system has been well calibrated to produce a desired single pulse, it is easy to reproduce it. Second, the differential stretching of the single input pulse at different diffraction orders can be calculated analytically [17]. In fact, the peak to peak stretching of a pulse at the m-th order for an N-slit planar grating will be $T_m = N\tau_m = Nm\lambda/c$. The experiment should first be calibrated with CW light to identify at what low intensity levels the different orders stop producing photoelectrons. This should then be compared with the results for pulsed light. We believe it might reveal whether photoelectrons require $h\nu$ quantity of field energy within its immediate vicinity for instantaneous ("wave function collapse") transition or it can accumulate energy from the traveling EM field over a finite period including the influence, if any, of the weak tails of pulses.

So far this N-slit grating experiment has been designed to validate that photon wave packets are classical and divisible. Then by the same classical model we should be able to synthesize a stronger field out of the many undetectable weak fields. Let us now propose another experiment using the same N-slit grating to establish our proposition. This experiment can be done with a CW light source assuming that each of the N-slits of the grating has identical opening and all the slits are illuminated with a uniform amplitude wave front. An array of identical detectors placed at the various orders with ample intensity in the beam would produce photoelectrons in all the detectors. Let us then place a broad opaque aperture with only one single slit matching that of the grating immediately after the grating on a translatable stage. This translatable single slit can now allow one to measure the photo count at selected places due to any one of the single slit out of the N-slits. Then one can reduce the input intensity to the minimum level that just stops the photoelectron production even after long integration time (except inevitable steady dark current). Then we remove the broad screen to allow all the N-diffracted wave fronts to arrive on the detection plane. The new intensity will now be $(N\sqrt{i_{\min}})^2$ or $N^2 i_{\min}$, where $i_{\min}$ is the intensity passing through one slit. With a typical 5 cm grating with $N = 3 \times 10^4$ slits one can enhance intensity by a factor of 9×10^8. We believe that under this new condition, photoelectrons can be counted again. The above two proposed experiments will establish that photons are classical wave packets that can both be split by optical components and recombined by detectors with proper experimental set up.

26.8 What Are the Possible Impacts if Photons Are Divisible Wave Packets?

First, the unnecessary claims that interference phenomenon is non-local can be replaced by a causal and local model without compromising any prediction of quantum mechanics [7]. Of course, we will have to give up the interpretation that each photoelectron implies the registration of a specific indivisible photon. We will have to give up the notion that no photons arrive at the location of the dark interference fringes. We also have to give up Dirac's statement, "Each photon then interferes only with itself. Interference between two different photons never occurs" [21]. And, of course, those conceived experiments that literally require the production, propagation, manipulation and detection of the same original indivisible photon, will have to be re-designed. EM field wave packets change constantly through incessant diffractive propagation. Also as a photon propagates through a material medium, it interacts with the dipoles of the medium and emerges as a different photon undergoing various changes in amplitude, phase, polarization and frequencies, depending upon the incident beam intensity and the polarizability χ_n of the medium. One should recognize that if photons were really indivisible and independent packets of energy and they can use their non-local properties to determine which place in an interferometer to appear or disappear from, then we should not have required any phase coherence property for superposition measurements (interferometry). The phase coherence is required by the detecting dipoles when they try to sum the induced dipole undulation amplitudes due to all the superposed fields at the same time. This is why the superposition effects necessarily have to be local (volume of the participating detecting molecules).

Acknowledgment

CR would like to acknowledge partial support from Nippon Sheet Glass Corporation. NT would like to acknowledge partial support from the Connecticut College of Technology through the Connecticut Business and Industry Association.

References

1. Arthur Zajonc, "Light reconsidered", p.S-2, in *The Nature of Light: What Is a Photon?* OPN trends, special issue, 2003, Eds. C. Roychoudhuri and R. Roy. http://www.osa-opn.org/abstract.cfm?URI=OPN-14-10-49.
2. Goulielmakis et al., "Direct measurement of light waves", *Science,* **305,** p. 1267, (2004).
3. W. E. Lamb, *Appl. Phys.* **B60,** p. 77–84 (1995); "Anti-photon".
4. Willis E. Lamb, Jr. and Marlan O. Scully, "The Photoelectric Effect without Photons", pp. 363–369, in *Polarization, matter and radiation;* Jubilee volume in honor of Alfred Kasler, Presses Universitaires de France, Paris (1969).

5. E. Panarella, *SPIE Proc.* Vol. **5866**, pp. 218–228, (2005), "Single Photons have not been detected. The alternative photon clump model". See also by E. Panarella, "Nonlinear behavior of light at very low intensities: the photon clump model", p. 105 in *Quantum Uncertainties – recent and future experiments and interpretations*, Eds. W. M. Honig, D. W. Kraft, & E. Panarella, Plenum Press (1987).
6. L. Mandel and E. Wolf, *Optical coherence and quantum optics*, Cambridge University Press (1995).
7. C. Roychoudhuri, Physics Essays, **19**(3), September 2006; "Locality of superposition principle is dictated by detection processes". [For many reprints related to this paper, go to the site: http://www.phys.uconn.edu/~chandra/]
8. C. Roychoudhuri and M. Tayahi, Intern. *J. of Microwave and Optics Tech.*, July 2006; "Spectral Super-Resolution by Understanding Superposition Principle & Detection Processes", manuscript ID# IJMOT-2006-5-46: http://www.ijmot.com/papers/papermain.asp
9. DongIk Lee and C. Roychoudhuri. "Measuring properties of superposed light beams carrying different frequencies", *Optics Express* **11**(8), 944-51, (2003); [http://www.opticsexpress.org/abstract.cfm?URI=OPEX-11-8-944].
10. C. Roychoudhuri, D. Lee, and P. Poulos, *Proc. SPIE* Vol. **6290**-02 (2006); "If EM fields do not operate on each other, how do we generate and manipulate laser pulses?"
11. C. Roychoudhuri and C. V. Seaver, *Proc. SPIE* Vol. **6285-**01, Invited, (2006) in *The nature of light: Light in nature*; "Are dark fringe locations devoid of energy of superposed fields?"
12. C. Roychoudhuri and N. Tirfessa, *Proc. SPIE* Vol. **6292**-01, Invited (2006); "A critical look at the source characteristics used for time varying fringe interferometry".
13. C. Roychoudhuri and V. Lakshminarayanan, *Proc. SPIE* Vol. **6285**-08 (2006) in *The nature of light: Light in nature;* "Role of the retinal detector array in perceiving the superposition effects of light".
14. C. Roychoudhuri, *Proc. SPIE* Vol. **6108**-50 (2006); "Reality of superposition principle and autocorrelation function for short pulses".
15. C. Roychoudhuri, *SPIE Conf. Proc.* **5866**, pp. 26–35 (2005); "If superposed light beams do not re-distribute each others energy in the absence of detectors (material dipoles), can an indivisible single photon interfere by/with itself?"
16. C. Roychoudhuri, *Proc. SPIE* Vol. **5866**, pp. 135–146 (2005); "What are the processes behind energy re-direction and re-distribution in interference and diffraction?"
17. C. Roychoudhuri, *Proc. SPIE* Vol. **5531**, 450–461 (2004); *Interferometry-XII: Techniques and Analysis*. "Propagating Fourier frequencies vs. carrier frequency of a pulse through spectrometers and other media".
18. B. E. A. Saleh and M. C. Teich, Ch. 17 in *Fundamentals of Photonics*, John Wiley (1991).
19. E. T. Jaynes, "Is QED Necessary?" in *Proceedings of the Second Rochester Conference on Coherence and Quantum Optics*, L. Mandel and E. Wolf (eds.), Plenum, New York, 1966, p. 21. See also: Jaynes, E. T., and F. W. Cummings, *Proc. IEEE.* **51,** 89 (1063), "Comparison of Quantum and Semiclassical Radiation Theory with Application to the Beam Maser". http://bayes.wustl.edu/etj/node1.html#quantum.beats.
20. J. W. Goodman, *Introduction to Fourier Optics*, McGraw-Hill (1996).
21. P. A. M. Dirac, *The Principles of Quantum Mechanics*, Clarendon Press (1974).

27

Direct Measurement of Light Waves

E. Goulielmakis,[1*] M. Uiberacker,[1*] R. Kienberger,[1] A. Baltuska,[1] V. Yakovlev,[1] A. Scrinzi,[1] Th. Westerwalbesloh,[2] U. Kleineberg,[2] U. Heinzmann,[2] M. Drescher,[2] F. Krausz[1,3†]

CONTENTS

The electromagnetic field of visible light performs ~10^{15} oscillations per second. Although many instruments are sensitive to the amplitude and frequency (or wavelength) of these oscillations, they cannot access the light field itself. We directly observed how the field built up and disappeared in a short, few-cycle pulse of visible laser light by probing the variation of the field strength with a 250-attosecond electron burst. Our apparatus allows complete characterization of few-cycle waves of visible, ultraviolet, and/or infrared light, thereby providing the possibility for controlled and reproducible synthesis of ultrabroadband light waveforms.

Although the wave nature of light has long been known, it has not been possible to measure directly the oscillating field of light. Radiation in the visible and higher frequency spectral ranges can so far only be characterized in terms of physical quantities averaged over the wave period. Nonlinear optical techniques now allow measurement of $\varepsilon_L(t)$, the amplitude envelope, and $\omega_L(t)$, the carrier frequency, as a function of time t, for light pulses with durations that approach the wave cycle (1, 2). The carrier-envelope phase φ, which determines the timing between $\varepsilon_L(t)$ and $\omega_L(t)$, can also be measured (3). These measurements rely on carrier-envelope decomposition, which is physically meaningful only as long as the frequency spectrum of the wave is

[1] Institut für Photonik, Technische Universität Wien, Gusshausstraβe 27, A-1040 Wien, Austria.
[2] Fakultät für Physik, Universität Bielefeld, D-33615 Bielefeld, Germany.
[3] Max-Planck-Institut für Quantenoptik, Hans-Kopfermann-Straβe 1, D–85748 Garching, Germany.
[*] These authors contributed equally to this work.
[†] To whom correspondence should be addressed. E-mail: ferenc.krausz@tuwien.ac.at

confined to less than one octave (4). If the radiation is composed of frequencies spanning a broader range (5–17), direct access to the field is required. Attosecond pulses of extreme ultraviolet (XUV) light were predicted to suit for this purpose (18, 19). We report the direct measurement of the buildup and disappearance of the electric field of a light pulse through the use of an attosecond probe.

The electric field is defined as the force exerted on a point charge of unit value. Its conceptually most direct measurement must therefore rely on measurement of this force. In a light wave, the electric field E_L, and hence the force $F = qE_L$ it exerts on a particle with charge q, are subject to rapid variations. Access to this force is possible only if the probe charge is instantly placed in the field, i.e., within a time interval τ_{probe} over which the temporal variation of the force is "frozen", i.e., $\tau_{probe} << T_0 = (2\pi)/\omega_L$, where T_0 is the wave period. The probe charge can be launched into the field by knocking electrons free from atoms or ions instantly. In a linearly polarized wave, the change of the electrons' momentum $\Delta p(\vec{r},t)$ at location $\vec{r}$ and time t along the direction of the electric field is given by

$$\Delta p(\vec{r},t) = e\int_t^{\infty} E_L(\vec{r},t')dt' = eA_L(\vec{r},t) \quad (27.1)$$

where e is the electron charge and $A_L(\vec{r},t)$ is the vector potential of the electric field $E_L(\vec{r},t) = E_0\varepsilon_L(\vec{r},t)\cos(kz - \omega_L t + \varphi)$, where E_0 is the maximum field amplitude, and k is the wave vector. In our analysis, we assumed the wave to propagate along the z direction, and $t = t_{real} - z/v_g$ was defined in a retarded frame to yield $t = 0$ as locked to the peak of the pulse travelling at the group velocity v_g.

The relation $E_L(\vec{r},t) = -\partial A_L(\vec{r},t)/\partial t$ implies that measuring the momentum boost $\Delta p(\vec{r},t)$ imparted to the freed electrons by the field at the location $\vec{r}$ at two instants differing in time by $\delta t << T_0/4$ will yield the electric field strength and direction directly as $E_L(\vec{r},t) = [\Delta p(\vec{r},t-\delta t/2) - \Delta p(\vec{r},t+\delta t/2)/e\delta t]$. This measurement procedure relies on a momentary release of the electrons within $\tau_{probe} \leq T_0/4$. For near infrared, visible, and ultraviolet light, this condition dictates that $\tau_{probe} < 1$ fs. Varying the timing of such a subfemto-second electron probe across the laser pulse provides complete information on the electric field of the light wave.

These considerations suggest that the electron probe needs to be localized not only in time to a tiny fraction of the wave period T_0, but also in space to a tiny fraction of the wavelength λ_L of the light wave to be measured. The latter requirement can be substantially relaxed if we trigger the electron release with an energetic photon pulse that copropagates with the laser wave in a collinear beam (Figure 27.1). Because the timing of the probe electrons relative to the light field is invariant to space in this case, in a gently focused laser beam they can be released and are subsequently allowed to move over distances substantially larger than λ_L, in a volume within which the

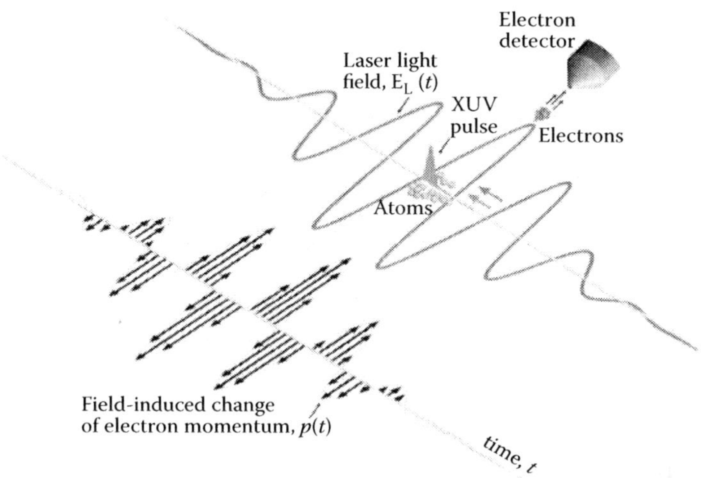

FIGURE 27.1
Schematic of the measurement principle. A few-cycle pulse of laser light, together with a synchronized subfemtosecond XUV burst, is focused into an atomic gas target. The XUV pulse knocks electrons free by photoionization. The light electric field $E_L(t)$ to be measured imparts a momentum change to the electrons (black arrows), which scales as the instantaneous value of the vector potential $A_L(t)$ at the instant of release of the probing electrons. The momentum change is measured by an electron detector, which collects the electrons ejected along the direction of the linearly polarized $E_L(\vec{r},t)$.

spatial variation of the field amplitude $\varepsilon_L(\vec{r},t)$ is negligibly small for a fixed value of t.

Putting the above concept into practice requires the electron probe to be scanned through the entire laser pulse. For each newly set timing t, measurement of the momentum shift $\Delta p(t)$ of the probing electrons requires the laser pulse to pass through the measurement apparatus again. Full characterization of the light waveform is therefore only feasible if it can be reproducibly generated for repeated measurements. Another equally important prerequisite for implementation of the above concept is the availability of an energetic instantaneous excitation (for launching the probing electrons) that is not only confined temporally to a fraction of 1 fs but is also synchronized to the light wave with similar accuracy. With the generation of waveform controlled, intense, few-cycle light pulses (20) and their successful application to producing single 250-as XUV pulses synchronized to the driver light wave (21), these preconditions are now fulfilled. The waveform-controlled pulses—after having produced the attosecond photon probe—allow through nonlinear optical frequency conversion the synthesis of reproducible, synchronized, ultrabroadband, few-cycle waveforms (5–17). These can be repeatedly sent into the measurement apparatus with exactly the same waveform, and the subfemtosecond XUV pulse is able to produce the electrons by photoionization for probing the oscillating light field with sufficient temporal resolution.

The electrons knocked free from the atoms by the XUV pulse can be most conveniently detected if the direction of their movement is left unchanged by the light field. This applies if electrons are detected within a narrow cone aligned with the electric field vector of the linearly polarized laser wave along the x direction and are ejected with a large-enough initial momentum p_i to fulfill $| p_i | > |\Delta p_{max}|$, where Δp_{max} is the maximum momentum shift induced by the field. A large initial momentum also benefits the measurement by enhancing the change of the electrons' kinetic energy ΔW, according to $\Delta W \approx (p_i|m)\Delta p$, and m is the electron's mass. This expression, together with Eq. 27.1, implies that the energy shift scales linearly with both the electric field and the wavelength of the light field to be probed (22). The importance of a large ΔW lies in the facts that the probing electrons are emitted with an inherent uncertainty $\delta W_{probe} \approx <= \hbar/\tau_{probe}$ (where $\hbar$ is Planck's constant h divided by 2π) and that the dynamic range over which the light field strength can be reliably measured scales with $\Delta W_{max}/\delta W_{probe}$ (ΔW_{max} is the maximum shift in the pulse).

Measurement of $E_L(t)$ over a substantial dynamic range requires a ΔW_{max} of several tens of electron volts. For an initial kinetic energy of $W_i \approx 100$ eV, this condition is satisfied for $E_0 < 10^8$ V/cm for near-infrared light and requires $E_0 \approx 3 \times 10^8$ V/cm for ultraviolet light (22). Noble gases with a low atomic number (such as helium and neon) safely resist ionization by a few-cycle field at these field strengths (23). The accuracy of definition of the location $\vec{r}$ is dictated by the size of the volume within which $\varepsilon(\vec{r}, t)$ is approximately independent of $\vec{r}$. If the field is probed in the beam focus, this condition requires the probing electrons to be confined—during their interaction with the laser field— laterally (xy) and longitudinally (z) to a small fraction of the diameter and to the confocal parameter of the beam, respectively.

In a proof-of-concept experiment, we directly measured the $E_L(t)$ of the few-cycle laser pulse used for producing the attosecond photon probe (Figure 27.1). Linearly polarized, waveform-controlled, < 5-fs, 0.4-mJ, 750-nm (T_0= 2.5 fs) laser pulses (20), with carefully optimized values of φ, and E_0, produce single 250-as XUV pulses at $(\hbar\omega_{xuv})_{mean}$ = 93 eV in a gas of neon atoms (21). The XUV pulse copropagates with the laser pulse in a collinear, laserlike beam to a second neon target placed in the focus of a spherical, two component, Mo/Si multilayer mirror (21). The mirror, of 120-mm focal length, reflects XUV radiation over a band of ~9 eV, centered at ~93 eV. Consequently, the XUV pulse sets electrons free by photoionization with an initial kinetic energy of $p_i^2/2m = h\omega_{xuv} - W$, (where W_b is the electron's binding energy) spread over an ~9-eV band, implying that $\delta W_{probe} \approx 9$ eV. The electrons' energy shift $\Delta W(t) \approx e(p_i/m)A_L(t)$ probes the laser vector potential. The volume of light-field probing is defined laterally by the < 10-μm diameter of the XUV beam at its waist and longitudinally by the <50-μm size of the neon jet, which is well confined within the focal volume of the laser beam (diameter, >60 μm; confocal parameter, >5 mm). For $p_i^2/2m \approx 100$ eV,

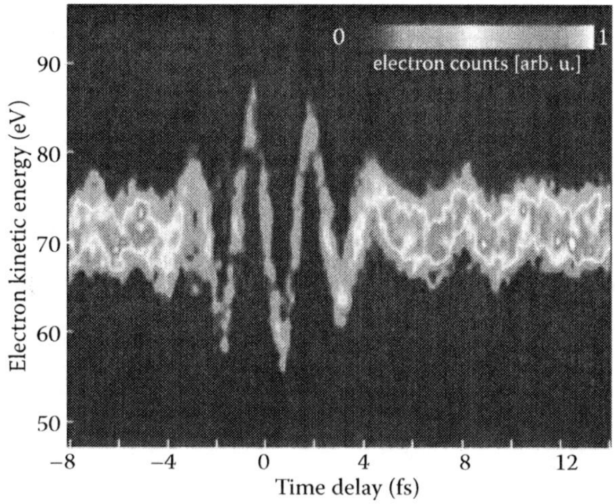

FIGURE 27.2
A series of kinetic energy spectra of electrons detached by a 250-as, 93-eV XUV pulse from neon atoms in the presence of an intense <5-fs, 750-nm laser field, in false-color representation. The delay of the XUV probe was varied in steps of 200 as, and each spectrum was accumulated over 100 s. The detected electrons were ejected along the laser electric field vector with a mean initial kinetic energy of $p_i^2/2m \approx \hbar\omega h_{xuv} - W_b = 93\ eV - 21.5\ eV = 71.5\ eV$. The energy shift of the electrons versus the timing of the XUV trigger pulse that launches the probing electrons directly represents $A_L(t)$. arb. u., arbitrary units.

the electrons traveled less than 1μm within 100 fs and hence remained safely confined to the region of constant laser field amplitude.

The field-induced variation of the final energy spectrum of the probe electrons versus delay between the XUV burst and the laser pulse (Figure 27.2) reveal, without the need of any detailed analysis, that probing is implemented by a single burst of subfemtosecond duration that is synchronized with subfemtosecond accuracy to the measured laser field. $E_L(t)$ can now be directly (i.e., without any iterative steps) obtained through the procedure outlined above (Figure 27.3). From the measured spectrum of the few-cycle laser pulse (Figure 27.3, inset), we calculated $E_L(t)$ by a simple Fourier transformation on the assumption of absence of spectral phase variations. The result, with E_0 and φ chosen to yield the best match to the measured values, is shown in gray. The excellent fit to the measured field evolution indicates a near-transform-limited pulse. Its duration was evaluated as 4.3 fs, in good agreement with the result of an autocorrelation measurement.

It has been predicted by theory that the few-cycle pulse pumping the XUV source has a "cosine" waveform ($\varphi \approx 0$) if a single subfemtosecond pulse emerges from the ionizing atoms (24). Our results (Figure 27.3) yield the experimental evidence. From this measurement, we also learn that the electric

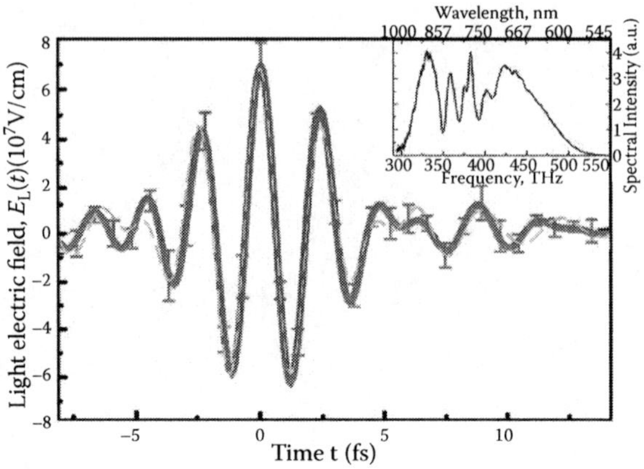

FIGURE 27.3
$E_L(t)$ reconstructed (solid line) from the data depicted in Fig. 2 and calculated (dashed line) from the measured pulse spectrum (inset) with the assumed absence of a frequency-dependent phase and with E_0 and φ chosen so as to afford optimum matching to the measured field evolution, a.u., arbitrary units.

field points toward the electron detector at the pulse peak and that its strength is ~7×10^7 V/cm. With the temporal evolution, strength, and direction of $E_L(t)$ measured, we have performed a complete characterization of a light pulse in terms of its classical electric field.

Direct probing of light-field oscillations represents what we believe to be a substantial extension of the basic repertoire of modern experimental science. The door to practical applications is opened by the creation of the key element of the demonstrated light-field detector, the synchronized attosecond electron probe, in a noninvasive manner. In fact, our intense <5-fs laser pulse appears to be capable of producing the necessary XUV trigger burst without suffering any noticeable back-action to its own temporal shape (Figure 27.3). After having produced the attosecond photon probe, this powerful few-femtosecond pulse is ideally suited for the synthesis of ultrabroadband, few-cycle, optical wave forms (5–17). Being composed of radiation extending from the infrared through the visible to the ultraviolet region, the resultant few-cycle, monocycle, and conceivably even subcycle waveforms will offer a marked degree of control over the temporal variation of electric and magnetic forces on molecular and atomic time scales.

These light forces, in turn, afford the promise of controlling quantum transitions of electrons in atoms and molecules and—at relativistic intensities—their center-of-mass motion. Reproducible ultrabroadband light wave synthesis, a prerequisite for these prospects to materialize, is inconceivable without subfemtosecond measurement of the synthesized waveforms. Beyond providing the subfemtosecond electron probe for these measurements, the substantial experimental efforts associated with the construction and reliable operation of a subfemtosecond photon source will pay off in yet another way. The envisioned control of electronic motion with light forces can only be regarded as accomplished once it has been measured. Owing to their perfect synchronism with the synthesized light waveforms, the subfemtosecond photon probe will allow us to test the degree of control achieved by tracking the triggered (and hopefully steered) motion in a time-resolved fashion.

References and Notes

1. R. Trebino et al., *Rev. Sci. Instrum.* **68**, 3277 (1997).
2. C. Iaconis, I. A. Walmsley, *Opt. Lett.* **23**, 792 (1998).
3. G. G. Paulus et al., *Phys. Rev. Lett.* **91**, 253004 (2003).
4. T. Brabec, F. Krausz, *Phys. Rev. Lett.* **78**, 3282 (1997).
5. T. W. Hänsch, *Opt. Commun.* **80**, 71 (1990).
6. S. Yoshikawa, T. Imasaka, *Opt. Commun.* **96**, 94 (1993).
7. A. E. Kaplan, P. L. Shkolnikov, *Phys. Rev. Lett.* **73**, 1243 (1994).
8. K. Shimoda, *Jpn. J. Appl. Phys.* **34**, 3566 (1995).
9. S. E. Harris, A. V. Sokolov, *Phys. Rev. Lett.* **81**, 2894 (1998).
10. A. Nazarkin, G. Korn, *Phys. Rev. Lett.* **83**, 4748 (1999).
11. O. Albert, G. Mourou, *Appl. Phys. B* **69**, 207 (1999).
12. M. Wittman, A. Nazarkin, G. Korn, *Phys. Rev. Lett.* **84**, 5508 (2000).
13. Y. Kobayashi, K. Torizuka, *Opt. Lett.* **25**, 856 (2000).
14. A. V. Sokolov, D. R. Walker, D. D. Yavuz, G. Y. Yin, S. E. Harris, *Phys. Rev. Lett.* **87**, 033402 (2001).
15. K. Yamane et al., *Opt. Lett.* **28**, 2258 (2003).
16. M. Y. Shverdin, D. R. Walker, D. D. Yavuz, G. Y. Yin, S. E. Harris, in *OSA Trends in Optics and Photonics Series (TOPS) Vol. 96, Conference on Lasers and Electro-Optics (CLEO)* (Optical Society of America, Washington, DC, 2004), Postdeadline paper CPDC1.
17. K. Yamane, T. Kito, R. Morita, M. Yamashita, in *OSA Trends in Optics and Photonics Series (TOPS)*, vol. 96, *Conference on Lasers and Electro-Optics (CLEO)*, (Optical Society of America, Washington, DC, 2004), Postdeadline paper CPDC2.
18. R. Kienberger et al., *Science* **297**, 1144 (2002).
19. A. D. Bandrauk, Sz. Chelkowski, N. H. Shon, *Phys. Rev.Lett.* **89**, 283903 (2002).
20. A. Baltuska et al., *Nature* **421**, 611 (2003).
21. R. Kienberger et al., *Nature* **427**, 817 (2004).
22. In the limit of $|\Delta p_{\max}| << |p_i|$, the change in the electrons' final kinetic energy is given by $\Delta W_{\max} \sim [8W_i \boldsymbol{U}_{p.\max}]^{1/2}$, where $U_{p,\max} = e^2E_0^2/4m_e\omega_L^2$ is the electrons' quiver energy averaged over an optical cycle at the peak of the light pulse. Increase of the excitation energy $\hbar\omega_{xuv}$ tends to reconcile the conflicting requirements of avoiding field ionization and ensuring a high dynamic range.
23. T. Brabec, F. Krausz, *Rev. Mod. Phys.* **72**, 545 (2000).

24. We are grateful to B. Ferus for creating the artwork. Sponsored by the fonds zur Forderung der Wissen-schafllichen Forschung (Austria, grant nos. Y44-PHY, P15382, and F016), the Deutsche Förschungsgemein-schaft and the Volkswagenstiftung (Germany), the European ATTO and Ultrashort XUV Pulses for Time-Resolved and Non-Linear Applications networks, and an Austrian Programme for Advanced Research and Technology fellowship to R.K. from the Austrian Academy of Sciences.

Index

G

H

X

Y

Z